T0183981

Lecture Notes in Computer Science   11451

Commenced Publication in 1973
Founding and Former Series Editors:
Gerhard Goos, Juris Hartmanis, and Jan van Leeuwen

Editorial Board

David Hutchison
  Lancaster University, Lancaster, UK
Takeo Kanade
  Carnegie Mellon University, Pittsburgh, PA, USA
Josef Kittler
  University of Surrey, Guildford, UK
Jon M. Kleinberg
  Cornell University, Ithaca, NY, USA
Friedemann Mattern
  ETH Zurich, Zurich, Switzerland
John C. Mitchell
  Stanford University, Stanford, CA, USA
Moni Naor
  Weizmann Institute of Science, Rehovot, Israel
C. Pandu Rangan
  Indian Institute of Technology Madras, Chennai, India
Bernhard Steffen
  TU Dortmund University, Dortmund, Germany
Demetri Terzopoulos
  University of California, Los Angeles, CA, USA
Doug Tygar
  University of California, Berkeley, CA, USA

More information about this series at http://www.springer.com/series/7407

Lukas Sekanina · Ting Hu ·
Nuno Lourenço · Hendrik Richter ·
Pablo García-Sánchez (Eds.)

Genetic Programming

22nd European Conference, EuroGP 2019
Held as Part of EvoStar 2019
Leipzig, Germany, April 24–26, 2019
Proceedings

Springer

Editors
Lukas Sekanina [iD]
Brno University of Technology
Brno, Czech Republic

Nuno Lourenço [iD]
University of Coimbra
Coimbra, Portugal

Pablo García-Sánchez [iD]
University of Granada
Granada, Spain

Ting Hu [iD]
Memorial University
St. John's, NL, Canada

Hendrik Richter [iD]
HTWK Leipzig University
of Applied Sciences
Leipzig, Germany

ISSN 0302-9743   ISSN 1611-3349   (electronic)
Lecture Notes in Computer Science
ISBN 978-3-030-16669-4   ISBN 978-3-030-16670-0   (eBook)
https://doi.org/10.1007/978-3-030-16670-0

Library of Congress Control Number: 2019936009

LNCS Sublibrary: SL1 – Theoretical Computer Science and General Issues

© Springer Nature Switzerland AG 2019
This work is subject to copyright. All rights are reserved by the Publisher, whether the whole or part of the material is concerned, specifically the rights of translation, reprinting, reuse of illustrations, recitation, broadcasting, reproduction on microfilms or in any other physical way, and transmission or information storage and retrieval, electronic adaptation, computer software, or by similar or dissimilar methodology now known or hereafter developed.
The use of general descriptive names, registered names, trademarks, service marks, etc. in this publication does not imply, even in the absence of a specific statement, that such names are exempt from the relevant protective laws and regulations and therefore free for general use.
The publisher, the authors and the editors are safe to assume that the advice and information in this book are believed to be true and accurate at the date of publication. Neither the publisher nor the authors or the editors give a warranty, expressed or implied, with respect to the material contained herein or for any errors or omissions that may have been made. The publisher remains neutral with regard to jurisdictional claims in published maps and institutional affiliations.

This Springer imprint is published by the registered company Springer Nature Switzerland AG
The registered company address is: Gewerbestrasse 11, 6330 Cham, Switzerland

Preface

The 22nd European Conference on Genetic Programming (EuroGP) took place at the Faculty of Electrical Engineering and Information Technology of the Leipzig University of Applied Sciences, Leipzig, Germany, 24–26 April, 2019.

Genetic programming (GP) is an evolutionary computation branch that has been developed to automatically solve design problems, in particular the computer program design, without requiring the user to know or specify the form or structure of the solution in advance. It uses the principles of Darwinian evolution to approach problems in the synthesis, improvement, and repair of computer programs. The universality of computer programs, and their importance in so many areas of our lives, means that the automation of these tasks is an exceptionally ambitious challenge with far-reaching implications. It has attracted a very large number of researchers and a vast amount of theoretical and practical contributions are available by consulting the GP bibliography[1].

Since the first EuroGP event in Paris in 1998, EuroGP has been the only conference exclusively devoted to the evolutionary design of computer programs and other computational structures. Indeed, EuroGP represents the single largest venue at which GP results are published. It plays an important role in the success of the field, by serving as a forum for expressing new ideas, meeting fellow researchers, and initiating collaborations. It attracts scholars from all over the world. In a friendly and welcoming atmosphere authors presented the latest advances in the field, also presenting GP-based solutions to complex real-world problems.

EuroGP 2019 received 36 submissions from around the world. The papers underwent a rigorous double-blind peer review process, each being reviewed by four members of the international Program Committee.

Among the papers presented in this volume, 12 were accepted for full-length oral presentation (33% acceptance rate) and six for short talks (50% acceptance rate for both categories of papers combined). Authors of both categories of papers also had the opportunity to present their work in poster sessions.

The wide range of topics in this volume reflects the current state of research in the field. With a special focus on real-world applications in 2019, the papers are devoted to topics such as the test data design in software engineering, fault detection and classification of induction motors, digital circuit design, mosquito abundance prediction, machine learning and cryptographic function design.

Together with three other co-located evolutionary computation conferences (EvoCOP 2019, EvoMusArt 2019, and EvoApplications 2019), EuroGP 2019 was part of the Evo* 2019 event. This meeting could not have taken place without the help of many people. The EuroGP Organizing Committee is particularly grateful to the following:

---

[1] http://liinwww.ira.uka.de/bibliography/Ai/genetic.programming.html.

- SPECIES, the Society for the Promotion of Evolutionary Computation in Europe and Its Surroundings, aiming to promote evolutionary algorithmic thinking within Europe and wider, and more generally to promote inspiration of parallel algorithms derived from natural processes.
- The high-quality and diverse EuroGP Program Committee. Each year the members give freely of their time and expertise, in order to maintain high standards in EuroGP and provide constructive feedback to help the authors improve their papers.
- Marc Schoenauer of Inria-Saclay, France, for his continued hosting and maintaining of the MyReview conference management system.
- Hendrik Richter and his local organizing team from the Leipzig University of Applied Sciences, Germany.
- Pablo García-Sánchez (University of Cádiz, Spain) for the Evo* 2019 publicity and website.
- The Faculty of Electrical Engineering and Information Technology of HTWK Leipzig University of Applied Sciences for supporting the local organization.
- HTWK Leipzig University of Applied Sciences for their patronage of the event.
- Our invited speakers, Risto Miikkulainen and Manja Marz, who gave inspiring, enlightening, and entertaining keynote talks.
- The Evo* coordinators: Anna I Esparcia-Alcázar, from Universitat Politècnica de València, Spain and Jennifer Willies.

April 2019

Lukas Sekanina
Ting Hu
Nuno Lourenço
Hendrik Richter
Pablo García-Sánchez

Organization

Program Co-chairs

Lukas Sekanina   Brno University of Technology, Czech Republic
Ting Hu   Memorial University, Canada

Publication Chair

Nuno Lourenço   University of Coimbra, Portugal

Local Chair

Hendrik Richter   HTWK Leipzig/Leipzig University of Applied
     Sciences, Germany

Publicity Chair

Pablo García-Sánchez   University of Cádiz, Spain

Conference Administration

Anna Esparcia-Alcazar   Evostar Coordinator

Program Committee

Ignacio Arnaldo	Massachusetts Institute of Technology, USA
R. Muhammad Atif Azad	Birmingham City University, UK
Wolfgang Banzhaf	Michigan State University, USA
Helio Barbosa	Federal University of Juiz de Fora, Brazil
Heder Bernardino	Federal University of Juiz de Fora, Brazil
Anthony Brabazon	University College Dublin, Ireland
Stefano Cagnoni	University of Parma, Italy
Mauro Castelli	Universidade Nova de Lisboa, Portugal
Ernesto Costa	University of Coimbra, Portugal
Antonio Della Cioppa	University of Salerno, Italy
Marc Ebner	Universität Greifswald, Germany
Anna Isabel Esparcia-Alcazar	Universitat Politècnica de València, Spain
Francisco Fernandez de Vega	Universidad de Extremadura, Spain
Gianluigi Folino	ICAR-CNR, Italy
James Foster	University of Idaho, USA
Christian Gagné	Université Laval, Canada
Jin-Kao Hao	University of Angers, France

Inman Harvey	University of Sussex, UK
Erik Hemberg	Massachusetts Institute of Technology, USA
Malcolm Heywood	Dalhousie University, Canada
Ting Hu	Memorial University, Canada
Domagoj Jakobović	University of Zagreb, Croatia
Colin Johnson	University of Kent, UK
Ahmed Kattan	Loughborough University, UK
Krzysztof Krawiec	Poznan University of Technology, Poland
Jiri Kubalik	Czech Technical University in Prague, Czech Republic
William B. Langdon	University College London, UK
Kwong Sak Leung	The Chinese University of Hong Kong, SAR China
Nuno Lourenço	University of Coimbra, Portugal
Penousal Machado	University of Coimbra, Portugal
Radek Matousek	Brno University of Technology, Czech Republic
James McDermott	University College Dublin, Ireland
Eric Medvet	University of Trieste, Italy
Julian Miller	University of York, UK
Xuan Hoai Nguyen	Hanoi University, Vietnam
Quang Uy Nguyen	Military Technical Academy, Vietnam
Miguel Nicolau	University College Dublin, Ireland
Julio Cesar Nievola	Pontificia Universidade Catolica do Parana, Brazil
Michael O'Neill	University College Dublin, Ireland
Una-May O'Reilly	Massachusetts Institute of Technology, USA
Ender Ozcan	University of Nottingham, UK
Gisele Pappa	Universidade Federal de Minas Gerais, Brazil
Andrew J. Parkes	University of Nottingham, UK
Tomasz Pawlak	Poznan University of Technology, Poland
Stjepan Picek	Delft University of Technology, The Netherlands
Clara Pizzuti	Institute for High Performance Computing and Networking, Italy
Thomas Ray	University of Oklahoma, USA
Denis Robilliard	University of Lille Nord de France, France
Peter Rockett	University of Sheffield, UK
Álvaro Rubio-Largo	Universidad de Extremadura, Spain
Conor Ryan	University of Limerick, Ireland
Lukas Sekanina	Brno University of Technology, Czech Republic
Marc Schoenauer	Inria, France
Sara Silva	University of Lisbon, Portugal
Moshe Sipper	Ben-Gurion University, Isreal
Lee Spector	Hampshire College, USA
Jerry Swan	University of York, UK
Ivan Tanev	Doshisha University, Japan
Ernesto Tarantino	ICAR-CNR, Italy
Leonardo Trujillo	Instituto Tecnológico de Tijuana, Mexico
Leonardo Vanneschi	Universidade Nova de Lisboa, Portugal

Zdenek Vasicek   Brno University of Technology, Czech Republic
David White   University of Sheffield, UK
Man Leung Wong   Lingnan University, Hong Kong, SAR China
Bing Xue   Victoria University of Wellington, New Zealand
Mengjie Zhang   Victoria University of Wellington, New Zealand

Contents

Long Presentations

Ariadne: Evolving Test Data Using Grammatical Evolution   3
  Muhammad Sheraz Anjum and Conor Ryan

Quantum Program Synthesis: Swarm Algorithms and Benchmarks   19
  Timothy Atkinson, Athena Karsa, John Drake, and Jerry Swan

A Genetic Programming Approach to Predict Mosquitoes Abundance   35
  Riccardo Gervasi, Irene Azzali, Donal Bisanzio, Andrea Mosca,
  Luigi Bertolotti, and Mario Giacobini

Complex Network Analysis of a Genetic Programming
Phenotype Network.   49
  Ting Hu, Marco Tomassini, and Wolfgang Banzhaf

Improving Genetic Programming with Novel Exploration - Exploitation
Control   64
  Jonathan Kelly, Erik Hemberg, and Una-May O'Reilly

Towards a Scalable EA-Based Optimization of Digital Circuits.   81
  Jitka Kocnova and Zdenek Vasicek

Cartesian Genetic Programming as an Optimizer of Programs Evolved
with Geometric Semantic Genetic Programming   98
  Ondrej Koncal and Lukas Sekanina

Can Genetic Programming Do Manifold Learning Too?   114
  Andrew Lensen, Bing Xue, and Mengjie Zhang

Why Is Auto-Encoding Difficult for Genetic Programming?   131
  James McDermott

Solution and Fitness Evolution (SAFE): Coevolving Solutions and Their
Objective Functions.   146
  Moshe Sipper, Jason H. Moore, and Ryan J. Urbanowicz

A Model of External Memory for Navigation in Partially Observable Visual
Reinforcement Learning Tasks   162
  Robert J. Smith and Malcolm I. Heywood

Fault Detection and Classification for Induction Motors Using Genetic
Programming   178
    Yu Zhang, Ting Hu, Xiaodong Liang, Mohammad Zawad Ali,
    and Md. Nasmus Sakib Khan Shabbir

Short Presentations

Fast DENSER: Efficient Deep NeuroEvolution   197
    Filipe Assunção, Nuno Lourenço, Penousal Machado,
    and Bernardete Ribeiro

A Vectorial Approach to Genetic Programming   213
    Irene Azzali, Leonardo Vanneschi, Sara Silva, Illya Bakurov,
    and Mario Giacobini

Comparison of Genetic Programming Methods on Design of Cryptographic
Boolean Functions.   228
    Jakub Husa

Evolving AVX512 Parallel C Code Using GP   245
    William B. Langdon and Ronny Lorenz

Hyper-bent Boolean Functions and Evolutionary Algorithms   262
    Luca Mariot, Domagoj Jakobovic, Alberto Leporati, and Stjepan Picek

Learning Class Disjointness Axioms Using Grammatical Evolution   278
    Thu Huong Nguyen and Andrea G. B. Tettamanzi

Author Index   295

Long Presentations

Oral Presentations

Ariadne: Evolving Test Data Using Grammatical Evolution

Muhammad Sheraz Anjum(✉) and Conor Ryan(iD)

Department of Computer Science and Information Systems, University of Limerick,
Castletroy, Limerick, Ireland
{sheraz.anjum,conor.ryan}@ul.ie

Abstract. Software testing is a key component in software quality
assurance; it typically involves generating test data that exercises all
instructions and tested conditions in a program and, due to its complexity, can consume as much as 50% of overall software development
budget. Some evolutionary computing techniques have been successfully
applied to automate the process of test data generation but no existing
techniques exploit variable interdependencies in the process of test data
generation, even though several studies from the software testing literature suggest that the variables examined in the branching conditions of
real life programs are often interdependent on each other, for example,
if (x == y), etc.

We propose the *Ariadne* system which uses Grammatical Evolution
(GE) and a simple Attribute Grammar to exploit the variable interdependencies in the process of test data generation. Our results show that
Ariadne dramatically improves both effectiveness and efficiency when
compared with existing techniques based upon well-established criteria,
attaining *coverage* (the standard software testing success metric for these
sorts of problems) of 100% on **all** benchmarks with far fewer program
evaluations (often between a third and a tenth of other systems).

Keywords: Automatic test case generation · Code coverage ·
Evolutionary testing · Grammatical Evolution ·
Variable interdependencies

1 Introduction

The primary goal of software testing is to uncover as many faults as possible.
In practice, a labor intensive testing is carried out in order to achieve a certain
level of confidence in a software system [1]. Studies have shown that manual
testing may consume as much as 50% of overall software development budget [2].
Researchers have been trying to automate software testing since as early as 1962,
when [3] proposed a random test data generator for COBOL, while more recently,
a variety of metaheuristic search techniques have successfully been applied to
automatically generate test data as surveyed by [4,5].

© Springer Nature Switzerland AG 2019
L. Sekanina et al. (Eds.): EuroGP 2019, LNCS 11451, pp. 3–18, 2019.
https://doi.org/10.1007/978-3-030-16670-0_1

The use of metaheuristic techniques to automatically generate test data is often referred as *Search Based Software Testing* (SBST). In SBST, in general, some metaheuristic technique is applied to search through the space of all possible inputs to a particular program to find a specific test set (set of inputs) that can then be used to satisfy a particular test adequacy criterion such as branch coverage.

Genetic Algorithms (GAs) [6] are the most commonly adopted search techniques in SBST [5,7] having been used with some success [8,9], while Memetic Algorithms (MAs), which are typically hybrids of GAs and some sort of Local Search Algorithms (LSAs), have also been shown to be useful in this case [10].

The most widely studied test adequacy criterion in literature is branch coverage [5], which aims at maximizing the number of branches executed during a test. An extended and more challenging form of branch coverage is *Condition-decision* coverage (details presented in Sect. 2); this paper is concerned with tackling this problem.

Variables in the branching conditions of real life programs are often interdependent on each other; for example, a branching condition may have a check if two variables have equal values. Further, there are many tests that appear in many programs, such as checking if a variable has a value equal to zero, or less than zero. This has been observed by several researchers; for example, a study of 50 COBOL programs [11] revealed that 64% of the total predicates were equality predicates and that 87% of them examined 2 or fewer variables. In [12], 120 production PL/I programs were analyzed and it was found that 98% of all expressions contained fewer than two operators while 62% of all operators were relational/comparison operators. To the best of our knowledge, to date no SBST technique has exploited these properties in the process of test data generation.

In this paper, we introduce a GE [13,14] based test data generator. GE is a grammar based genetic algorithm which enables the use of a simple grammar to exploit simple relationships between input variables, including the sorts of properties mentioned above. As condition-decision coverage involve efficiently searching the space of paths in a program to make sure that the program has been thoroughly explored, we call our system *Ariadne*, after the mythological figure who helped Theseus find his way out of the Minotaur's Labyrinth.

We apply Ariadne to the problem of condition-decision coverage, although demonstrate that the technique can be deployed for other test adequacy criteria. Our results suggest that Ariadne significantly improves both effectiveness and efficiency when compared with well-known results from the literature [8–10], showing that across 11 popular benchmark programs, Ariadne achieved 100% coverage for all while reducing the search budget up to multiple times.

2   Background and Related Work

Software testing is a key component in software quality assurance and it can be broadly categorized as structural/white-box testing and specification-based/black-box testing. Structural testing inspects program *structure* while specification-based testing examines *functionality*. Structural testing is labor intensive and, as a result, costs more time and money compared to other software development activities [2]. Significant research has been conducted on ways to automate the testing process to help reduce costs.

A test adequacy criterion is a certain property that a program must satisfy to gain confidence about the absence of certain types of errors. For example, branch coverage is a structure-based adequacy criterion which requires that every branching condition must take each possible outcome at least once. If the following piece of code is under test:

If ((x < y) and (x < z)) { some program statement(s)}
If (y < z) { some program statement(s)}
else { some program statement(s)}

to achieve 100% branch coverage (often referred as complete coverage), this program must be executed with a test suite such that both if-conditions evaluate to both TRUE and FALSE outcomes at least once each. To manually achieve complete branch coverage in this case, a tester must generate a set of test inputs (test data) that execute every branch of the program under test.

Condition coverage is a more refined criterion than branch coverage. It requires that each single **condition** in the program must get both TRUE and FALSE values at least once. In case of a compound branching condition, such as in the first if-condition in the above example, all single conditions must individually get both TRUE and FALSE values. Another criterion, condition-decision coverage, is essentially a combination of branch coverage and condition coverage as it requires that all branching conditions as well as single conditions must get both TRUE and FALSE values at least once.

2.1 Related Work

The automation of test data generation has been the subject of increasing research interest [15] and it has been an area of investigation for many years [3,16,17]. One of the most straightforward methods of test data generation is to simply deploy a random search mechanism and use it to repeatedly generate input values until the required input is found. Sander's work [3] is one of the earliest random test data generators reported in the literature, as he was the first to consider the space of all possible inputs (of COBOL programs) as a search space. *Random test data generation* can be inefficient as the test generation is not guided in any way and, in general, becomes increasingly more inefficient as the search space increases.

Another paradigm for test data generation presented in the literature is *static test data generation*. This paradigm does not require executing the program under test as a mathematical system is deployed to find the required input values. One such technique is symbolic execution, in which program variables are assigned symbolic values and the resulting mathematical expression is solved to generate test data [16,18–20]. Major challenges associated with symbolic execution include handling complexity of constraints, procedure calls, loops and pointers. Techniques such as domain reduction [21] and dynamic domain reduction [22] have been proposed to address some of the issues associated with symbolic execution but dealing with loops and pointers still remains a problem.

Dynamic test data generation is based on the idea of actually executing the program under test to observe its behavior. Data recorded during this observation is then used to direct the search of required test data. This idea was first presented in [17] where numerical maximization techniques were employed to generate floating point test data. This idea was later extended by various researchers [23–26]. All these dynamic test data generation techniques were based on LSAs and consequently had a risk of getting stuck in local minima.

To overcome some inherent problems associated with local search, global search techniques including simulated annealing [27,28] and GAs [8,9,29–32] have been employed for test data generation. MAs have also been successfully deployed for test data generation [10,33,34]. GA based test data generation is detailed in the next section.

The conventional test data generation techniques target one coverage goal at a time. For example, in case of branch coverage, the coverage of one branch is targeted at one time. However, Whole Test Suite Generation [34,35] and Many-Objective Optimization [36] techniques target multiple coverage goals simultaneously.

GA Based Test Data Generation. GA-based test data generation, like other SBST techniques (often referred as evolutionary testing), considers the space of all possible inputs of the program under test a search space. The individuals in the GA population are generally vectors of input parameters which serve as the test data (test cases) for the program under test. The code is usually instrumented to monitor the execution of the program under test and the fitness value is assigned according to the execution of the program. The test adequacy criterion is implemented as fitness function which, in general, measures how far or close an individual is from covering the current target. Different techniques presented in the literature use different definitions of fitness functions but, in general, fitness functions can be broadly categorized into *branch distance based* and *control flow based* fitness functions.

The concept of branch distance was introduced by [23] and it simply describes how close an individual is from satisfying the target predicate. For example in Fig. 1, if the TRUE branch from node 2 is the target branch then the predicate to be satisfied is *if (i == j)*. The fitness function in this case of equality operator will be *absolute(i − j)* and the fitness value of all the individuals reaching the predicate can be measured accordingly. If the following individuals are evaluated for the above mentioned branch: <1, 2, 5>, <0, 4, 29>, <5, 250, 251>, where in each the three values are assigned to i, j and k respectively, then the respective fitness values would be 1, 4 and 245 where lower is better. That is, their fitnesses are computed using *absolute(i − j)* in each case, giving us $|1 - 2|$, $|0 - 4|$ and $|5 - 250|$, respectively, so the individual <1, 2, 5> will get the best fitness being

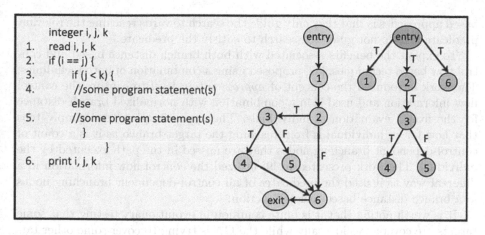

Fig. 1. An example program on the left, its Control Flow Graph in the center, and its Control Dependence Graph on the right.

closet from satisfying the target predicate. A list of fitness functions for different types of predicates is presented in [23].

GAs were first employed in SBST by [29] where a branch distance based fitness function was used. The apparent weakness of branch distance based fitness functions is that they don't guide the search towards reaching the concerned branching predicate and only effectively evaluate the individuals already reaching the concerned predicate. All the individuals missing the concerned predicate get a similar low fitness irrespective to the number of nodes they miss to reach it. This problem was partially handled by [8] where those branches were targeted first whose relevant predicates were already reached in some previous fitness evaluations. This problem can be addressed by bringing control flow information into consideration.

Control flow based fitness functions, on the other hand, solely rely on control flow information to measure the fitness. The list of nodes traversed during the execution of the program is compared with the list of critical/control-dependent branching nodes. Control-dependent nodes are the ones which must be traversed in order to reach the target. The individuals traversing more number of control dependent nodes get a better fitness value. For example if the TRUE branch from node 3, in the example presented as Fig. 1, is the target branch then the list of control dependent branches are {entryT, 2T, 3T}. If the following individuals are evaluated for this branch; <6, 6, 4>, <241, 241, 12>, <5, 0, 1>, then the first two individuals will get the equal higher fitness as both of them traverse 2 control dependent branches.

The work presented in [30] primarily used a branch distance based fitness functions but it also utilized control flow information as the fitness function considered the number of required loop iterations for loop testing. Later on [31], used the number of traversed control dependent predicates as a fitness measure for statement and branch coverage. The problem associated with control flow based approaches is that they only guide the search towards reaching the relevant predicate and do not guide the search to satisfy the predicate.

To exploit the benefits associated with both branch distance based and control flow based techniques, [32] proposed using a combination of both techniques. The work introduced the concept of *approximation level* to capture the control flow information and used it in a combination with normalized branch distance for the fitness evaluation of individuals. The approximation level simply tells that how far an individual is from reaching the target branch as it is a count of control-dependent branching nodes that are missed in the path executed by the individual. The work presented in [9] utilized the control flow information in a different way as it used the predicates of all control-dependent branching nodes in a branch distance based fitness function.

It is worth noting that it is quite common in evolutionary testing that some targets are covered accidentally while the GA is trying to cover some other targets. This phenomenon is called accidental (serendipitous) coverage and it significantly shortens the execution time as the GA does not need to be executed for the already covered targets. The effectiveness of any evolutionary testing technique is measured according to the selected adequacy criterion e.g. percentage of branches covered will be the effectiveness measure in case of branch coverage.

2.2   Grammatical Evolution

GE is a grammar-based GA that uses a grammar-based mapping process that separates search space from solution space. The evolutionary processes are applied to the genotype (search space) while the generated phenotype (solution domain) is generated by means of a problem specific grammar and is used for the purpose of fitness evaluation.

A grammar is composed of four elements i.e. terminals, non-terminals, productions rules and a start symbol. The terminals represent the constructs from the solution domain and the non-terminals are associated with a set production rule each. The production rules direct the mapping process for the expansion of non-terminals into one or more terminals and non-terminals. The start symbol is a selected non-terminal and the mapping process starts from applying an associated production rule for the start symbol. A sample grammar is shown in Fig. 2.

The genotype is a binary string where each 8 bit codon represents an integer value. These integer values are consumed one at a time in the mapping process for the selection of appropriate production rules. The production rules are selected by the following formula:

Rule = (Codon Integer Value) MOD (number of total production rules for the current non-terminal)

For example, if the non-terminal <op> is to be expanded by selecting a production rule from the set of these four rules:

<op> ::= +   (0)
        | -   (1)

$$| \ / \quad (2)$$
$$| \ * \quad (3)$$

And the next genotype integer to be consumed is 51, then 51 MOD 4 = 3. So the # 3 (<op> ::= *) is selected. The use of the MOD operator ensures that only relevant rules will be chosen. A complete example of genotype to phenotype mapping is presented in Fig. 2. While standard GE typically uses context free grammars (CFGs), it is a simple matter to change to Attribute Grammars (AGs). AGs are grammars in which some of the non-terminals have attributes that can be used to pass contextual information around a derivation tree.

3 GE Based Test Data Generation-Ariadne

In this section, we propose a GE based evolutionary test data generator named Ariadne that automatically detects and exploits variable dependencies using a simple attribute grammar designed based on the observations discussed in Sect. 1. We test Ariadne on a set of benchmark problems with condition-decision coverage as the test adequacy criterion, but the technique can also be used for other test adequacy criteria. Grammars allow GE to impose constraints on its individuals; typically this involves available instructions or preventing the use of certain sequences of instructions [37]. We use this power to impose dependencies between the variables being generated.

3.1 Overview

Ariadne, like other evolutionary testing techniques, considers the space of all possible inputs of the program under test as a search space. It deploys GE as a search algorithm to automate the process of test data generation, using the grammar presented below in Sect. 3.2, and produces a set of input variables for the program under test. We use the **same** grammar for all numeric benchmark problems.

To fulfil the adequacy criterion of condition-decision coverage, both TRUE and FALSE outcomes of all the branching nodes and condition predicates are considered as test objectives. Ariadne targets these objectives one by one and for every selected target the GE is initialized and an attempt to achieve the target, via the evolutionary process, is made. Ariadne also allows accidental/ serendipitous coverage as it records whenever a condition or decision outcome is executed for the first time, regardless of whether or not it is the current target. Once the objective at hand is achieved, the next objective is selected from the pool of currently unachieved test objectives and it continues until all the objectives are either achieved or had a failed try i.e. a run of GE. The objectives which stay unachieved can be considered either infeasible or simply unreachable by the applied technique. The success of any SBST technique can be measured in terms of percentage of test objectives achieved by that technique.

Recall from the Sect. 2.1 that in dynamic test data generation, the total number of Fitness Evaluations (FE) is the same as total number of target program executions (because the target program is executed once for every fitness evaluation). This count FE was used as the efficiency measures by [8–10] and we also used the same metric measure to compare our results with the ones presented in the literature.

3.2 Grammar

We use the following simple grammar to capture observed characteristics of conditions commonly found in code. The start symbol of the grammar has only one production rule and it produces a non-terminals for each input variable in the target program.

$$<start> :: = <var_1><var_2><var_3> \cdots <var_N> \qquad (1)$$

where N is the number of input variables. Each of these non-terminals is then expanded using a set of production rules of the form:

$$<var_M> :: = 0|1| - 1| <rand> | <dep_{var_1}> | <dep_{var_2}>$$
$$|\ldots| <dep_{var_{M-2}}> | <dep_{var_{M-1}}> \qquad (2)$$

The first three rules give grammar the ability to generate input values that can satisfy commonly used checks for zero, positive and negative integer values represented by 0, 1 and −1 respectively. The next rule, i.e. <rand>, produces a 32 bit signed random number which is generated from a seed value taken from the individual's genome as shown in genotype to phenotype mapping example in Fig. 2. This ensures that not only each time the individual is evaluated it will produce the same set of random numbers, but that an offspring that inherits this seed will also generate the same set. These seeds are subject to mutation, so the sets of random numbers available can change during evolution.

The remaining non-terminals essentially simple synthesized attributes in which the value of the variable is calculated based on the value of a previously generated variable. Each variable M can use the value of any previously declared variable and, when a production rule of this form is chosen, has its value calculated as follows:

$$<dep_{var_X}> := var_X|(var_X + 1)|(var_X - 1) \qquad (3)$$

where var_X is the value of a previously generated input variable. This set of production rules is responsible for exploiting variable dependencies as it generates values which are dependent on previously generated input variables. Recall from Sect. 1, that it is very common for the conditions to have comparison operators (dependencies) between two variables.

The dependencies are very simple; either the new variable has the same value or is ±1 the value. Similarly, because of the way in which the start symbol produces a single non-terminal for each input variable in a fixed order, the flow of dependencies is always from left to right and there can be no circular dependencies. These dependencies facilitate the generation of input data that is likely to satisfy conditions with comparison operators between variables. A complete example of a grammar-based genotype to phenotype mapping for a program having three input variables is presented in Fig. 2. Note that we use the **same** grammar for all experiments in this paper; the only difference is the number of input variables.

The grammar in Fig. 2 doesn't have explicit attributes, rather contextual information is passed from left to right using grammar rules that can interrogate the derived value of another variable as in rule 3.

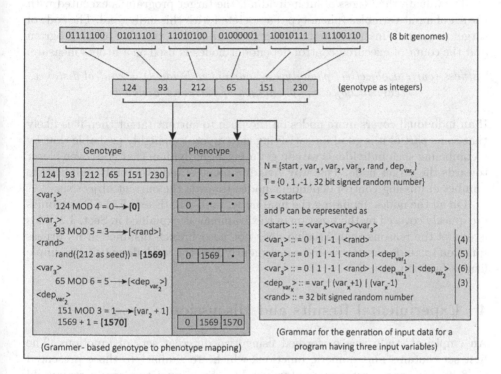

Fig. 2. An example with the genotype on the top, grammar on the right and the mapping sequence on the left. Note that the only required changes to the grammar for different problems are the number of variables and the number of dependency rules, but these are all of the same form.

3.3   Fitness Function

A fitness function, in general, measures how far or close an individual is from achieving the current target. The fitness function deployed in Ariadne is similar to [31] and approximation level of [32] as it directs the search towards the current objective based on control dependencies. The Control Dependence Graph (CDG) of the program is used to identify the sequences of control dependent nodes for all the test objectives. Figure 1 presents an example program and it's CDG.

Control dependent nodes are the nodes that must be executed to reach the current objective. For example, node 3 is control dependent for the execution of node 4, as node 4 cannot be executed without executing node 3. It is also worth noting that, for any given input the execution path of the program depends on the results of branching conditions on intermediate nodes. For example, in the program presented as Fig. 1, the branching condition on node 3 must be evaluated as TRUE in order to execute node 4 whereas it must be evaluated as FALSE for the execution of node 5.

To evaluate the fitness of an individual, the target program is executed with the set of input variables (phenotype) associated with this individual. The code of target program is instrumented in order to monitor the execution of the program and the count of executed control dependent nodes is used as a fitness measure:

$$fitness\ (current\_objective,\ phenotype\_of\_current\_individual) = control\ distance \\ (current\_objective,\ phenotype\_of\_current\_individual)$$

If an individual covers more nodes on the path to current target then it is likely that some part of it is good for satisfying the target. It is highly probable that by recombining such individuals we can get a better individual that goes even closer towards the target node. This way Ariadne reduces and eventually nullifies the number of missing control dependent nodes towards the current objective.

On all the nodes, including the final node, both TRUE and FALSE outcomes are quickly covered by the function of the grammar as explained in Sect. 3.2. This is one of the reasons that Ariadne does not need branch distance in its fitness function to guide the search and consequently gets the work done using a simple fitness function.

4   Experimental Results and Discussion

An empirical study was performed using two different sets of functions. The first set contains nine numeric functions which are similar[1] to those presented in [8,9]. The second set contains two validity-check functions from a real-world program **bibclean-2.08** [38].

The validity check functions are taken from a study presented in [10]. That study generalized and validated the Holland's schema theory [39] and the *Royal Road* theory [40] using a set of condition coverage problems. The generalized theory essentially predicts that GAs should perform well for problems that exhibit Royal Road property. For example, a branch is said to exhibit Royal Road property if the fitter individuals for the coverage of that branch contain some building blocks which independently can participate towards improving the fitness value. In this case it will be highly probable for crossover to generate fitter offspring

---

[1] Where possible we used the same functions, but where the actual source code isn't available we have tried to reproduce the code as closely as possible and in order to facilitate future comparisons we have made available the source code at http://bds.ul.ie/?page_id=390/.

as compared to their parents. The validity check functions were selected for the purpose of above mentioned study because some of their branches exhibit Royal Road property.

For the purpose of this study, we used Ariadne to generate test data for the selected functions and we compared our results with the results of GA based techniques presented in [8–10]. A brief description of the selected functions is presented below in next section.

4.1 Test Functions

The function sets we use vary in complexity. The functions in the Set 1 represent different types of test-data generation problems. These include Binary Search, Bubble Sort, *Days*, Greatest Common Divisor (GCD), Insertion Sort, Median, Quadratic Formula, Warshall's Algorithm and *Triangle Classification*.

Most of these are well known and self-explanatory, but the lesser known ones are as follows. *Days* calculates the total number of days between two dates, Warshall's Algorithm finds the shortest path in a weighted graph and Triangle Classification, which is one of the most widely studied functions in SBST [8, 9,26,31,41] etc., classifies a triangle based on the lengths of three sides of the triangle.

Function Set 2 contains check_ISBN and check_ISSN functions from an open source program bibclean-2.08 [38]. Both check_ISBN and check_ISSN take a string of 30 characters as input and perform a sequential search to find valid characters for an International Standard Book Number (ISBN) and an International Standard Serial Number (ISSN) respectively. ISBN and ISSN are used to uniquely identify publications while they comprise of 10 and 8 characters respectively.

4.2 Experimental Setup

A small set of initial experiments were conducted to identify reasonable run parameters. While we assumed a population of 50 would be reasonable for the functions in Set 1, we quickly discovered that several of them were solved with very small populations of just ten individuals. These include Binary Search, Bubble Sort, GCD, Insertion Sort, Median, Quadratic Formula and Warshall's Algorithm; complete coverage was achieved on all very quickly due to a combination of the constraints imposed by the grammar and the fact that multiple targets were covered in different iterations of the loops in those functions.

For Triangle Classification and Days, both maximum number of generations and population size were left at the standard GE setting of 50, while the more complex functions contained in Set 2 (check_ISBN and check_ISSN) were given 300 generations with a population size of 100. The probabilities of crossover and mutation were set as 0.9 and 0.05 and were kept the same during all the experiments while the methods of One Point Crossover & Flip Mutation were employed. It is worth noting that Ariadne produces similar results even when these genetic parameters are set high because the terminating criteria causes the search to stop once the target at hand is achieved, so even if we had left the population at 50 the runs would terminate early.

The automatically generated data via our system contain values including 0, 1, −1 and 32-bit signed integer values (random) in the range of −2,147,483,648 to 2,147,483,647. These generated values directly work as input values for most of the selected functions as they take integer inputs. For the functions with more complex inputs, specifically Days, check_ISBN and check_ISSN an extra mapping step is needed as the input type for Days is date and check_ISBN and check_ISSN take character strings as input. A mod based mapping function is used to convert six integer values (generated as a result of genotype to phenotype mapping) into two valid dates each consisting of a year, month and day. Similarly, for check_ISBN and check_ISSN, a mod based mapping function is used to convert thirty integer values into the ASCII codes which represent the characters of the input string.

4.3   Detailed Analysis of Experiments

In order to illustrate both the effectiveness and the efficiency of Ariadne, the results of our experiments are reported here. For all the nine functions from the first set, thirty runs were performed separately for each function and their mean performance is presented in Table 1. A comparison of our results with the best results of [8] is also presented in Table 1. Since the source code of only the Triangle Classification function was provided, we created our own versions of the other functions based on the literature. It is worth noting that [8] reported the **highest** performance among five runs in contrast to our **mean** performance over of thirty runs. It can be seen that at least one of their best performers was not able to achieve 100% coverage for each of Binary Search, Insertion Sort, Quadratic Formula and Triangle Classification, while Ariadne was able to achieve a 100% coverage in all thirty runs for all the functions.

The coverage for Binary Search, Bubble Sort, GCD, Insertion Sort, Median, Quadratic Formula and Warshall's Algorithm was immediately achieved in our experiments as either the structure of the function was simple or the condition predicates were quickly satisfied by grammar. The function Days, on average, took around 300 FE to achieve 100% coverage as the comparison conditions in its nested structure were quickly satisfied by the function of grammar. We were not able to compare these results as, among all nine functions, [8] reported the number of fitness evaluations for Triangle Classification only.

For Triangle Classification, [8] reported the best coverage of 94.29% with the search cost of about 8,000 FE as presented in Table 2. [9] reported a 100% coverage using the Program Dependence Graphs(PDG) based approach as discussed in Sect. 2.1. However, it can be clearly seen that [9] improved the coverage with a huge cost in terms of FE. On the other hand, Ariadne achieved 100% coverage on a multiple times smaller search cost. The reason Ariadne was so effective and efficient for Triangle Classification is that it was able to quickly generate input data of the form $i = j = k$ which is extremely difficult for a GA to generate otherwise.

Table 1. A comparison of Ariadne with GADGET [8] on nine benchmark functions

Program	GADGET [8]		Ariadne	
	GA	Differential GA	Ariadne	Avg. FE
Binary Search	70%	100%	100%	3
Bubble Sort	100%	100%	100%	1
Days	100%	100%	100%	300
GCD	100%	100%	100%	6
Insertion Sort	92.9%	100%	100%	1
Median	100%	100%	100%	4
Quadratic Formula	75%	75%	100%	15
Warshall's Algorithm	100%	100%	100%	1
Triangle Classification	94.29%	84.3%	100%	935

Table 2. Results on Triangle Classification for condition-decision coverage

Method	Coverage	Avg. FE
GADGET [8]	94%	8000
TDGen [9]	100%	97300
Ariadne	100%	935

For both of the functions from the Function Set 2, i.e. check_ISBN and check_ISSN, experiments are carried out on the same lines as that of [10] in order to have a fair comparison with their best results. [10] reported the GA-based technique's results for non-trivial branches only i.e. the branches which were not covered by random testing in their experiments. We computed our results based on sixty independent runs, for each of the non-trivial branches and compared them with the best results of [10] as Table 3.

The results are reported based on two metrics i.e. Success Rate (SR) and average number of FE where SR is defined as percentage of times that a particular branch was covered when targeted. Table 3 shows that Ariadne exhibited a 100% SR for all the branches in comparison to 95% and 98% SRs reported in [10] while the average number of FE was reduced up to an order of magnitude in many cases, but by two thirds at least.

The reason Ariadne performed even better for Royal Road functions as compared to standard GAs is that it encourages the generation of variables similar to the previously generated variables. So if the parents contain any valid characters for check_ISBN/check_ISSN, it will be highly probable that the offspring will not only retain these characters like standard GAs but also generate additional similar valid characters by the function of the grammar. The results presented in this section show that Ariadne outperforms other GA based SBST techniques and improves the effectiveness as well as efficiency by a wide margin.

Table 3. A comparison of Ariadne with [10] on non-trivial branches of check_ISBN & check_ISSN functions from **bibclean-2.08**.

Branch ID	Harman and McMinn [10]		Ariadne	
	SR	Avg. FE	SR	Avg. FE
B3-ISBN	95%	7986	100%	745
B4-ISBN	95%	7986	100%	747
B6-ISBN	95%	8001	100%	708
B7-ISBN	95%	9103	100%	3313
B3-ISSN	98%	5273	100%	655
B4-ISSN	98%	5273	100%	550
B6-ISSN	98%	5324	100%	542
B7-ISSN	98%	6380	100%	2662

5   Conclusion and Future Work

We have presented Ariadne, a GE-based tool to automate the process of test data generation. The work proposes the use of a simple grammar to exploit the variable interdependencies present in the branching conditions of real life programs. We have conducted our experiments using two sets of functions representing different types of test-data generation problems. Results of our experiments show that Ariadne clearly outperforms existing techniques both in terms of effectiveness and efficiency which are measured in terms of percentage (of condition-decision) coverage and number of fitness evaluations (function executions) respectively.

This paper serves an introduction to GE based test data generation and we believe that there is a lot of potential to further improve the technique. We are actively working towards improving Ariadne in a number of ways including the optimization of the grammar to make Ariadne even more efficient, as well as the extension of the grammar to accommodate more constructs such as constant values present in condition predicates.

Acknowledgments. The authors would like to thank Muhammad Hamad Khan for his help with the graphic designs. This work is supported by Lero, the Irish Software Research Centre, and the Science Foundation of Ireland.

References

1. Myers, G.J., Sandler, C., Badgett, T.: The Art of Software Testing. Wiley, Hoboken (2011)
2. Beizer, B.: Software Testing Techniques, 2nd edn. Van Nostrand Reinhold Inc., New York (1990). ISBN 0-442-20672-0
3. Sauder, R.L.: A general test data generator for COBOL. In: Proceedings of the 1–3 May 1962, Spring Joint Computer Conference, pp. 317–323. ACM (1962)

4. McMinn, P.: Search-based software test data generation: a survey. Softw. Test. Verif. Reliab. **14**(2), 105–156 (2004)
5. Ali, S., Briand, L.C., Hemmati, H., Panesar-Walawege, R.K.: A systematic review of the application and empirical investigation of search-based test case generation. IEEE Trans. Softw. Eng. **36**(6), 742–762 (2010)
6. Holland, J.H.: Genetic algorithms. Sci. Am. **267**(1), 66–73 (1992)
7. Aleti, A., Buhnova, B., Grunske, L., Koziolek, A., Meedeniya, I.: Software architecture optimization methods: a systematic literature review. IEEE Trans. Softw. Eng. **39**(5), 658–683 (2013)
8. Michael, C.C., McGraw, G., Schatz, M.A.: Generating software test data by evolution. IEEE Trans. Softw. Eng. **12**, 1085–1110 (2001)
9. Miller, J., Reformat, M., Zhang, H.: Automatic test data generation using genetic algorithm and program dependence graphs. Inf. Softw. Technol. **48**(7), 586–605 (2006)
10. Harman, M., McMinn, P.: A theoretical and empirical study of search-based testing: local, global, and hybrid search. IEEE Trans. Softw. Eng. **36**(2), 226–247 (2010)
11. Cohen, E.I.: A finite domain-testing strategy for computer program testing. Ph.D. thesis, The Ohio State University (1978)
12. Elshoff, J.L.: An analysis of some commercial PL/I programs. IEEE Trans. Softw. Eng. **2**, 113–120 (1976)
13. Ryan, C., Collins, J.J., Neill, M.O.: Grammatical evolution: evolving programs for an arbitrary language. In: Banzhaf, W., Poli, R., Schoenauer, M., Fogarty, T.C. (eds.) EuroGP 1998. LNCS, vol. 1391, pp. 83–96. Springer, Heidelberg (1998). https://doi.org/10.1007/BFb0055930
14. O'Neill, M., Ryan, C.: Grammatical evolution. IEEE Trans. Evol. Comput. **5**(4), 349–358 (2001)
15. Harman, M., Jia, Y., Zhang, Y.: Achievements, open problems and challenges for search based software testing. In: 2015 IEEE 8th International Conference on Software Testing, Verification and Validation (ICST), pp. 1–12. IEEE (2015)
16. Boyer, R.S., Elspas, B., Levitt, K.N.: Selecta formal system for testing and debugging programs by symbolic execution. ACM SIGPLAN Not. **10**(6), 234–245 (1975)
17. Miller, W., Spooner, D.L.: Automatic generation of floating-point test data. IEEE Trans. Softw. Eng. **3**, 223–226 (1976)
18. Clarke, L.A.: A system to generate test data and symbolically execute programs. IEEE Trans. Softw. Eng. **3**, 215–222 (1976)
19. Ramamoorthy, C.V., Ho, S.B., Chen, W.: On the automated generation of program test data. IEEE Trans. Softw. Eng. **4**, 293–300 (1976)
20. Offutt, A.J.: An integrated automatic test data generation system. In: Yeh, R.T. (ed.) Case Technology, pp. 129–147. Springer, Boston (1991). https://doi.org/10.1007/978-1-4615-3644-4_7
21. DeMilli, R., Offutt, A.J.: Constraint-based automatic test data generation. IEEE Trans. Softw. Eng. **17**(9), 900–910 (1991)
22. Offutt, A.J., Jin, Z., Pan, J.: The dynamic domain reduction procedure for test data generation. Softw. Pract. Exp. **29**(2), 167–193 (1999)
23. Korel, B.: Automated software test data generation. IEEE Trans. Softw. Eng. **16**(8), 870–879 (1990)
24. Korel, B.: Dynamic method for software test data generation. Softw. Test. Verif. Reliab. **2**(4), 203–213 (1992)
25. Korel, B.: Automated test data generation for programs with procedures. In: ACM SIGSOFT Software Engineering Notes, vol. 21, pp. 209–215. ACM (1996)

26. Ferguson, R., Korel, B.: The chaining approach for software test data generation. ACM Trans. Softw. Eng. Methodol. (TOSEM) 5(1), 63–86 (1996)
27. Tracey, N., Clark, J., Mander, K., McDermid, J.: An automated framework for structural test-data generation. In: ASE, p. 285. IEEE (1998)
28. Tracey, N., Clark, J.A., Mander, K.: The way forward for unifying dynamic test-case generation: the optimisation-based approach. In: Proceedings of the IFIP International Workshop on Dependable Computing and Its Applications (DCIA), York (1998)
29. Xanthakis, S., Ellis, C., Skourlas, C., Le Gall, A., Katsikas, S., Karapoulios, K.: Application of genetic algorithms to software testing. In: Proceedings of the 5th International Conference on Software Engineering and Applications, pp. 625–636 (1992)
30. Jones, B.F., Sthamer, H.H., Eyres, D.E.: Automatic structural testing using genetic algorithms. Softw. Eng. J. 11(5), 299–306 (1996)
31. Pargas, R.P., Harrold, M.J., Peck, R.R.: Test-data generation using genetic algorithms. Softw. Test. Verif. Reliab. 9(4), 263–282 (1999)
32. Wegener, J., Baresel, A., Sthamer, H.: Evolutionary test environment for automatic structural testing. Inf. Softw. Technol. 43(14), 841–854 (2001)
33. Arcuri, A., Yao, X.: Search based software testing of object-oriented containers. Inf. Sci. 178(15), 3075–3095 (2008)
34. Fraser, G., Arcuri, A., McMinn, P.: A memetic algorithm for whole test suite generation. J. Syst. Softw. 103, 311–327 (2015)
35. Fraser, G., Arcuri, A.: Whole test suite generation. IEEE Trans. Softw. Eng. 39(2), 276–291 (2013)
36. Panichella, A., Kifetew, F.M., Tonella, P.: Reformulating branch coverage as a many-objective optimization problem. In: 2015 IEEE 8th International Conference on Software Testing, Verification and Validation (ICST), pp. 1–10. IEEE (2015)
37. Karim, M.R., Ryan, C.: Sensitive ants are sensible ants. In: Proceedings of the 14th Annual Conference on Genetic and Evolutionary Computation, pp. 775–782. ACM (2012)
38. bibclean.c (1995). http://www.cs.bham.ac.uk/~wbl/biblio/tools/bibclean.c. Accessed 09 Nov 2018
39. Reeves, C.R., Rowe, J.E.: Genetic Algorithms Principles and Presentation: A Guide to GA Theory. Springer, New York (2002). https://doi.org/10.1007/b101880
40. Mitchell, M., Forrest, S., Holland, J.H.: The royal road for genetic algorithms: fitness landscapes and GA performance. In: Proceedings of the First European Conference on Artificial Life, pp. 245–254 (1992)
41. DeMillo, R.A., Offutt, A.J.: Experimental results from an automatic test case generator. ACM Trans. Softw. Eng. Methodol. (TOSEM) 2(2), 109–127 (1993)

Quantum Program Synthesis: Swarm Algorithms and Benchmarks

Timothy Atkinson[1]([⊠]), Athena Karsa[1], John Drake[2], and Jerry Swan[1]

[1] Department of Computer Science, University of York, York, UK
tja511@york.ac.uk
[2] School of Electronic Engineering and Computer Science,
Queen Mary University of London, London, UK

Abstract. In the two decades since Shor's celebrated quantum algorithm for integer factorisation, manual design has failed to produce the anticipated growth in the number of quantum algorithms. Hence, there is a great deal of interest in the automatic synthesis of quantum circuits and algorithms. Here we present a set of experiments which use Ant Programming to automatically synthesise quantum circuits. In the proposed approach, ants choosing paths in high-dimensional Cartesian space are analogous to transformation of qubits in Hilbert space. In addition to the proposed algorithm, we introduce new evaluation criteria for searching the space of quantum circuits, both for classical simulation and simulation on a quantum computer. We demonstrate that the proposed approach significantly outperforms random search on a suite of benchmark problems based on these new measures.

Keywords: Quantum algorithms · Ant Programming

1 Introduction

As traditional computational models approach their physical limits, so the importance of quantum computation is increasing. In contrast to classical models of computation, in which the two possible states of a bit are mutually exclusive, the qubits of quantum systems are able to represent both states simultaneously. Through the quantum phenomena of superposition and entanglement, quantum computers can potentially process information at an exponentially faster rate than classical computation.

However, devising algorithms that are able to leverage this power has proved elusive. Indeed, even designing the gates and circuits required to execute such algorithms is a nontrivial task. Despite notable successes, such as Shor's and Grover's algorithms, human ingenuity has so far failed to produce many quantum alternatives to classical algorithms. In the literature on quantum algorithms, heuristic methods such as Genetic Programming (GP) [14] have therefore been popular methods for the automatic design of quantum circuits and algorithms. GP is a heuristic search method, inspired by the natural process of evolution.

© Springer Nature Switzerland AG 2019
L. Sekanina et al. (Eds.): EuroGP 2019, LNCS 11451, pp. 19–34, 2019.
https://doi.org/10.1007/978-3-030-16670-0_2

Some early work, due to Williams and Gray, used linear GP to evolve quantum circuits [33]. They searched for more efficient implementations of known quantum circuits, successfully evolving a system for quantum teleportation. They found a quantum teleportation algorithm with a more efficient 'receive' sub-circuit than that previously presented by Brassard [2]. Around the same time, Spector et al. used the gamut of popular GP representations (tree, stack and linear) to evolve quantum algorithms automatically [26,27]. Their linear GP system was able to evolve quantum algorithms that outperformed any classical algorithm in terms of error probability, on the two-bit "AND/OR" Oracle problem in which only a single access to the Oracle is permitted. Further work by Spector and Klein [28] used the stack-based PushGP system to evolve circuits for an extension of the same problem that allows the Oracle to be accessed twice. Massey et al. [18,19] used GP with the domain-specific Q-PACE language to evolve generalizing n-qubit circuits for problems such as the quantum Fourier transformation. Leier et al. [15] used a linear-tree GP scheme to evolve quantum circuits capable of using measurement 'gates', evolving a scalable quantum algorithm for 1-SAT. In addition to GP, a variety of other evolutionary-based approaches have been proposed [5,16,34]. For a more detailed overview of the existing literature, and in particular the experimental settings and variety of representations used, we refer the interested reader to Stepney and Clark [30] and Gepp and Stocks [9].

As will be evident from the above, a significant share of the existing literature is made up of GP-based approaches. *Ant Programming* is an alternative metaheuristic based on Ant Colony Optimisation (ACO) [7], that has previously been used for automated programming in other domains [13,23]. In this paper we introduce Quantum Ant Programming (QAP), an ACO-based approach to the generation of quantum circuits.

A significant limitation of the existing literature for evolving quantum circuits is the methods used to compare and evaluate the performance of different methods. Many of the existing fitness functions wrongly classify correct answers in terms of physical distinguish-ability. This is because they use a raw Euclidean measure of distance, rather taking into account the actual similarity between two states. To overcome this we introduce 'Mean Square Fidelity' (MSF) as a fitness measure, which correctly ignores global phase differences between target and produced outputs. Additionally, as the literature lacks a consistent baseline, we propose a benchmark suite for evaluating heuristic search for quantum circuits, evaluating the proposed approach in comparison with random search.

The rest of this paper is as follows: in Sect. 2 we introduce basic concepts regarding the circuit model of quantum computation. In Sect. 3 we discuss related work on Ant Programming. In Sect. 4, we describe our proposed approach, Quantum Ant Programming, and its parameters. In Sects. 5 and 6 we describe our new fitness function MSF and proposed benchmark suite, respectively. We experimentally compare our proposed approach to random search in Sect. 7. Finally in Sect. 8 we offer conclusions and areas for future work.

2 Quantum Computation Using the Circuit Model

The pure states of a quantum system are described by the unit vectors, up to some overall immeasurable phase, of a Hilbert space. The most basic example is that of a *qubit*—a two-level quantum system whose state space is spanned by the basis vectors $|0\rangle = \begin{bmatrix} 1 & 0 \end{bmatrix}^T$ and $|1\rangle = \begin{bmatrix} 0 & 1 \end{bmatrix}^T$. While analogous to classical bits with mutually-exclusive states of 0 or 1, qubits residing in a space *spanned* by these states may also at any given time be in any superposition of them. That is, they may take the form

$$|q\rangle = \alpha|0\rangle + \beta|1\rangle, \tag{1}$$

where $\alpha, \beta \in \mathbb{C}$ such that $|\alpha|^2 + |\beta|^2 = 1$. The significance of this is such that measurement of the quantum state $|q\rangle$ in the computational basis $\{|0\rangle, |1\rangle\}$ will certainly yield as outcome one of either of the two possible physical states. The measurement process changes the state of the system, collapsing it from the superposition into the observed outcome state. This will be either $|0\rangle$ with probability $|\alpha|^2$ or the state $|1\rangle$ with probability $|\beta|^2$. Thus the unknown quantum state in fact forms a probability distribution over all available states. Similar procedures may be used to describe quantum systems of arbitrary dimension d, referred to as qudits.

Manipulation of qubits in particular ways gives rise to quantum algorithms and can be done either discontinuously, through previously described measurement procedures, or smoothly during unitary evolution. In the circuit model, *quantum logic gates* carry out unitary operations on input qubits over a given period of time, corresponding to a series of rotations and reflections of the Bloch sphere. Some of the most important single-qubit gates are:

$$X = \begin{pmatrix} 0 & 1 \\ 1 & 0 \end{pmatrix}, \quad Z = \begin{pmatrix} 1 & 0 \\ 0 & -1 \end{pmatrix}; \quad \text{and} \quad H = \frac{1}{\sqrt{2}} \begin{pmatrix} 1 & 1 \\ 1 & -1 \end{pmatrix}. \tag{2}$$

Respectively, these are the Pauli-X (NOT) gate, the Pauli-Z gate, which applies a phase to the $|1\rangle$ state while leaving the $|0\rangle$ state unchanged, and the Hadamard gate, which when applied to a state in the computational basis transforms it to the X-basis. The Pauli-Z gate can be generalised to an arbitrary phase gate:

$$\Phi = \begin{pmatrix} 1 & 0 \\ 0 & e^{-i\phi} \end{pmatrix}, \tag{3}$$

a form of which the T-gate, defined as $T = \Phi(\phi = -\pi/4)$, will be used later in this work. It is important to note here that, by the definition of unitarity all unitary matrices are reversible by the direct application of their Hermitian conjugates. That is, for any unitary matrix U, we have its Hermitian conjugate $U^\dagger$ and that $U^\dagger U = \mathbb{1}$, the identity matrix.

Composite quantum systems over n individual qubits reside in $\{0, 1\}^n$ tensor product Hilbert spaces. Considering a 2-qubit system with four possible outcomes when measured in the computational basis, we have that:

$$|q_1 q_1\rangle = \alpha|00\rangle + \beta|01\rangle + \gamma|10\rangle + \delta|11\rangle, \tag{4}$$

where $|\alpha|^2 + |\beta|^2 + |\gamma|^2 + |\delta|^2 = 1$ and, in a similar manner to before, the arbitrary 2-qubit state can be written in column vector form $|v_1 v_1\rangle = \begin{bmatrix} \alpha \ \beta \ \gamma \ \delta \end{bmatrix}^T$. Evolution of 2^n-dimensional composite systems is carried out via $2^n \times 2^n$ complex unitary matrices which act on the entire n-qubit register. These are built through the ordered tensor product of single-qubit operations acting on each of the individual component subspaces, whose action can be computed though direct matrix multiplication.

Separability of composite states can be directly affected by the action of quantum gates. A quantum state is separable if it can be decomposed into the tensor product of its composite qubit states, as in the 2-qubit state:

$$|q\rangle = \frac{1}{\sqrt{2}} \left(|00\rangle + |01\rangle \right) \equiv \frac{1}{\sqrt{2}} |0\rangle_1 \otimes \left(|0\rangle_2 + |1\rangle_2 \right). \tag{5}$$

This means that application of a single-qubit gate to one of the composite qubits does not affect the state of the other. On the other hand, the 2-qubit state:

$$|q\rangle = \frac{1}{\sqrt{2}} \left(|00\rangle + |11\rangle \right) \tag{6}$$

cannot be decomposed in such a manner. As consequence, the desire to perform a unitary operation on any one of the composite qubits will always affect the other since their Hilbert spaces are intrinsically combined. When a quantum state is not separable it is entangled.

3   Ant Programming

Stochastic search has long been used as a method of generating programs, whether represented as finite state automata [8], expression trees [14], or graphs [20]. While the prevalent stochastic approach has historically been Genetic Programming (GP) [14], a competitive approach of increasing popularity is Ant Programming (AP) [23], a variant of Dorigo's Ant Colony Optimisation (ACO) [7]. ACO is motivated by the foraging behavior of ants. A notable aspect of this approach is that the environment is modified to facilitate search (a process known as 'stigmergy'), in this case by the deposition of pheromone trails by the ants. Dorigo's initial application was to the well-known Traveling Salesperson Problem (i.e. obtaining the minimal weight Hamiltonian circuit of a graph [3]). In Dorigo's 'Ant System' approach, artificial ants construct solutions incrementally, in a probabilistic manner.

Ant Programming was first proposed by Roux and Fonlupt [23]. Their algorithm was applied to symbolic regression using an alternative to the traditional mutation operation of Genetic Programming: pheromone values are associated with terminals and functions at given positions, and the artificial ants perform mutations by walking the expression trees. Subsequent work by Green et al. [10] represents a further move towards ACO from traditional GP, in which all of the "genetic" operators are instead performed by artificial ants.

4   Quantum Ant Programming

We present an extension of the Ant Colony System (ACS) [6] designed to generate Quantum Circuits. In our approach, 'Quantum Ant Programming' (QAP), a team of ants choose paths through a graph, generating a sub-graph. This sub-graph then represents a Quantum Circuit.

For example, Fig. 1 shows a graph that QAP might be used to explore. A collection of ants start at the left-hand-side and moves rightwards. Each ant in the team corresponds to a qubit. The ants build a circuit by choosing edges that correspond to wires of a quantum circuit. Where an ant passes through a 1-qubit gate, its respective qubit is transformed by that gate. Where multiple ants pass through an n-qubit gate, their respective qubits are collectively transformed by that gate. If an insufficient number of ants pass through an n-qubit gate, then each ant is treated as passing through a wire and the gate is effectively ignored.

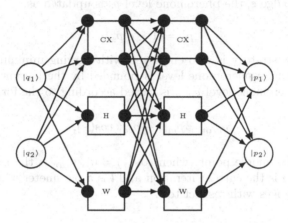

Fig. 1. An example graph for QAP. This example may be used to generate 2-qubit circuits of maximum depth 2. Each node in each column is fully connected with the next column.

A team of ants move rightwards in unison; the first ant chooses its next location, and then the second ant and so forth. Any two ants cannot choose to visit the same location; this constraint ensures that the constructed quantum circuit has a consistent number of qubits. To counteract the limitations this may have on generated routes, priorities are assigned to the ants at random before the circuit is generated, with ants choosing their next location in priority order. In contrast to the traditional behaviour of Ant Systems, ants do not share pheromone values: each ant effectively moves on its own graph. This means that if an ant visits a previously unexplored node, it will not be biased towards the movements taken by another ant on that node. The rationale for this departure from the traditional approach is that, since each ant (qubit) may perform a different function, such bias is highly likely to be unhelpful. An example of a generated circuit and its Hilbert space representation is given in Fig. 2. This Figure shows routes chosen by a pair of ants generating a 2-qubit circuit implementing an Bell-Pair generator, with the first qubit's path given in red and the second qubit's path given in blue. Unused edges are colored gray.

In each iteration, a different circuits are generated and evaluated. A pheromone update is then performed, before moving on to the next iteration. This update lays a pheromone trail on all of the edges taken in a set of routes. An edge e's pheromone p_e is updated as follows:

$$p_e = p_e + \alpha.f, \tag{7}$$

where α is the learning rate and f is the fitness associated with the set of routes. The routes that actually contribute to pheromone updates are controlled with elitism and pseudo-elitism described in Sect. 4.1 below. As part of the pheromone update, the pheromone levels on unused edges are reduced using the evaporation rate ϵ. For each edge e, the pheromone level p_e is updated as:

$$p_e = (1 - \epsilon)p_e. \tag{8}$$

A MAX-MIN strategy is also employed with varying minimum pheromone levels. The maximum pheromone level is bounded by the parameter p_{max}, and the minimum pheromone level p_{min} is varied according to the formula:

$$p_{min} = p_{mid} + p_{var}.cos(\frac{i}{l}), \tag{9}$$

where p_{mid} is the middle point (when $cos(\frac{i}{l}) = 0$), p_{var} is the maximum $+/-$ change in p_{min}, i is the current iteration and l is a parameter which varies how quickly p_{min} changes with respect to i.

4.1   Elitism and Pseudo-elitism

QAP employs both elitism and what we refer to as *pseudo-elitism*. The elitist approach lays pheromones on the routes of the best solution found so far with learning rate α. Variation is also encouraged by laying pheromones for sets of routes which are 'close' in fitness to the elite route. The pseudo-elitism parameter β describes how close a solution's fitness must be to the best solution found so far for its component routes to be updated with pheromone. Specifically, if f_{el} is the fitness of the best solution found so far, and f is the fitness of a candidate solution, the candidate solution only contributes to pheromone levels if

$$f \geq f_{el} - \beta. \tag{10}$$

4.2   Controllable Gates

We create a special case for controllable functions. If an n-qubit function f is listed as controllable in the function set F, then it is present in the column as a

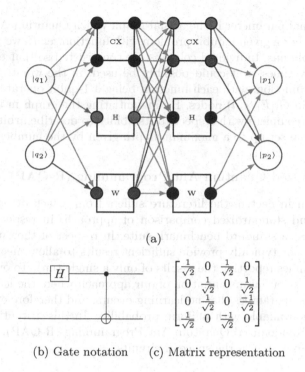

(a)

$$H \text{---}\bullet$$

$$\text{---}\oplus$$

$$\begin{bmatrix} \frac{1}{\sqrt{2}} & 0 & \frac{1}{\sqrt{2}} & 0 \\ 0 & \frac{1}{\sqrt{2}} & 0 & \frac{1}{\sqrt{2}} \\ 0 & \frac{1}{\sqrt{2}} & 0 & \frac{-1}{\sqrt{2}} \\ \frac{1}{\sqrt{2}} & 0 & \frac{-1}{\sqrt{2}} & 0 \end{bmatrix}$$

(b) Gate notation   (c) Matrix representation

Fig. 2. (a) Paths chosen by a team of ants through the graph given in Fig. 1, (b) the corresponding quantum circuit described and (c) that solution's matrix representation. (Color figure online)

$n + 1$ qubit gate with one of its inputs specifically corresponding to the control qubit. When f is used and the control qubit is used, it is treated as a controlled version of f. However, if only f is used then f is used as normal, rather than treated as an identity (e.g. if the controlled form of f were included as its own function).

4.3 Parameters

Quantum Ant Programming has the following parameters:

- a. The number of teams of ants (circuits) to use in each iteration.
- p_{mid}. The middle point of the varying p_{min}.
- p_{var}. The change in p_{min} over time.
- l. How quickly p_{min} changes over time.
- p_{max}. The maximum pheromone level.
- α. The learning rate.
- ϵ. The evaporation rate.
- β. The degree to which a circuit's fitness can deviate from the elite solution's fitness to qualify for pseudo-elitism.

An additional parameter is the actual graph which Quantum Ant Programming explores. For a given n qubit problem with function set F, we generally use $10 \times (n-1)$ columns. Individual columns are constructed so that each function occurs as many times as it could possibly be used; in the worst case this is n times for each function, with each function being a 1-qubit operator. Hence our columns contain $O(|F| \times n)$ nodes. By instantiating the graph in this way, our search space is complete with respect to its ability to describe arbitrary circuits over the function set up to a maximum depth given by the number of columns.

4.4 Randomized Quantum Ant Programming (R-QAP)

As will be seen in Sect. 6, the literature suffers from a lack of both standard benchmarks and standardized comparison of approach. In respect of the former, we propose a standard benchmark suite. In respect of the latter, existing approaches do not typically provide sufficient results to allow meaningful comparison (sometimes reporting the results of only a single run). To overcome this, we compare to a special instantiation of our approach where the learning rate α is set to 0. This guarantees that no learning occurs, and therefore every possible set of routes is available with uniform probability. In this instantiation, which we refer to as Randomized Quantum Ant Programming (R-QAP), we therefore perform random search in the space of circuits.

5 Fitness of a Quantum Circuit: Mean Square Fidelity

To guide any metaheuristic search in the space of quantum circuits, a measure of correctness is required. For unitary maps, the measure of *fidelity* as a real numerical value for the closeness of two quantum states can be used. For two density matrices ρ and σ, the fidelity is defined as:

$$\mathcal{F}(\rho, \sigma) \equiv tr\sqrt{\rho^{\frac{1}{2}} \sigma \rho^{\frac{1}{2}}}. \tag{11}$$

In our case, $\rho = |\psi\rangle\langle\psi|$ and $\sigma = |\phi\rangle\langle\phi|$ are the density matrices of vector states $|\psi\rangle$ and $|\phi\rangle$ respectively, where $|\psi\rangle$ and $|\phi\rangle$ commute and are diagonal in the same basis. The fidelity becomes:

$$\mathcal{F}(\rho, \sigma) = \sqrt{\langle\psi|\phi\rangle\langle\phi|\psi\rangle} = \sqrt{|\langle\psi|\phi\rangle|^2} = |\langle\psi|\phi\rangle| = ||\psi\rangle^{\dagger}|\phi\rangle|. \tag{12}$$

When $\mathcal{F}(\rho, \sigma) = 1$ it is impossible to distinguish $|\phi\rangle$ and $|\psi\rangle$ in any basis. In the context of fitness for metaheuristics, we design an example-based fitness function 'Mean Square Fidelity' (MS$\mathcal{F}$). Circuits are trained against a dataset (X, Y) consisting of a set of n inputs $X = \{|x_1\rangle, |x_2\rangle \ldots |x_n\rangle\}$ and n corresponding target outputs $Y = \{|y_1\rangle, |y_2\rangle \ldots |y_n\rangle\}$, where the objective given input $|x_i\rangle$ is to produce target output $|y_i\rangle$. Let A be the matrix representation of some circuit to be evaluated against a dataset (X, Y). Then the Mean Square Fidelity is defined as:

$$MS\mathcal{F}(A, X, Y) = \frac{\sum_{i=1}^{|X|} \mathcal{F}(|y_i\rangle, A|x_i\rangle)^2}{|X|}, \tag{13}$$

yielding the mean squared fidelity between the target outputs and the outputs produced by the circuit. An appealing property of fidelity as a measure of correctness is that it effectively ignores phase; global phase can be trivially corrected but ignoring this fact yields a misleading measure of fitness. As an example we consider the fitness measure used in [18] which, in our notation, takes the form:

$$f(A, X, Y) = \sum_{i=1}^{|X|} \| |y_i\rangle - A|x_i\rangle \|. \tag{14}$$

Consider the following input and target output given by: $|x_1\rangle = \begin{bmatrix} 1 & 0 \end{bmatrix}^T$ and $|y_1\rangle = \begin{bmatrix} 0 & 1 \end{bmatrix}^T$ respectively. Then the candidate circuit A which may produce as a solution $A|x_0\rangle = \begin{bmatrix} 0 & -1 \end{bmatrix}^T$. This is indistinguishable from $|y_0\rangle$ to an outside observer as it differs only by a global phase as:

$$A|x_0\rangle = -1|y_0\rangle, \tag{15}$$

That is, experimentally speaking they are identical up to an overall, unobservable, phase and, as a result A should have a perfect score for fitness. Despite this, the fitness function from Eq. 14 gives the worst possible fitness for this single example as:

$$f(A, \{|x_1\rangle\}, \{|y_1\rangle\}) = \| \begin{bmatrix} 0 & 2 \end{bmatrix}^T \| = 2. \tag{16}$$

In contrast, the MS$\mathcal{F}$ indicates that A is a perfect solution as:

$$MS\mathcal{F}(A, \{|x_1\rangle\}, \{|y_1\rangle\}) = \frac{\mathcal{F}(|y_1\rangle, A|x_1\rangle)^2}{1} = 1 \tag{17}$$

As noted by Stepney and Clark [30], when quantum computers and particularly universal quantum computers [4] become readily available, it may then be possible to evaluate candidate solutions while avoiding the exponential cost of simulation. One approach could then be to execute solutions repeatedly to approximate the probability distribution over their outcomes. These approximated probability distributions readily lend themselves to the MS$\mathcal{F}$ measure, and clearly cannot contain knowledge of global phase. We therefore argue that MS$\mathcal{F}$ provides a suitable basis for subsequent work in this area.

6   Benchmark Problems

It is common in the literature on quantum program synthesis to tackle problems by simply running the proposed approach and reporting the results. In some works only a single run was performed [18,19,26,28], whereas others have reported on the general or best behavior of their approach over multiple runs [15,29,33,34]. Additionally, the set of logic gates varies significantly throughout the literature; e.g. in [29] solutions may contain rotation gates of 128 different angles, whereas in [34] solutions may only be constructed from 3 fixed gates (CNOT and two variants of a Hadamard).

With metaheuristics evaluated on diverse problems under different experimental conditions, without statistical comparison, it is not possible to ascertain whether one approach outperforms another. In fact, it is not clear from the existing literature whether *any* existing approach outperforms random search. We argue that it is difficult for progress to be made in the absence of meaningful comparison. We therefore propose a suite of benchmark problems, taken from various sources in both the metaheuristic and quantum computation literature. We hope that future work can use this benchmark suite as a basis for future principled comparison. The problems we identify as motivating benchmarks are as follows:

Bell-Pair Generator (BP). Bell-Pair generation is included as a relatively easy problem for baseline validation of a proposed approach. Used to prepare the entanglement resource for the quantum teleportation protocol [1], Bell-Pair generation involves the transformation of some pure state into a maximally entangled Bell-Pair, for example, $|00\rangle \rightarrow \frac{|00\rangle+|11\rangle}{\sqrt{2}}$. A Bell-Pair generator can be constructed from a CNOT and a Hadamard gate shown in Fig. 3(a). Figure 3(b) shows its expected behaviour in generating each of the four maximally entangled Bell-Pairs which will be used as the dataset for learning them.

(a) Bell-pair Generator (b) Input-Output Examples

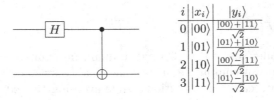

i	$|x_i\rangle$	$|y_i\rangle$
0	$|00\rangle$	$\frac{|00\rangle+|11\rangle}{\sqrt{2}}$
1	$|01\rangle$	$\frac{|01\rangle+|10\rangle}{\sqrt{2}}$
2	$|10\rangle$	$\frac{|00\rangle-|11\rangle}{\sqrt{2}}$
3	$|11\rangle$	$\frac{|01\rangle-|10\rangle}{\sqrt{2}}$

Fig. 3. (a) A Bell-Pair generator and (b) its expected behavior.

Quantum Fourier Transformation. The Quantum Fourier Transformation (QFT) [12] is the quantum analog of the classical Fourier transform. The QFT is described for some quantum state $|x\rangle$, decomposed as a linear superposition over its N basis states, $|x\rangle = \sum_{j=0}^{N-1} \alpha_j |j\rangle$, according to the transformation

$$|y\rangle = \sum_{k=0}^{N-1} \left(\sum_{j=0}^{N-1} \alpha_j.(e^{\frac{2\pi i}{N}})^{j.k} \right) |k\rangle. \tag{18}$$

The QFT is included as it is utilized in many prominent quantum algorithms, including Shor's algorithm for Prime Factorization [25]. The QFT is well-defined for any number of qubits, but for the purposes of the benchmark set, we include the 2-qubit (QFT-2), 3-qubit (QFT-3) and 4-qubit (QFT-4).

We highlight a separate problem, studied by Massey et al. [19], of generating an n-qubit QFT (QFT-n) by learning some expandable circuit over n qubits. The accessibility of this problem depends on a system being able to learn expandable circuits, which is in general not the case either here or in much of the literature. For this reason, we do not include QFT-n in our own set of benchmark problems but encourage the comparison of systems which are capable of learning expandable circuits on this problem.

Grover's Diffusion Operator (GDO). Grover's diffusion operator is the amplitude amplification component of Grover's algorithm [11], which offers a quadratic speedup w.r.t. classical approaches for searching an unordered database. If $|s\rangle$ denotes the uniform superposition over N basis states:

$$|s\rangle = \frac{1}{N} \sum_{i=0}^{N-1} |i\rangle, \tag{19}$$

then the diffusion operator is defined as $U_s = 2\,|s\rangle\langle s| - I_n$, where I_n is an $N \times N$ identity matrix.

We propose the 2-qubit (GDO-2), 3-qubit (GDO-3) and 4-qubit (GDO-4) Grover's diffusion operators as benchmark problems, and again note the possibility of searching for an n-qubit Grover's operator (GDO-n) when using an approach capable of learning expandable circuits.

Toffoli Gate (TOF). The Toffoli Gate originates as a reversible logic gate for classical (rather than quantum) computing [31]. However its quantum analog has a number of quantum applications, e.g. in quantum error correction [22]. The Toffoli Gate effectively implements a controlled version of a CNOT gate, and is therefore sometimes referred to as a CCNOT gate. Although the Toffoli gate is simple to describe, its actual implementation using CNOT gates and single-qubit gates is surprisingly non-trivial; at least 6 correctly arranged CNOT gates are required to implement a Toffoli gate [24].

Teleportation Circuit (TP). A Teleportation Circuit is the circuit model interpretation of the Teleportation Protocol [1]. We use a simplified isolated-qubit version of the Teleportation Circuit, in which it is assumed that the qubit to be teleported is not entangled with any other qubits. A simple model of the Teleportation Circuit is given in Fig. 4.

The purpose of the Teleportation protocol is to send an arbitrary qubit from the sender (Alice) to the receiver (Bob) using only local operations and classical communication, relying on the presence of a shared entangled resource.

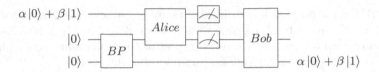

Fig. 4. An abstract Teleportation Circuit.

The 'standard' Teleportation Circuit generates a Bell-Pair (the BP component) which is shared between Alice and Bob. Alice then performs a quantum *joint* measurement on both the qubit she wishes to send and her portion of the entangled Bell-Pair. Her measurement result is then sent to Bob via classical means, who then uses them to apply a correction (a single-qubit gate) to his own portion of the Bell-Pair to recover the sent qubit.

For a given input qubit $|\alpha\rangle$, there are four valid outcomes:

$$Y(|\alpha\rangle) = \{|00\rangle\,|\alpha\rangle\,,|01\rangle\,|\alpha\rangle\,,|10\rangle\,|\alpha\rangle\,,|11\rangle\,|\alpha\rangle\}. \qquad (20)$$

We devise a modified form of the $MF\mathcal{F}$ measure for Teleportation Circuits which do not bias towards certain orderings of these outcomes. For a given input $|\alpha\rangle\,|00\rangle$ to a candidate circuit four outcomes may be produced (the results of different measurements after the 'Alice' component of the circuit has been performed). Referring to these four outcomes as $A(|\alpha\rangle)$ for a candidate circuit A, we construct a fitness function for Teleportation Circuits with X as the set of example inputs:

$$MS\mathcal{F}_{TP}(A, X) = \sum_{|\alpha\rangle\in X} \frac{\displaystyle\sum_{|o\rangle\in A|\alpha\rangle} max_{|y_i\rangle\in Y(|\alpha\rangle)}(\mathcal{F}(|y_i\rangle\,,|o\rangle)).}{4}. \qquad (21)$$

7   Experiments

In this section, we describe the experimental evaluation of our proposed approach on benchmark suite which is listed in Table 1. For all experiments, we use a function set consisting of 1-qubit Hadamard, Pauli-X, T and $T^\dagger$ gates and the 2-qubit Swap gate. Additionally, we allow every 1-qubit gate to be controllable (see Sect. 4.2). Hence the overall function set we propose for use with these benchmarks consists of 9 gates. Additionally, we allow search procedures to use identity operators by including wires in the function set. It is well-known that the subset of T, H and CNOT gates are capable of approximating any arbitrary circuit [21].

For each benchmark problem, we run each search algorithm 100 times, allowing them to execute for up to $1,000,000$ circuit evaluations or until a correct solution is found. A 'correct' solution is one which scores a fitness of ≥ 0.98. For each independent run we record the best discovered solution.

Table 1. Benchmark problems for metaheuristics synthesizing quantum circuits.

Problem	Description	Fitness function
BP	Bell-Pair Generator	$MS\mathcal{F}$
QFT-2	2-Qubit Quantum Fourier Transformation	$MS\mathcal{F}$
QFT-3	3-Qubit Quantum Fourier Transformation	$MS\mathcal{F}$
QFT-4	4-Qubit Quantum Fourier Transformation	$MS\mathcal{F}$
GDO-2	2-Qubit Grover's Diffusion Operator	$MS\mathcal{F}$
GDO-3	3-Qubit Grover's Diffusion Operator	$MS\mathcal{F}$
GDO-4	4-Qubit Grover's Diffusion Operator	$MS\mathcal{F}$
TOF	Toffoli Gate	$MS\mathcal{F}$
TP	Teleportation Circuit	$MS\mathcal{F}_{TP}$

The parameters used for QAP are: $a = 50$, $p_{mid} = \frac{0.3}{q^3}$, $p_{var} = \frac{0.3}{q^3}$, $l = 500$, $p_{max} = 10.0$, $\alpha = 1.0$, $\epsilon = 0.1$, $\beta = 0.05$, where q is the number of qubits associated with a given problem. We use the same parameters for R-QAP, except that $\alpha = 0$. For the TP problem, we treat the BP, Alice and Bob components (Fig. 4) as separate graphs with their own pheromone levels. All 3 graphs are sampled simultaneously to generate a candidate circuit, and all 3 generated sub-circuits share fitness scores. As BP and Alice are 2-qubit circuits, whereas Bob is a 3-qubit circuit, their corresponding graphs have different numbers of rows, columns and differing p_{min} levels. Additionally, as the TP problem is much 'deeper' than other 3-qubit problems with respect to the total number of columns (40 vs. 20), we change both p_{mid} and p_{var} to $\frac{0.2}{q^3}$ to reduce the overall level of exploration correspondingly.

7.1 Results

Results are given in Table 2. We list the success rate (SR), median fitness (MF), best fitness (BF) and inter-quartile range in fitness (IQR). We test for statistical differences between results using the two-tailed Mann-Whitney U test [17] and where significant differences are found, we use a Vargha-Delaney A test [32] to determine effect size.

Quantum Ant Programming was able to find correct solutions to all problems except the 3-qubit and 4-qubit Grover's Diffusion Operator problems. In contrast, R-QAP was only able to find correct solutions for the simpler 2-qubit problems. On every problem except BP and GDO-2 (where both algorithms almost always found correct solutions), the Mann-Whitney U test reveals a statistically significant ($p < 0.05$) improvement when using a non-zero learning rate. Additionally, in each of these cases, the Vargha-Delaney A test confirms that the effect size is large ($A > 0.71$).

Table 2. Results from benchmark problems for QAP and R-QAP. SR is the success rate of each algorithm. MF is median globally best fitness observed across each run of each algorithm, with *(BF)* indicating the best fitness observed in any run. IQR is the inter-quartile range in the globally best fitness across each run of each algorithm. The p value is from the two-tailed Mann-Whitney U test comparing the globally best fitness observed across each run of the algorithm. Where $p < 0.05$, the effect size from the Vargha-Delaney A test is shown; large effect sizes ($A > 0.71$) are shown in **bold**.

Problem	QAP			R-QAP			p	A
	SR	MF *(BF)*	IQR	SR	MF *(BF)*	IQR		
BP	100%	1.0 *(1.0)*	0.0	100%	1.0 *(1.0)*	0.0	1.0	–
QFT-2	100%	1.0 *(1.0)*	0.0	55%	1.0 *(1.0)*	0.07	10^{-14}	**0.73**
QFT-3	68%	1.0 *(1.0)*	0.06	0%	0.69 *(0.81)*	0.05	10^{-35}	**1.0**
QFT-4	2%	0.84 *(0.98)*	0.07	0%	0.37 *(0.48)*	0.05	10^{-34}	**1.0**
GDO-2	100%	1.0 *(1.0)*	0.0	99%	1.0 *(1.0)*	0.0	0.32	–
GDO-3	0%	0.67 *(0.85)*	0.06	0%	0.67 *(0.68)*	0.05	10^{-32}	**0.96**
GDO-4	0%	0.90 *(0.90)*	0.0	0%	0.57 *(0.77)*	0.11	10^{-38}	**1.0**
TOF	9%	0.83 *(1.0)*	0.06	0%	0.63 *(0.69)*	0.04	10^{-34}	**1.0**
TP	57%	1.0 *(1.0)*	0.18	0%	0.63 *0.81*	0.04	10^{-35}	**1.0**

8   Conclusion

We have introduced the application of Ant Programming to the synthesis of quantum circuits, also proposing an evaluation criterion with increased discriminatory power and an associated benchmark suite.

Our results offer a number of interesting conclusions: firstly, the use of ACO is superior or equal to to random search on our benchmark suite in every circumstance. Secondly, the fact that ACO was able to achieve this level of performance strongly indicates that the proposed MSF measure is a helpful heuristic for searching the space of quantum circuits, since this was the only guiding measure during the ACO search. Finally, our benchmark suite is non-trivial: in over $200,000,000$ total evaluations using QAP or R-QAP we were not able to find a correct solution to the GDO-3 or GDO-4 problems. This indicates that these circuits are not trivially constructible from our function set, implying that these benchmarks can indeed be of value to subsequent researchers.

There are a number of potential areas for future work. An investigation into the local optima visited by QAP on the unsolved GDO-3 and GDO-4 problems may reveal limitations in the current approach. It may be worth investigating sub-systems for the generation of useful single-qubit gates as linear combinations of the function set to help approximate larger circuits. In addition the extension of QAP to n-qubit problems [19] would allow the learning of quantum algorithms that generalize beyond concrete circuit instances. Further experiments varying each of QAP's parameters may clarify their roles with respect to performance. Alongside this, it may be useful to compare the performance of various approaches (e.g. [16,18,28]) from the literature, using the Mean Square Fidelity fitness function, on our newly proposed benchmark suite.

Acknowledgements. T. Atkinson and J. Swan acknowledge the support of EPSRC grant EP/J017515/1.

References

1. Bennett, C.H., Brassard, G., Crépeau, C., Jozsa, R., Peres, A., Wootters, W.K.: Teleporting an unknown quantum state via dual classical and Einstein-Podolsky-Rosen channels. Phys. Rev. Lett. **70**(13), 1895–1899 (1993)
2. Brassard, G., Braunstein, S.L., Cleve, R.: Teleportation as a quantum computation. In: Proceedings of the Fourth Workshop on Physics and Computation, PhysComp 1996, pp. 43–47. Elsevier, Amsterdam (1998)
3. Cormen, T.H., Leiserson, C.E., Rivest, R.L., Stein, C.: Introduction to Algorithms, 3rd edn. MIT Press, Cambridge (2009)
4. Deutsch, D.: Quantum theory, the Church-Turing principle and the universal quantum computer. Proc. Roy. Soc. Lond. A **400**(1818), 97–117 (1985)
5. Ding, S., Jin, Z., Yang, Q.: Evolving quantum circuits at the gate level with a hybrid quantum-inspired evolutionary algorithm. Soft Comput. **12**(11), 1059–1072 (2008)
6. Dorigo, M., Gambardella, L.M.: Ant colony system: a cooperative learning approach to the traveling salesman problem. IEEE Trans. Evol. Comput. **1**(1), 53–66 (1997)
7. Dorigo, M., Maniezzo, V., Colorni, A.: Ant system: optimization by a colony of cooperating agents. IEEE Trans. Syst. Man Cybern. Part B (Cybern.) **26**(1), 29–41 (1996)
8. Fogel, L.J.: Autonomous automata. Ind. Res. **4**, 14–19 (1962)
9. Gepp, A., Stocks, P.: A review of procedures to evolve quantum algorithms. Genet. Program. Evolvable Mach. **10**(2), 181–228 (2009)
10. Green, J., Whalley, J., Johnson, C.: Automatic programming with ant colony optimization. In: Proceedings of the 2004 UK Workshop on Computational Intelligence, pp. 70–77 (2004)
11. Grover, L.K.: Quantum mechanics helps in searching for a needle in a haystack. Phys. Rev. Lett. **79**(2), 325–328 (1997)
12. Hales, L., Hallgren, S.: An improved quantum Fourier transform algorithm and applications. In: Proceedings of 41st Annual Symposium on Foundations of Computer Science, pp. 515–525. IEEE (2000)
13. Keber, C., Schuster, M.G.: Option valuation with generalized ant programming. In: Proceedings of the 4th Annual Conference on Genetic and Evolutionary Computation, pp. 74–81. Morgan Kaufmann Publishers Inc. (2002)
14. Koza, J.R.: Genetic Programming: On the Programming of Computers by Means of Natural Selection. MIT Press, Cambridge (1992)
15. Leier, A., Banzhaf, W.: Evolving Hogg's quantum algorithm using linear-tree GP. In: Cantú-Paz, E., et al. (eds.) GECCO 2003. LNCS, vol. 2723, pp. 390–400. Springer, Heidelberg (2003). https://doi.org/10.1007/3-540-45105-6_48
16. Lukac, M., et al.: Evolutionary approach to quantum and reversible circuits synthesis. Artif. Intell. Rev. **20**(3–4), 361–417 (2003)

17. Mann, H.B., Whitney, D.R.: On a test of whether one of two random variables is stochastically larger than the other. Ann. Math. Stat. **18**(1), 50–60 (1947)

18. Massey, P., Clark, J.A., Stepney, S.: Evolving quantum circuits and programs through Genetic Programming. In: Deb, K. (ed.) GECCO 2004. LNCS, vol. 3103, pp. 569–580. Springer, Heidelberg (2004). https://doi.org/10.1007/978-3-540-24855-2_66

19. Massey, P., Clark, J.A., Stepney, S.: Human-competitive evolution of quantum computing artefacts by Genetic Programming. Evol. Comput. **14**(1), 21–40 (2006)

20. Miller, J.F.: An empirical study of the efficiency of learning Boolean functions using a Cartesian Genetic Programming approach. In: Proceedings of the 1st Annual Conference on Genetic and Evolutionary Computation - Volume 2, GECCO 1999, pp. 1135–1142. Morgan Kaufmann Publishers Inc., San Francisco (1999)

21. Nielsen, M.A., Chuang, I.: Quantum Computation and Quantum Information. Cambridge University Press, Cambridge (2002)

22. Reed, M., et al.: Realization of three-qubit quantum error correction with superconducting circuits. Nature **482**(7385), 382–385 (2012)

23. Roux, O., Fonlupt, C.: Ant programming: or how to use ants for automatic programming. In: Dorigo, M. (ed.) ANTS 2000 From Ant Colonies to Artificial Ants: 2nd International Workshop on Ant Algorithms (2000)

24. Shende, V.V., Markov, I.L.: On the CNOT-cost of TOFFOLI gates. Quantum Inf. Comput. **9**(5), 461–486 (2009)

25. Shor, P.W.: Polynomial-time algorithms for prime factorization and discrete logarithms on a quantum computer. SIAM J. Comput. **26**(5), 1484–1509 (1997)

26. Spector, L., Barnum, H., Bernstein, H.J., Swamy, N.: Finding a better-than-classical quantum AND/OR algorithm using Genetic Programming. In: Proceedings of the IEEE Congress on Evolutionary Computation, CEC 1999, vol. 3, pp. 2239–2246. IEEE (1999)

27. Spector, L., Barnum, H., Bernstein, H.J., Swamy, N.: Quantum computing applications of Genetic Programming. Adv. Genet. program. **3**, 135–160 (1999)

28. Spector, L., Klein, J.: Machine invention of quantum computing circuits by means of Genetic Programming. AI EDAM **22**(3), 275–283 (2008)

29. Stadelhofer, R., Banzhaf, W., Suter, D.: Evolving blackbox quantum algorithms using Genetic Programming. AI EDAM **22**(3), 285–297 (2008)

30. Stepney, S., Clark, J.A.: Searching for quantum programs and quantum protocols. J. Comput. Theor. Nanosci. **5**(5), 942–969 (2008)

31. Toffoli, T.: Reversible computing. In: de Bakker, J., van Leeuwen, J. (eds.) Automata, Languages and Programming. LNCS, vol. 85, pp. 632–644. Springer, Heidelberg (1980). https://doi.org/10.1007/3-540-10003-2_104

32. Vargha, A., Delaney, H.D.: A critique and improvement of the CL common language effect size statistics of McGraw and Wong. J. Educ. Behav. Stat. **25**(2), 101–132 (2000)

33. Williams, C.P., Gray, A.G.: Automated design of quantum circuits. In: Williams, C.P. (ed.) QCQC 1998. LNCS, vol. 1509, pp. 113–125. Springer, Heidelberg (1999). https://doi.org/10.1007/3-540-49208-9_8

34. Yabuki, T., Iba, H.: Genetic algorithms for quantum circuit design - evolving a simpler teleportation circuit. In: Late Breaking Papers at the 2000 Genetic and Evolutionary Computation Conference, pp. 421–425. ACM (2000)

A Genetic Programming Approach
to Predict Mosquitoes Abundance

Riccardo Gervasi[1,2] , Irene Azzali[1] , Donal Bisanzio[3,4] , Andrea Mosca[5] ,
Luigi Bertolotti[1] , and Mario Giacobini[1(✉)]

[1] DAMU - Data Analysis and Modeling Unit, Department of Veterinary Sciences,
University of Torino, Turin, Italy
riccardo.gervasi@edu.unito.it,
{irene.azzali,luigi.bertolotti,mario.giacobini}@unito.it
[2] Department of Management and Production Engineering (DIGEP),
Politecnico di Torino, Turin, Italy
[3] RTI International, Washington, DC, USA
dbisanzio.epi@gmail.com
[4] Division of Epidemiology and Public Health, School of Medicine,
University of Nottingham, Nottingham, UK
[5] Istituto per le Piante da Legno e l'Ambiente (IPLA), regional government-owned
corporation of Regione Piemonte, Turin, Italy
mosca@ipla.org

Abstract. In ecology, one of the main interests is to understand species
population dynamics and to describe its link with various environmental
factors, such as habitat characteristics and climate. It is especially important to study the behaviour of animal species that can hosts pathogens,
as they can be potential disease reservoirs and/or vectors. Pathogens of
vector borne diseases can only be transmitted from an infected to a susceptible individual by a *vector*. Thus, vector ecology is a crucial factor
influencing the transmission dynamics of vector borne diseases and their
complexity. The formulation of models able to predict vector abundance
are essential tools to implement intervention plans aiming to reduce the
spread of vector-borne diseases (e.g. West Nile Virus). The goal of this
paper is to explore the possible advantages in using Genetic Programming (GP) in the field of vector ecology. In this study, we present the
application of GP to predict the distribution of *Culex pipiens*, a mosquito
species vector of West Nile virus (WNV), in Piedmont, Italy. Our modelling approach took into consideration the ecological factors which affect
mosquitoes abundance. Our results showed that GP was able to outperform a statistical model that was used to address the same problem in
a previous work. Furthermore, GP performed an implicit feature selection, discovered automatically relationships among variables and produced fully explorable models.

Keywords: Genetic Programming · Ecological modeling · Prediction ·
West Nile virus

© Springer Nature Switzerland AG 2019
L. Sekanina et al. (Eds.): EuroGP 2019, LNCS 11451, pp. 35–48, 2019.
https://doi.org/10.1007/978-3-030-16670-0_3

1   Introduction

West Nile virus (WNV) is a zoonotic virus belonging to the *Flaviviridae* family, genus *Flavivirus*, eventually neuropathogen for birds and mammals including humans [1]. West Nile virus is transmitted by the bite of infected mosquitoes, mainly belonging to the genus *Culex* [2]. The natural cycle (enzootic cycle) of the virus involves the pathogen passing from mosquitoes to wild bird species which can be reservoirs. The virus can infect several vertebrate species (mammals, birds, reptiles) and, among mammals, humans and horses are considered dead-end hosts [3] and can show clinical symptoms [4]. In humans, most WNV infections occur asymptomatically: about 20% of infected individuals develop a febrile disease, commonly called *West Nile fever* (WNF). In less than 1% of cases the disease manifests itself in neuro-invasive form (usually encephalitis, meningoencephalitis or flaccid paralysis) [5].

In Italy, the first WNV outbreak occurred in 1998 involving several horses in the Tuscany region [6]. After the first outbreak of 1998, Italian public health authorities implemented a national surveillance plan to detect WNV circulation (in vectors and host) and quantify vector abundance [7]. Given the complexity of the biological cycle of WNV, the control of its circulation requires the integration of surveillance systems in different areas: entomological, veterinary and human. The main objective of integrated surveillance is to detect early, through targeted programmes, the circulation of WNV in birds, insects or mammals on the national territory. The early detection allows to assess the risk of transmission of the disease to humans and to implement all available measures to prevent transmission [8]. *Culex pipiens* is the most common mosquito species belonging to the genus *Culex* in the northern hemisphere. *Cx. pipiens* mosquitoes are the principal vector for WNV in Europe, since they have ornithophilic and anthropophilic subspecies, often inbreeded, and their populations are usually large [9,10]. For this reason, it is important to keep their abundance under control in order to avoid the outbreak of cases of West Nile Disease (WND).

The main goal of our study is to explore the potential application of genetic programming (GP) [23] for an ecological problem, in particular to create a predictive model for the abundance of *Cx. pipiens* in an eastern area of Piedmont region. The obtained GP models are compared with a statistical model developed to address the same problem in a previous work [11].

In the first section we present the available data and how they were collected. Next, we describe the statistical model used to compare the performance of GP and the results obtained using it. The third section contains the results obtained by GP and the comparison with the statistical model via hypothesis tests. Last section concerns conclusions and possible improvements.

2   Data Set

Mosquito data used in this study are based on entomological collections performed during the surveillance Piedmont program started by "Regione

Piemonte" through local Municipalities Agreements, and subsequently carried out by "Istituto per le Piante da Legno e l'Ambiente" (IPLA) from 2002 to 2006. The area of study covered a territory of 987 km^2 in the eastern part of the Piedmont region, where the Municipalities Agreement of Casale Monferrato operated. This territory offers favourable habitats for the reproduction of the local *Cx. pipiens* mosquitoes [11], therefore it is suitable for the introduction and amplification of WNV. From now on when we mention mosquitoes we will refer to the ones of *Cx. pipiens* species. Half of the territory is made up of hills, with an average elevation of 268 m, and the other half of plains where the landscape is mainly composed of mixed agricultural patches (72.2%, mainly in the northeast), rice fields (14.2%, mainly concentrated in the north), deciduous tree forests (8.6%, in the south), urban areas (3.1%) and the river Po with its tributaries (1.9%, in the north). The climate is characterized by cold winters (0.4 °C on average) and hot-warm summers (24.0 °C on average), with heavy rainfall in spring and autumn (about 600 mm/yr) [12]. To acquire mosquitoes abundance, 36 sampling location were selected on the territory, at a minimum distance of 5 km from each other, in areas suitable for mosquito proliferation (near rice fields, forests, urban suburbs). Captures of mosquitoes were taken at night weekly using CO_2-baited trap from the beginning of May to late September, which is the main period of mosquitoes activity in the area, for a total of 20 collections per year. In addition, various environmental and ecological parameters were collected as influential variables of mosquitoes abundance. The previous work of *Bisanzio et al.* [11] selected among all the variables the ones deemed to be the most effective in predicting the abundance and the spatial distribution of mosquitoes. Table 1 lists the final predictors after processing data as in [11].

TWEEK and *RAIN* refer to time windows that capture respectively the impact of the temperature and the rain on the abundance of mosquitoes. The parameter *NDVI* reflects environmental changes due to human agricultural activities that influence the distribution of the vector. *ELEV* takes into account altitude differences within the study area. The effect of the surrounding environment is considered through the variables *DISTU, DISTR, DISTW, DISTF* and *RICEA*. *SIN* represents the seasonal pattern of mosquitoes abundance as a positive sinusoidal curve with a peak in period the 12th week of collection (end of July-beginning of August), when mosquitoes are usually more numerous. For our experiments, the years 2002–2006 had been taken into account. The main reasons for this choice are the fact that during this period there was the highest spatial coverage of CO_2baited-traps and that mosquitoes control actions in the region were minimal. Our dataset was composed by the number of captured mosquitoes in each trap for each of the 20 periods, joined with the related predictors values.

3   Classical Statistical Approach

As mentioned in Sect. 2, mosquitoes abundance in eastern Piedmont had already been investigated in *Bisanzio et al.* [11]. The authors used a spatio-temporal

Table 1. Description of variables chosen as mosquitoes predictors.

Predictor variable	Description
TWEEK	The average land surface temperature 8–15 days prior to trapping [13]
NDVI	16-days average of normalized difference vegetation index [13]
RAIN	Cumulative rainfall 10–17 days prior to trapping [12]
ELEV	Elevation of the sample location [13]
DISTU	Distance of the sampling location from the nearest urban area
DISTR	Distance of the sampling location from the nearest rice field
DISTW	Distance of the sampling location from the nearest water source
DISTF	Distance of the sampling location from the nearest forest
RICEA	Area of the nearest rice field
SIN	Value of a sinusoidal curve with phase 1 year

Generalized Linear Mixed Model (GLMM) [14] to study the association between the abundance of *Cx. pipiens* and environmental and ecological variables. They formulated and tested 169 GLMM models and selected the best through the Deviance Information Criterion (DIC) [15]. The parameters of the derived model were tuned according to data using a Bayesian approach based on the Integrated Nested Laplace Approximation (INLA) [16]. For any further detail refer to [11]. In order to investigate the predictive behaviour of a GP approach, we borrowed the selected best model and the INLA technique and we fitted again the statistical model dividing the data in a training and a test dataset. We chose data of weekly collection from 2002 to 2005 as the training set and the 2006 trapping as the test set, which contains unseen data used to validate model performances. This choice follows the natural order of the years, therefore we learn from the past and we test on the future what we have learned. The resulting model is:

$$y = I + \beta_1 * RAIN + \beta_2 * TWEEK + \beta_3 * SIN + \beta_4 * ELEV + \beta_5 * DISTU + RNDtrap \quad (1)$$

where the variables are those described in Sect. 2, *RNDtrap* is an unstructured spacial random effect to represent unaccountable differences in each trap, and I is the intercept of the model. The estimated parameters according to INLA technique are reported in Table 2.

The GLMM model obtained a training Root Mean Square Error (RMSE) of 70.32851 and a test RMSE of 88.58552. We can notice that:

– Cumulative rain (*RAIN*) does not seem to have a significant effect in predicting abundance, as the credible interval contains zero.
– The average temperature of about a week before (*TWEEK*) has a significant negative effect on the amount of mosquitoes. This result, although opposite

Table 2. Fitted terms of the GLMM applied to predict mosquitoes abundance.

Predictors	Coefficients (mean)	Credible intervals
Intercept	1.4831	(0.7400, 2.2303)
RAIN	0.0018	(−0.0006, 0.0043)
TWEEK	−0.1285	(−0.1441, −0.1129)
SIN	7.8351	(7.4870, 8.1846)
ELEV	−0.0069	(−0.0104, −0.0035)
DISTU	−0.1920	(−0.3260, −0.0583)
Random effects	Mean	St. deviation
RNDtrap	2.7468	0.7143

to that of a previous work [11], is confirmed in a more recent article [17] in which it is pointed out that high temperatures lead to a decrease of *Cx. pipiens* abundance.

- The seasonality variable (*SIN*) is quite important and has a positive effect on the abundance. As expected, the closer to the period of mosquitoes high activity, the greater the abundance.
- The elevation (*ELEV*) has a negative effect on the expected amount of *Cx. pipiens*. At high altitudes, environmental conditions would not support their proliferation.
- Distance from the nearest urban center (*DISTU*) has a significant negative effect on the amount of *Cx. pipiens*. Thus, the closer to towns, the higher the abundance. This could be explained by the possible presence of stagnant water, which is an important resource for mosquitoes development. Indeed, in an urban environment there is a large number of small breeding sites, mainly underground, particularly adapted for the development of *Cx. pipiens*. The number of potential hosts is large, too.

4   Application of GP

4.1   Experimental Setting

In this work we used a tree-based GP where the individuals are represented in a tree structure. The set of terminal nodes is composed by the 10 variables described in Sect. 2 and a random constant between 0 and 1 generated during the initialization phase, while the function set is $F = \{\texttt{plus}, \texttt{minus}, \texttt{times}, \texttt{kozadivide}\}$ where `plus`, `minus`, `times` are the classical binary addition, subtraction and multiplication operators and `kozadivide` is the division protected as in [18] that returns 1 if the second argument (i.e. the divisor) is 0. The chosen fitness function was the Root Mean Square Error (RMSE), transforming the problem into a minimization one, since lower values represent better solutions. As the aim of this investigation was to explore the possible application of GP in ecological modeling, we prioritized simplicity considering the parameters default setting proposed in GPLab [19], the GP toolbox implemented in MATLAB that we used for our experiments. The experimental parameters setting is provided in Table 3.

Table 3. GP parameters used for experiments.

Parameter	Setting
Population size	500
Maximum number of generations	100
Initialization method	Ramped Half-and-Half [18]
Selection method	Lexicographic Parsimony Pressure [20]
Elitism	Survival of the best ('keepbest')
Subtree Crossover rate	0.9
Subtree Mutation rate	0.1
Maximum tree depth	17

Each experiment was repeated 60 times, to ensure a large enough set of simulations in order to investigate the algorithm performance. In each run the algorithm was trained using mosquitoes collection data from 2002 to 2005 and the best model found in each run was tested on mosquitoes abundance prediction of 2006.

4.2   Experimental Results

Figure 1 shows the median training and test RMSEs at each generation over the 60 runs. We chose the median value due to its robustness to outliers which can emerge in such a stochastic algorithm. The evolution of the error reveals the ability of GP in learning the relationship between variables. Moreover, the constant decrease of the test error indicates that no overfitting is occurring, therefore the algorithm is learning with generalization.

At the end of each of the 60 runs of the experiment, we obtained a best solution. The median test RMSE obtained by considering these 60 models is 83.63461, which is less than test RMSE of the GLMM (see Sect. 5 for the statistical comparison of the performances of the two approaches). After removing possible introns from each best individual, we decided to observe the median frequency with which the variables appear in the terminal nodes of the best solutions. This consideration could give a general idea of which variables play a key role in achieving the best results, while also highlighting variables that could be not so important for explaining the variability of the dependant variable. Table 4 contains the median frequencies obtained for each variable considering the 60 best solutions.

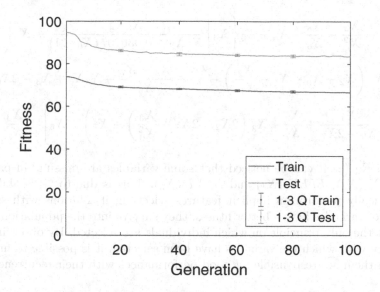

Fig. 1. GP evolution plots. The blue line represents the median RMSE on the training set, while the green line represents the median RMSE on the test set. The bars describe the range between the 1st and 3rd quartiles. (Color figure online)

Table 4. Median frequency of each variable in the best 60 solutions.

Variable	Median frequency
X_1 - RAIN	0
X_2 - TWEEK	4
X_3 - SIN	27.5
X_4 - ELEV	0
X_5 - NDVI	1
X_6 - DISTU	4
X_7 - DISTR	7
X_8 - DISTW	4.5
X_9 - DISTF	12
X_{10} - RICEA	0

SIN and DISTF are the most frequent variables on median, suggesting that these variables could be relevant predictors for mosquito abundance. RAIN, ELEV and RICEA seem to be not so relevant, since their median frequency is 0.

Among the 60 best solutions, we chose as a prediction model the expression that produced the lowest RMSE on the test set. The selected solution obtained a training RMSE of 65.99388 and a test RMSE of 81.41473. Equation (2) represents the selected model for mosquitoes prediction.

$$Y = X_9 + \frac{X_3}{X_6^2} - \frac{X_8}{X_6} - X_3 + X_3 \Bigg\{ X_3 \Bigg[\frac{X_3 X_6^2}{X_7 X_6 + 0.036707 \left(X_6 - 1 \right)} + X_3 \Bigg(X_3 \cdot$$

$$\Bigg(X_3 \Bigg(X_2 + X_9 - X_7 - \frac{X_2}{X_{10}} \Bigg) - X_7 + \frac{X_3 - X_7}{X_6} - X_3 X_6 + X_3 - 2 X_6 \Bigg) +$$

$$- \frac{X_9}{X_9 - 2 X_7} + \frac{X_3}{X_7} + X_3^3 \Bigg(2 X_2 - 2 X_7 + \frac{X_3}{X_6^2} \Bigg) - X_7 \Bigg) - X_6 \Bigg] + \frac{X_3}{X_6^2} - 1 \Bigg\}$$

$$(2)$$

From Eq. (2) it can be noticed that some variables are missing, in particular $RAIN$ (X_1), $ELEV$ (X_4) and $NDVI$ (X_5). This is due to GP's ability to automatically perform an implicit features selection; if solutions with smaller number of variables have a better fitness, they survive into the population, since fitness is the only principle on which individuals are selected. By observing the frequency with which the variables have been selected, it is possible to find out which of them are responsible for good performances with their recurrence.

Table 5. Frequency of each variable in the best model.

Variable	Frequency
X_1 - RAIN	0
X_2 - TWEEK	4
X_3 - SIN	17
X_4 -ELEV	0
X_5 - NDVI	0
X_6 - DISTU	14
X_7 - DISTR	10
X_8 - DISTW	1
X_9 - DISTF	4
X_{10} - RICEA	1

Table 5 shows that SIN is the most frequent variable, and the fact that it was a significant predictor also in the previous GLMM gives greater credence to its importance. Although the model (2) found by GP is very complex, it is still possible to interpret it and highlight the effect of some variables:

- SIN (X_3) has a general positive effect on the amount of trapped mosquitoes: as expected, as you approach the peak of the mosquito season, the abundance increases. This effect is also confirmed by the previous GLMM.
- $DISTF$ (X_9) appears in the model as a standalone term, directly influencing the abundance of mosquitoes. This fact may lead us to believe that it could play an important role in the correct prediction. Moreover, it has a general positive effect on the abundance of $Cx.$ $pipiens$: the further a place is from a

forest, the more mosquitoes are expected to be trapped. This could be justified by the fact that in the forest there are few habitats suitable as breeding sites for this species. Moreover, another explanation could be that CO_2-baited traps close to woodlands may have less attraction than those ones placed far, due to abundance of potential hosts for mosquitoes for *Cx. pipiens*, such as various bird species.

- The average temperature of a week before, *TWEEK* (X_2), has a general positive effect on the abundance of collected mosquitoes. Mosquitoes need to accumulate heat so that they can hatch from eggs, and consequently develop.
- The distance from a urban centre, *DISTU* (X_6), appears often negative at numerator or positive at denominator, having basically a negative effect on the abundance of *Cx. pipiens*. Indeed, the closer a place is to an urban area, the more mosquitoes there are. In a urban area there are many breeding sites for mosquitoes (e.g. catch basins and plant pot saucers).
- The distance from a water source, *DISTW* (X_8), has a negative effect on the abundance of *Cx. pipiens*. Water sources (e.g. rivers, lakes) provide higher humidity rates that positively effect the presence and the survival of mosquitoes. Thus, the closer a place is to a water source, the more mosquitoes is likely to be captured.
- The distance from the nearest rice field, *DISTR* (X_7), has in general a negative effect: the closer a place is to a rice field, the more mosquitoes are collected. The dimension of the rice field, *RICEA* (X_{10}), has a positive effect: at the same distance from a rice field, it is expected that more mosquitoes will be trapped in location with the largest rice field. Rice fields are indeed a suitable habitat for mosquitoes.

The study area allows to evaluate the role of environmental variables at fine scale. In general, it can be said that differences in habitat features have a greater impact than differences in climate (temperature or rainfall) when studying a small area.

To validate the ability of the GP model in predicting mosquitoes abundance, we generated a vector suitability map for August 2006, a period that includes the peak of abundance. We represented the median numbers of predicted mosquitoes against the median numbers of truly collected mosquitoes for each trap. Figure 2 shows a good prediction of high populated patterns, information that can be used to identify areas at high risk of exposure to the virus.

5   Comparison with the Statistical Model

To further explore GP ability in prediction we compared the best models found by GP over the 60 runs with the statistical model described in Sect. 3. We assumed test RMSEs of best solutions followed a normal distribution, since this hypothesis was not rejected by Lilliefors test (p-value: 0.8137). Moreover, looking at the Q-Q plot (Fig. 3), the test RMSEs follow pretty well the line and are almost all contained in the grey stip, supporting the normality hypothesis.

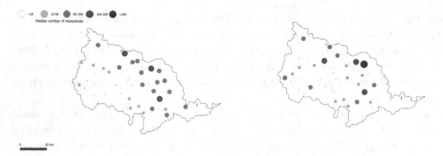

(a) Median number of mosquitoes pre-   (b) Median number of mosquitoes col-
dicted by GP.   lected.

Fig. 2. Prediction maps for mosquitoes in August 2006. Circles indicates the abundance
in each trap. Size and darkness increase with the number of mosquitoes

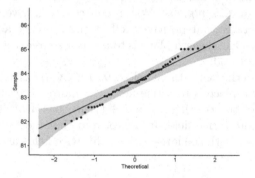

Fig. 3. Q-Q plot for normality of the sample of GP test RMSEs.

We tested whether test RMSEs of the best GP solutions were significantly
different from test RMSE of the GLMM, performing a lower-tailed one-sample t-
test, with $\alpha = 0.01$. We obtained an extremely significant p-value ($<2.2 \cdot 10^{-16}$),
which indicates a significant difference between the performances of the two
methods, therefore GP outperforms the statistical model. Moreover we depicted
the best errors on 2006 prediction of both models using a boxplot representation.
From Fig. 4 emerges that the models selected by GP have better results rather
than the GLMM model.

To give an idea of how much is the difference we used the *A statistics* (also
called *measure of stochastic superiority*). It is a non-parametric effect size measure that quantifies the difference between two populations in terms of the probability that a score sampled at random from the first population will be greater
than a score sampled at random from the second [21]. Choosing as first population the 60 test RMSEs of the best GP solutions and as second one the test
RMSE of the GLMM, we obtained an A measure equal to 0, indicating that GP
greatly outperforms the GLMM. From an ecological point of view, the relevance

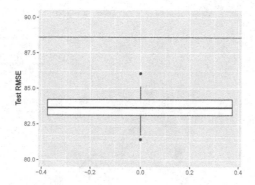

Fig. 4. Boxplot of GP test RMSEs against test RMSE of the GLMM (red line). (Color figure online)

of the performance difference between the two methods may be marginal, but GP has proved to be consistent in obtaining numerically better results.

Focusing on the GP model with the lowest test RMSE (Eq. 2), Fig. 5 shows the median predicted values of abundance compared to the median observed values for each of the 20 periods of 2006. We can notice that the GP model was able to detect quite well the moments of greatest mosquito abundance, while the GLMM missed the right peak of abundance.

6   Discussion

Some observations arise from the results achieved by GP and their comparison with the statistical model performance. It is interesting to notice that GP gave some importance to a variable not particularly significant in the GLMM, that is the distance from the nearest woodland (*DISTF*). This result has highlighted how the proximity to forests may be a disturbing element for collection of mosquitoes by means of CO_2-baited traps. Furthermore, this discovery is an example of how certain statistical models could limit the interaction between variables with their predefined structure. The exploration of other types of regression models, with not-only-linear terms, is obviously possible, but we based our work on the state-of-art of the predictive models for the specific problem addressed here.

The *SIN* variable seemed to be quite important in predicting the abundance of *Cx. pipiens* in both methods. This artificial variable was the only one that allowed to model seasonal peaks and periods of total absence of mosquitoes. We thought it would be interesting if this pattern could be learned by the model directly from ecological variables, such as temperature and rainfall. However, to recognize interesting patterns in these time series variables the algorithm needs to receive them in their entire sequence and not splitted into different fitness cases. This point made us think of a possible application of a vector approach of GP [22] to predict mosquitoes abundance. Using this technique we would have

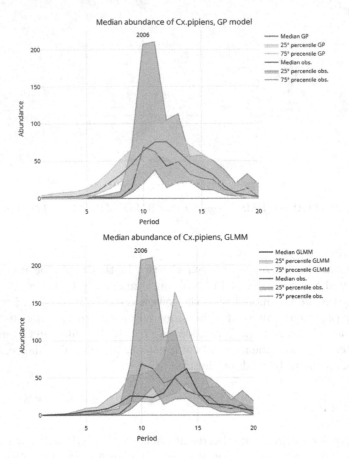

Fig. 5. Predicted median abundance of mosquitoes for 2006 by the GP model (upper figure) and the GLMM (lower figure) compared with median observed values (in green). (Color figure online)

the opportunity to avoid artificial predictors and let GP use only environmental variables to discover the model underlying data.

7   Conclusion

The goal of our work was to explore the possibility of using GP to tackle ecological problems, possibly highlighting the advantages of this technique. In our investigation, we used GP to create a predictive model for the abundance of *Cx. pipiens* in the eastern area of Piedmont region, to detect suitable locations for the entry of WNV. We evaluated GP performances using a statistical model produced in a previous work [11] as threshold. GP was able to outperform the statistical model in predicting the abundance on unseen data, while also finding out interesting relationships among predictors.

In conclusion, GP has proven to be a powerful tool that opens up important perspectives for the creation of forecasting models, particularly in the ecological field. The ability of not requiring a prefixed model structure represent a huge advantage over typical statistical models. In conjunction with the explorability of the model, GP allows to automatically discover rather complex relationships among variables, which is a really attractive feature in the ecological field. Future development could be the application of a vector based GP, as some of the predictors involve time series variables.

References

1. Diamond, M.S. (ed.): West Nile Encephalitis Virus Infection: Viral Pathogenesis and the Host Immune Response. Emerging Infectious Diseases of the 21st Century. Springer, New York (2009). https://doi.org/10.1007/978-0-387-79840-0
2. Chambers, T., Monath, T.: The Flaviviruses: Detection, Diagnosis and Vaccine Development. Advances in Virus Research. Elsevier Science, Amsterdam (2003). https://doi.org/10.1016/S0065-3527(03)61017-1
3. Sfakianos, J.N.: West Nile Virus. Chelsea House Publications, Langhorne (2005)
4. Kramer, L.D., Styer, L.M., Ebel, G.D.: A global perspective on the epidemiology of West Nile virus. Ann. Rev. Entomol. **53**, 61–81 (2008). https://doi.org/10.1146/annurev.ento.53.103106.093258
5. Istituto Superiore di Sanità. http://old.iss.it/
6. Autorino, G.L., et al.: West Nile virus epidemic in horses, Tuscany region, Italy. Emerg. Infect. Dis. **8**(12), 1372–1378 (2002). https://doi.org/10.3201/eid0812.020234
7. Ministero della Salute: Piano di sorveglianza nazionale per la encefalomielite di tipo West Nile (West Nile Disease). Gazzetta Ufficiale della Repubblica Italiana, N. 113, 16 May 2002
8. EpiCentro - Portale di epidemiologia. http://www.epicentro.iss.it/
9. Zeller, H.G., Schuffenecker, I.: West Nile virus: an overview of its spread in Europe and the Mediterranean basin in contrast to its spread in the Americas. Eur. J. Clin. Microbiol. Infect. Dis.: Off. Publ. Eur. Soc. Clin. Microbiol. **23**(3), 147–156 (2004). https://doi.org/10.1007/s10096-003-1085-1
10. Becker, N., Jöst, A., Weitzel, T.: The Culex pipiens complex in Europe. J. Am. Mosq. Control Assoc. **28**(4 Suppl.), 53–67 (2012). https://doi.org/10.2987/8756-971X-28.4s.53
11. Bisanzio, D., et al.: Spatio-temporal patterns of distribution of West Nile virus vectors in eastern Piedmont region, Italy. Parasites Vectors **4**, 230 (2011). https://doi.org/10.1186/1756-3305-4-230
12. Arpa Piemonte. http://www.arpa.piemonte.it
13. NASA MODIS Web. https://modis.gsfc.nasa.gov/
14. Stroup, W.: Generalized Linear Mixed Models: Modern Concepts, Methods and Applications. CRC Press, Boca Raton (2012)
15. Spiegelhalter, D.J., Best, N.G., Carlin, B.P., Van-der Linde, A.: Bayesian measures of model complexity and fit. J. Roy. Stat. Soc. **64**(4), 583–639 (2002). https://doi.org/10.1111/1467-9868.02022
16. Rue, H., Martino, S., Chopin, N.: Approximate Bayesian inference for latent Gaussian models by using integrated nested Laplace approximations. J. Roy. Stat. Soc.: Ser. B (Stat. Methodol.) **71**(2), 319–392 (2009). https://doi.org/10.1111/j.1467-9868.2008.00700.x

17. Rosà, R., et al.: Early warning of West Nile virus mosquito vector: climate and land use models successfully explain phenology and abundance of Culex pipiens mosquitoes in north-western Italy. Parasites Vectors **7**, 269 (2014). https://doi.org/10.1186/1756-3305-7-269
18. Koza, J.R.: Genetic Programming: On the Programming of Computers by Means of Natural Selection. MIT Press, Cambridge (1992)
19. Silva, S.: GPLAB - A Genetic Programming Toolbox for MATLAB. http://gplab.sourceforge.net/index.html
20. Luke, S., Panait, L.: Lexicographic parsimony pressure. In: Proceedings of the 4th Annual Conference on Genetic and Evolutionary Computation, GECCO 2002, pp. 829–836. Morgan Kaufmann Publishers Inc., San Francisco (2002)
21. Vargha, A., Delaney, H.D.: A critique and improvement of the "CL" common language effect size statistics of McGraw and Wong. J. Educ. Behav. Stat. **25**(2), 101–132 (2000). https://doi.org/10.2307/1165329
22. A vectorial approach to genetic programming. Submitted to EuroGP 2019
23. Poli, R., Langdon, W., McPhee, N., Koza, J.: A Field Guide to Genetic Programming. Lulu.com, Morrisville (2008). https://doi.org/10.1007/s10710-008-9073-y

Complex Network Analysis of a Genetic Programming Phenotype Network

Ting Hu[1](✉), Marco Tomassini[2], and Wolfgang Banzhaf[3]

[1] Department of Computer Science, Memorial University,
St. John's, NL, Canada
`ting.hu@mun.ca`
[2] Faculty of Business and Economics, Information Systems Department,
University of Lausanne, Lausanne, Switzerland
`marco.tomassini@unil.ch`
[3] Department of Computer Science and Engineering, and BEACON Center,
Michigan State University, East Lansing, MI, USA
`banzhafw@msu.edu`

Abstract. The genotype-to-phenotype mapping plays an essential role in the design of an evolutionary algorithm. Since variation occurs at the genotypic level but fitness is evaluated at the phenotypic level, this mapping determines how variations are effectively translated into quality improvements. We numerically study the redundant genotype-to-phenotype mapping of a simple Boolean linear genetic programming system. In particular, we investigate the resulting phenotypic network using tools of complex network analysis. The analysis yields a number of interesting statistics of this network, considered both as a directed as well as an undirected graph. We show by numerical simulation that less redundant phenotypes are more difficult to find as targets of a search than others that have much more genotypic abundance. We connect this observation with the fact that hard to find phenotypes tend to belong to small and almost isolated clusters in the phenotypic network.

Keywords: Complex networks · Genetic programming ·
Genotype-phenotype mapping · Phenotype networks · Evolvability

1 Introduction

In evolutionary algorithms, the quality of a candidate solution is assessed based on its phenotype, i.e., how well the phenotype is able to produce a desired outcome judged by a fitness measure. Yet, the actual EA search occurs in genotype space, where the encoding of candidate solutions is modified by mutation or recombination operations. Thus, how genotypes are mapped to phenotypes will substantially influence the search effectiveness of an evolutionary algorithm [1,2].

Redundant genotype-to-phenotype mappings are common in both natural [3,4] and computational evolution [5–8], where multiple genotypes can map to

© Springer Nature Switzerland AG 2019
L. Sekanina et al. (Eds.): EuroGP 2019, LNCS 11451, pp. 49–63, 2019.
https://doi.org/10.1007/978-3-030-16670-0_4

the same phenotype. Such a *redundancy* is often unevenly distributed among phenotypes, where some phenotypes are over-represented by many genotypes, and some are under-represented by only a few [7,9]. When the target phenotype is under-represented, its evolutionary search is often more difficult than having a genotypically over-represented target. This is intuitive since it can be more difficult to find one of the few genotypes that map to an under-represented target phenotype.

If the genotype-to-phenotype mapping is redundant, a mutation to a genotype may not change the phenotype it encodes, a phenomenon defined as *neutrality* [10,11], and such mutations are called neutral mutations [12–15]. Neutrality is facilitated by redundancy, but not guaranteed. For instance, there are cases where genotypes map to the same phenotype but are not mutationally connected, i.e., one genotype cannot be reached from the other through single point mutations, thus mutations that need to occur on the way from one to the other will need to alter the phenotype.

In contrast to neutral mutations, non-neutral mutations connect genotypes of distinct phenotypes. Such non-neutral mutational connections among phenotypes might also be heterogeneous [9,16], i.e., a phenotype may not have the same likelihood of mutating to other phenotypes and thus may tend to "prefer" some phenotypes over others. The difficulty of finding a target phenotype is thus influenced not only by its genotypic abundance, but also by how mutational connections are distributed among different phenotypes.

In this contribution, we quantitatively measure the genotypic redundancy of phenotypes and the mutational connections among them, and take a network approach to analyze how these properties correlate with the difficulty of finding a target phenotype. We use a linear genetic programming (LGP) algorithm for Boolean optimization, and numerically characterize its genotype, phenotype, and fitness space. Using random sampling and random walks, we construct a phenotype network to depict the mutational connections among different phenotypes. Once a specific target phenotype is chosen, this changes the connectivity of the phenotype network since only non-deleterious mutations, i.e. mutations that do not decrease fitness, are allowed. We show that such changes can significantly influence the difficulty of finding a target.

2 Methods

2.1 A Boolean Linear Genetic Programming Algorithm

We use a linear genetic programming (LGP) algorithm for our empirical analysis. LGP is a branch of genetic programming and employs a sequential representation of computer programs to encode an evolutionary individual [17]. Such an LGP program is often comprised of a set of imperative instructions, which are executed sequentially. Registers are used to either read input variables (input registers) or to enable computational capacity (calculation register). One or more registers can be designated as the output register(s) such that the final stored value(s) after the program is executed will be the program's output.

In this study, we use an LGP algorithm for a three-input, one-output Boolean function modeling application. Each instruction has one return, two operands and one Boolean operator. The operator set has four Boolean functions {AND, OR, NAND, NOR}, any of which can be selected as the operator for an instruction. Three registers R_1, R_2, and R_3 receive the three Boolean inputs, and are write-protected in an LGP program. That is, they can only be used as an operand in an instruction. Registers R_0 and R_4 are calculation registers, and can be used as either a return or an operand. Register R_0 is also the designated output register, and the Boolean value stored in R_0 after an LGP program's execution will be the final output of the program. All calculation registers are initialized as FALSE before execution of a program. An LGP program can have any number of instructions, however, for the ease of simulation in this study, we determine that an LGP program has a fixed length of six instructions. An example LGP program is given as follows.

$$I_1 : R_4 = R_2 \ \text{AND} \ R_3$$
$$I_2 : R_0 = R_1 \ \text{OR} \ R_4$$
$$I_3 : R_4 = R_4 \ \text{NAND} \ R_0$$
$$I_4 : R_4 = R_3 \ \text{AND} \ R_2$$
$$I_5 : R_0 = R_1 \ \text{NOR} \ R_1$$
$$I_6 : R_0 = R_3 \ \text{AND} \ R_0$$

2.2 Genotype, Phenotype, and Fitness

The *genotype* in our evolutionary algorithm is a unique LGP program. Since we have a finite set of registers and operators, as well as a fixed length for all programs, the genotype space is finite. Specifically, considering an instruction, two registers can be chosen as the return, all five registers can be used as the two operands, and the operator is picked from the set of four possible Boolean functions. Thus, there are $2 \times 5 \times 5 \times 4 = 200$ unique instructions. Given the fixed length of six instructions for all LGP programs, we have a total number of $200^6 = 6.4 \times 10^{13}$ possible different programs. Although finite, the genotype space is enormous and is not amenable to exhaustive enumeration. Therefore, we conduct a simulation by randomly generating one billion LGP programs (≈ 15.6 ppm $= 0.00156\%$ of the genotype space) to approximate the genotype space.

The *phenotype* in our evolutionary algorithm is a Boolean relationship that maps three inputs to one output, represented by an LGP program, i.e., $f : \mathbf{B}^3 \to \mathbf{B}$, where $\mathbf{B} = \{\text{TRUE, FALSE}\}$. There are thus a total of $2^{2^3} = 256$ possible Boolean relationships. Having 6.4×10^{13} genotypes to encode 256 phenotypes, our LGP algorithm must have a highly redundant genotype-to-phenotype mapping. We define the *genotypic redundancy* of a phenotype as the total number of genotypes that map to it.

The *fitness* of an LGP program is dependent on the target Boolean relationship, and it is defined as the dissimilarity of the presented and the target

Boolean relationships. Given three inputs, there are $2^3 = 8$ combinations of Boolean inputs. The Boolean relationship encoded by an LGP program can be seen as a 8-bit string representing the outputs that correspond to all 8 possible combinations of inputs. Fitness is defined as the Hamming distance of this 8-bit output and the target output. For instance, if the target relationship is $f(R_1, R_2, R_3) = R_1$ AND R_2 AND R_3, represented by the 8-bit output string of 00000001, the fitness of an LGP program encoding the FALSE relationship, i.e., 00000000, is 1. Fitness falls into the range of $[0, 8]$ where 0 is the perfect fitness and 8 is the worst, and is to be minimized.

2.3   Phenotype Networks

Point mutations to genotypes may change the encoded phenotypes from one to another. In the context of our LGP algorithm, a point mutation is to replace any one of the four elements, i.e., return, two operands, and operator, of an instruction in an LGP program. The mutational connections among pairs of phenotypes can be modeled using a *phenotype network*. In such a network, each node represents one of the 256 phenotypes that can be possibly encoded by the LGP genotypes. Two nodes (phenotypes) are directly connected by an edge if there exist at least one pair of underlying genotypes, one from each phenotype, that can be transitioned from one to the other through a single point mutation.

Since it is infeasible to enumerate all possible genotypes, sampling the mutational connections among phenotypes is also necessary. We assemble one million randomly generated LGP programs and allow each to take a 1000-step random walk in genotype space. All the phenotypes each random walker encountered are recorded in order to estimate the number of point mutations that can transition one phenotype to another. This random walk simulation yields a *undirected, weighted* phenotype network, where the weight of an edge is proportional to the number of sampled point mutations that can change the genotypes of one phenotype to that of the other phenotype.

Assigning a fitness to each phenotype and preventing deleterious mutations changes the reversible feature of point mutations and further transforms the weighted phenotype network into a *directed* graph. We pick two target phenotypes with a considerable difference in their genotypic redundancies, given the consideration that whether a target phenotype is over- or under-represented by genotypic encodings may influence the difficulty level of finding that target [7]. The first target is phenotype 11110000 (decimal 240) which has a genotypic redundancy of 46,729,920, i.e., 4.673% of the one billion sampled genotypes. The second target is phenotype 10110100 (decimal 180) with only a genotypic redundancy of 86. Setting such different targets will render the corresponding directed, weighted phenotype networks different. Thus, we investigate a variety of network properties to compare these two networks.

2.4   Complex Network Analysis

Since we need a few concepts and methods from the field of network science [18, 19], we here collect some useful definitions to be referenced and used later.

Strength. This term refers to the generalization of the vertex degree to weighted networks. It is defined as the sum of weights of the edges from node i to its neighbors $\mathcal{N}(i)$,

$$s_i = \sum_{j \in \mathcal{N}(i)} w_{ij},$$

where w_{ij} is the weight of the edge connecting nodes i and j.

Disparity. A given value of a node's strength can be obtained with very different values of edge weights. The contributing weights could be of about the same size or they could be very different. To measure the degree of heterogeneity of a node's edges *disparity* can be used. It is defined as follows:

$$Y_2(i) = \sum_{j \in \mathcal{N}(i)} \left(\frac{w_{ij}}{s_i} \right)^2.$$

If all the connections are of the same order then Y_2 is small and of order $1/k$ where k is the vertex degree. On the other hand, if there is a small number of high weight connections Y_2 is larger and may approach unity.

Average Shortest Paths. We use weighted and unweighted shortest paths between pairs of vertices. The average values of all two-point shortest paths in a graph give an idea of the typical distances between nodes.

Clustering Coefficient. The clustering coefficient $C(i)$ of a node i is defined as the ratio between the e edges that actually exist between the k neighbors of i and the number of possible edges between these nodes:

$$C(i) = \frac{e}{\binom{k}{2}} = \frac{2e}{k(k-1)}.$$

The clustering coefficient can be interpreted intuitively as the likelihood that two of node i's neighbors are also neighbors. The *average clustering coefficient* $\bar{C}$ is the average of $C(i)$ over all N vertices in the graph G, $i \in V(G)$: $\bar{C} = (1/N) \sum_{i=1}^{N} C(i)$.

Degree, Strength, and Weights Distribution Functions. These discrete distributions give, respectively, the frequency of a given node degree, node strength, or edge weight in the network. These distributions are useful for evaluating whether they are, for instance, homogeneous or heterogeneous, unimodal or multimodal.

3   Results

3.1   Sampled Genotype Space and Mutational Connections

When we decode the one billion randomly generated genotypes, we find that 17 of the total 256 phenotypes are never sampled. The distribution of the genotypic redundancy of the remaining 239 sampled phenotypes is highly heterogeneous (see Fig. 1a). The most over-represented phenotype is 0, i.e., FALSE, which has over 108 million genotypes, while phenotype 255, i.e., TRUE, its symmetric counterpart, is the second most abundant with over 93 million genotypes. The asymmetry in count is due to the initialization of calculation registers, including the output register R_0, to FALSE in all LGP programs prior to execution. In addition to the 17 phenotypes never sampled, under-represented phenotypes include 105, 231, 24, 219, 189, 36, and 66, none of which has more than 40 sampled genotypic encodings.

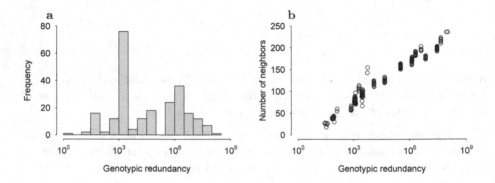

Fig. 1. (a) Distribution of the genotypic redundancy of sampled phenotypes using one billion randomly generated LGP programs. (b) Number of neighbors in relation to the genotypic redundancy of a phenotype. Their linear-log correlation has a coefficient of 0.9642 $(p < 2.2 \times 10^{-16})$.

Using the assembly of one million 1,000-step random walkers, the mutational connections among pairs of phenotypes can be approximated. 16 out of 256 phenotypes are never encountered, i.e., they are isolated nodes in the phenotype network, 15 of which belong to the 17 never-sampled phenotypes discussed previously. This also suggests that under-represented phenotypes are hard to reach by random walks. Figure 1b shows the correlation of node degree, i.e., the number of distinct phenotypes accessible from a phenotype through point mutations, and the genotypic redundancy of a phenotype. We observe a strong and highly significant positive correlation.

3.2   Properties of the Undirected Weighted Phenotype Network

We first construct an undirected, weighted phenotype network using the sampled mutational connections among pairs of phenotypes. The network has 240 nodes

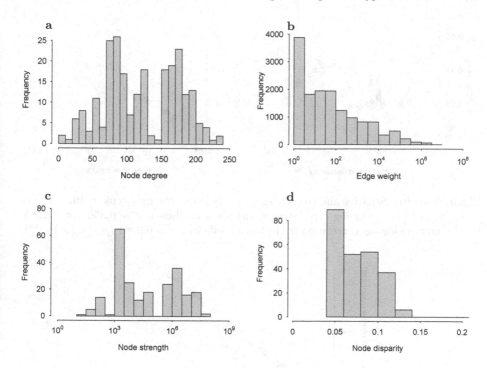

Fig. 2. Distribution of (a) node degree, (b) edge weight, (c) node strength, and (d) node disparity in the undirected weighted phenotype network.

representing 240 unique phenotypes, since 16 phenotypes were never encountered in the random walk sampling. See Sect. 2.4 for definitions of the various measures used. Ignoring edge weights, the network has 14,663 edges, which yields an average node degree of 122. There is only one connected component in this network. Its average shortest path is only 1.5 and its diameter is as short as 3, which means that any pair of phenotypes can be reached from one to another by point mutations through no more than 3 hops in the phenotype network. The clustering coefficient is high at 0.75; this is due to the fact that many nodes have neighbors that are themselves connected, giving rise to many closed triangles.

The degree and edge weight distributions are shown in Fig. 2(a) and (b). Phenotypes 0 (**FALSE**) and 255 (**TRUE**) have the highest node degree of 236, and phenotypes 22 and 104 only have a degree of 2. The distribution of edge weights is roughly monotonic, with the majority of edges having a weight of less than 50, while the edge connecting phenotypes 0 and 255 has the highest weight of 5,673,803.

Figure 2(c) and (d) show node strength and node disparity distributions, respectively. Strength, being a generalization of degree for weighted networks, has a shape that is qualitatively similar to the degree histogram, with a bi-modal distribution. The node disparity shows that most phenotypes have a low disparity, i.e. their links tend to have similar weights, and the distribution decays

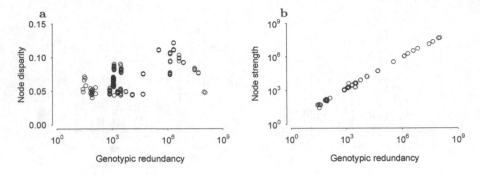

Fig. 3. Node (a) disparity and (b) strength in relation to the genotypic redundancy of a phenotype. The linear-log correlation in (a) has a coefficient $r^2 = 0.3568$ ($p < 2.2 \times 10^{-16}$), and the log-log correlation in (b) has a coefficient $r^2 = 0.9981$ ($p < 2.2 \times 10^{-16}$).

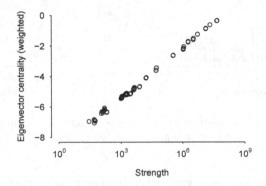

Fig. 4. Correlation of node strength and weighted eigenvector centrality. This positive correlation has a coefficient $r^2 = 0.9982$ ($p < 2.2 \times 10^{-16}$).

quickly. Since we were only interested in the qualitative aspects of these distributions, we did not attempt to fit any particular functions to them. Finally, Fig. 3 shows that genotypic redundancy is strongly correlated with node strength.

An important property of phenotypes is their *evolvability* which is the ability to generate novel and adaptive phenotypes. Evolvability can be defined quantitatively and has been studied in detail in previous work (see, e.g., [9,20,21]). In particular, it was found in [21] that a phenotypic network centrality measure called *weighted eigenvector centrality* was a good predictor for phenotypic evolvability. In that respect, we note that the simpler node strength is highly correlated with this centrality measure, as can be seen in Fig. 4. This means that a good proxy for phenotypic evolvability is the easily computable phenotype strength.

3.3   Communities in the Undirected Weighted Phenotype Network

Communities in a complex network can be loosely defined as collections of vertices that are more strongly linked among themselves than with the rest of the network. A precise and unique definition cannot be given, which makes community detection a hard and somewhat ill-defined task. Nevertheless, several community detection algorithms have been proposed that work well in practice.

Here we use the methods implemented in the *igraph* R package [22], which also cover weighted networks. Before submitting our phenotypic network G to a community detection algorithm some manipulations are necessary. In fact, the graph has a mean degree of about 122 which makes it a very dense network. Community detection algorithms typically do not work well, or at all, on such graphs. However, we note that edge weights in G span seven orders of magnitude (see Fig. 2b), which means that many links are comparatively very weak. Thus we have discarded weak network connections by cutting all edges with weights below a threshold of $w_{ij} < 10^5$. As a consequence, some of the original nodes also become disconnected but we have ensured that all edges of the target phenotypes are kept, especially for target node 180, which would have become isolated otherwise, since all its edges have weights lower than the threshold.

Modularity is a measure that estimates the cohesiveness of a partition found by a community detection algorithm with respect to a graph with the same degree distribution but with edges placed at random [23]. The community partition found with several community detection algorithms from *igraph* has a modularity value of about two, which is not very high but still significantly different from random. Figure 5 shows the communities found by the *Louvain* algorithm. It is important to note that the small community to which vertex 180 belongs is almost always found identically by all the different algorithms tried. Figure 5 clearly shows that node 180, together with its neighbors belonging to the same community, appears to be extremely difficult to reach, all the more taking into account that the intra-community and extra-community edges are weak. On the other hand, phenotype 240 is at the intersection of two bigger and well connected communities and thus it is intuitively reasonable that it should be easier to find. These indications will find a numerical confirmation in the next section where we shall use random walks to traverse the network.

3.4   Random Walks

Although the GP system searches the genotype space and not the much smaller phenotype space, it is still interesting to simulate random walk search in the latter to numerically confirm the above idea that some phenotypes are easy to find while others are hard. Random walks on networks are reviewed in [24]. In an unweighted network, the probability for going from node i to node j is $p_{ij} = a_{ij}/k_i$, where a_{ij} is the corresponding entry in the graph adjacency matrix being 1 if nodes i and j are connected and 0 if they are not. The random walk we are interested in is biased, since edges of the undirected phenotype network are weighted. So we have to modify the probabilities accordingly, but the changes are

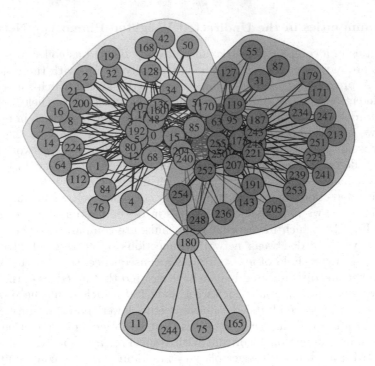

Fig. 5. Community structure of the edge-filtered undirected phenotypic network. Phenotype 180 clearly belongs to a small community that is very weakly connected to the rest of the network, while phenotype 240 is located in the center of the network and belongs to a larger and well connected community.

minor: the transition probability from node i to node j through an edge $\{ij\}$ with weight $w_{ij} \geq 0$ now becomes $p_{ij} = w_{ij}/s_i$, where s_i is the *strength* of node i and is defined as the sum of the weights of the edges from i to its neighbors $\mathcal{N}(i)$ (see definitions in Sect. 2.4). These probabilities are well behaved since a connected node must have a positive finite strength, $p_{ij} \geq 0$, and $\sum_{j \in \mathcal{N}(i)} p_{ij} = 1$.

For each of the two target phenotypes 180 and 240, we numerically simulate biased random walks in the original unfiltered network starting from all network nodes except the target nodes themselves. For each starting node we perform 10^5 random walk steps, for a total of $(N-1) \times 10^5 = 237 \times 10^5$ steps. For each of the phenotypes 180 and 240 we record the number of times it is found, i.e., the number of hits, and the mean number of steps to the first hit, when the node is found.

Results are shown in Table 1. From the number of hits and the first hit times, it is apparent that phenotype 180 is much harder to find than phenotype 240. Furthermore, if we exclude the first neighbors of the target node as starting nodes in the random walk, it becomes even more difficult, comparatively, to find phenotype 180 (figures after the comma in Table 1). We can also see that phenotype 240 is very often found directly from a starting node that is a first neighbor given that 240 has 223 neighbors while 180 only has 41 connections.

Table 1. Average number of hits and first hitting times for random walks having nodes 180 and 240 as targets. 237×10^5 random walks steps are performed in total. The figures after the commas refer to the same quantities when the first neighbors of nodes 180 and 240 are excluded as starting nodes for the walk.

	Target phenotype 180	Target phenotype 240
Number of hits	2, 0	97465, 5862
First hitting time	910306, –	10, 10

3.5   In-degree and Out-degree in the Directed Phenotype Networks

When a fitness is assigned to each phenotype based on its Hamming distance to the target phenotype, the phenotype network becomes oriented since we only allow non-deleterious point mutations. Note that we only consider simple graphs in the current study, i.e., self-loops are excluded in our network analysis, in order to focus on the mutational connections among distinct phenotypes. A phenotype/node now has edges with two directions, pointing to its neighbors (out edges) and being pointed from its neighbors (in edges). Subsequently, in-degree and out-degree can be used to depict how many unique phenotypes can access or can be reached from a reference phenotype.

Figure 6 shows the correlations of in- and out-degrees with the fitness of a phenotype in two directed phenotype networks with different targets. Using both targets, in-degrees are negatively correlated with fitness while out-degrees are positively correlated with fitness. Note that fitness is to be minimized. Phenotypes with better fitness will have less edges going out but more edges coming in, i.e., fitter phenotypes are easier to reach and harder to leave, which is intuitive and desirable since we hope reaching fitter phenotypes will be more likely leading to the path to the target. However, when we compare the correlations using different targets, it can be seen that using a relatively harder target (i.e., phenotype 180) results in weaker correlations of in-/out-degrees and fitness. This indicates that some targets are difficult to find not only because they under-represented by genotypes, but also because they render the guidance of the fitness gradient less effective. That is, reaching fitter phenotypes at a current stop does not necessarily lead to better paths to finding the target.

3.6   Fitness Correlation of Neighboring Phenotypes

Fitness correlation can give statistical information about the fitness assortativity of neighboring nodes in the network. A practical way for evaluating fitness correlation is given by the average fitness of neighbors $\bar{f}_{neighbor}(i)$ of a node i

$$\bar{f}_{neighbor}(i) = \frac{1}{|\mathcal{N}(i)|} \sum_{j \in \mathcal{N}(i)} f_j,$$

where f_i is the fitness of phenotype/node i.

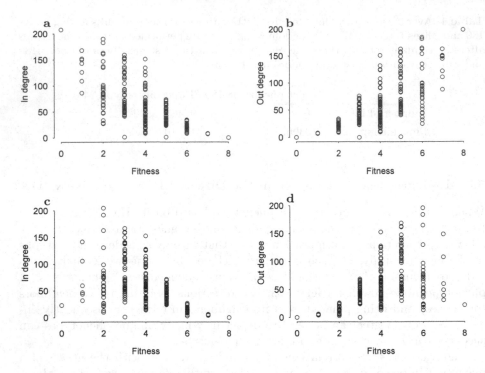

Fig. 6. Correlation of node in-degree (a, c) and out-degree (b, d) with fitness in the directed phenotype networks using an over-represented target 240 (a, b) and under-represented target 180 (b, d). In (a) and (c), the negative correlations are with a coefficient of $R^2 = 0.4623$ ($p < 2.2 \times 10^{-16}$), and $R^2 = 0.2519$ ($p < 4.6 \times 10^{-16}$), respectively. In (b) and (d), the positive correlations are with a coefficient of $R^2 = 0.4611$ ($p < 2.2 \times 10^{-16}$), and $R^2 = 0.2622$ ($p < 4.6 \times 10^{-16}$), respectively.

From this quantity one can compute the average fitness of the neighbors $\bar{f}_{neighbor}(f)$ for nodes of the phenotypic network having fitness value f which is a good approximation to the fitness-fitness correlation:

$$\bar{f}_{neighbor}(f) = \frac{1}{N_f} \sum_i \bar{f}_{neighbor}(i),$$

where N_f is the number of nodes with fitness f.

Remembering that a low fitness value is better in our context, one can see from Fig. 7 that the neighboring fitness correlations are quite different when different phenotypes are used as the search target. Specifically, when the over-represented phenotype 240 is set as the target (see Fig. 7a), the fitness values of neighboring phenotypes do not correlate, meaning that a phenotype can be connected to phenotypes with any fitness values. However, as shown in Fig. 7b, when the target is under-represented, phenotypes having fitnesses greater than four tend to have neighbors with lower (better) fitness, while good phenotypes

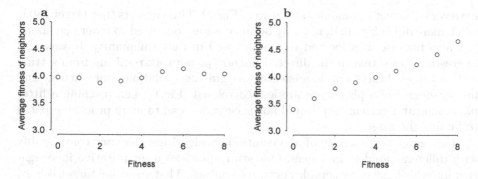

Fig. 7. Average neighbor fitness versus node fitness in the phenotypic network with the target phenotype (a) 240 and (b) 180.

below four tend to have as neighbors phenotypes with higher (worse) fitness. This situation indicates that, in the average, good phenotypes, i.e., those that have a fitness lower than the average of four, are surrounded by less good ones.

4   Discussion

The genotype-to-phenotype mapping plays a central role in enabling a system to be evolvable, since the variations characterizing the search occur in the genotype space, but the quality or behavior of a system can only be observed and evaluated at the phenotypic level. We argue that some target phenotypes are more difficult to find than others, not only because they are most likely underrepresented in genotypic space, but also because the mutational connections between phenotypes are altered through setting different target phenotypes.

In this study, we took a network approach and quantitatively analyzed the distribution of mutational connections among phenotypes and how this distribution is changed with different phenotypes set as target. Using a Boolean LGP algorithm, we sampled the genotypic and phenotypic spaces and constructed a phenotype network to characterize the distribution of mutational connections among phenotypes. By setting two different phenotypes as the target, one genotypically over-represented and one under-represented, we compared the properties of the resulting directed, weighted phenotype networks.

Similar to many GP systems, our Boolean LGP algorithm has a highly redundant mapping from genotypes (6.4×10^{13}) to phenotypes (256). Such redundancy is heterogeneously distributed among phenotypes, with the most abundant phenotype possessing about 10% of the entire genotype space while some other phenotypes never appeared in our samples (Fig. 1a). By examining the undirected, weighted phenotype network, we found that more abundant phenotypes have more access to different phenotypes (Fig. 1b) and more tendency to mutate into certain neighboring phenotypes (Fig. 3a).

We chose two phenotypes, 180 and 240, as targets, and observed that in addition to having considerably different degrees, 41 and 233, the two phenotypes have very different community structures (Fig. 5). This suggests that target 180 is much more difficult to find, not only because it was connected to fewer neighbors, but also because it is located in a small and distant community. It was also interesting to see that in the directed phenotype network resulting from setting 180 as target, fitness was less effective at guiding evolution, since fitness and in-/out-degree of a phenotype are less correlated (Fig. 6), i.e., reaching a fitter phenotype at a current stop would not necessarily lead to more promising paths to finding the target.

The search performance of an evolutionary algorithm can vary considerably with different problem instances. Our study provides a quantitative investigation into this issue using complex network analysis. That a specific target is hard to reach can have multiple explanations: (1) the target is under-represented in genotype space; (2) the target is connected to only a few phenotypes in phenotype space; (3) the target belongs to a small community distant from the rest of the phenotypes in that space; and (4) setting the target has wired the connections among phenotypes in a way that renders following fitter phenotypes in order to reach the target a less effective strategy. We hope our observations can be found useful to inspire more intelligent search mechanisms that are able to overcome these challenges.

Acknowledgements. This research was supported by the Natural Sciences and Engineering Research Council (NSERC) of Canada Discovery Grant RGPIN-2016-04699 to T.H., and the Koza Endowment fund provided to W.B. by Michigan State University.

References

1. Kell, D.B.: Genotype-phenotype mapping: genes as computer programs. Trends Genet. **18**(11), 555–559 (2002)
2. de Visser, J.A.G.M., Krug, J.: Empirical fitness landscapes and the predictability of evolution. Nat. Rev. Genet. **15**, 480–490 (2014)
3. Schaper, S., Louis, A.A.: The arrival of the frequent: how bias in genotype-phenotype maps can steer populations to local optima. PLoS One **9**(2), e86635 (2014)
4. Catalan, P., Wagner, A., Manrubia, S., Cuesta, J.A.: Adding levels of complexity enhances robustness and evolvability in a multilevel genotype-phenotype map. J. R. Soc. Interface **15**(138), 20170516 (2018)
5. Banzhaf, W.: Genotype-phenotype-mapping and neutral variation—a case study in Genetic Programming. In: Davidor, Y., Schwefel, H.-P., Männer, R. (eds.) PPSN 1994. LNCS, vol. 866, pp. 322–332. Springer, Heidelberg (1994). https://doi.org/10.1007/3-540-58484-6_276
6. Smith, T., Husbands, P., O'Shea, M.: Neutral networks and evolvability with complex genotype-phenotype mapping. In: Kelemen, J., Sosík, P. (eds.) ECAL 2001. LNCS (LNAI), vol. 2159, pp. 272–281. Springer, Heidelberg (2001). https://doi.org/10.1007/3-540-44811-X_29
7. Rothlauf, F., Goldberg, D.E.: Redundant representations in evolutionary computation. Evol. Comput. **11**(4), 381–415 (2003)
8. Hu, T., Banzhaf, W., Moore, J.H.: The effect of recombination on phenotypic exploration and robustness in evolution. Artif. Life **20**(4), 457–470 (2014)

9. Hu, T., Payne, J., Banzhaf, W., Moore, J.H.: Evolutionary dynamics on multiple scales: a quantitative analysis of the interplay between genotype, phenotype, and fitness in linear genetic programming. Genet. Program. Evolvable Mach. **13**(3), 305–337 (2012)
10. Newman, M.E.J., Engelhardt, R.: Effects of selective neutrality on the evolution of molecular species. Proc. R. Soc. B **265**(1403), 1333–1338 (1998)
11. Wagner, A.: Robustness, evolvability, and neutrality. Fed. Eur. Biochem. Soc. Lett. **579**(8), 1772–1778 (2005)
12. van Nimwegen, E., Crutchfield, J.P., Huynen, M.A.: Neutral evolution of mutational robustness. Proc. Natl. Acad. Sci. **96**(17), 9716–9720 (1999)
13. Galvan-Lopez, E., Poli, R.: An empirical investigation of how and why neutrality affects evolutionary search. In: Cattolico, M. (ed.) Proceedings of the Genetic and Evolutionary Computation Conference, pp. 1149–1156 (2006)
14. Hu, T., Banzhaf, W.: Neutrality and variability: two sides of evolvability in linear genetic programming. In: Proceedings of the 18th Genetic and Evolutionary Computation Conference (GECCO), pp. 963–970 (2009)
15. Hu, T., Banzhaf, W.: Neutrality, robustness, and evolvability in genetic programming. In: Riolo, R., Worzel, B., Goldman, B., Tozier, B. (eds.) Genetic Programming Theory and Practice XIV. GEC, pp. 101–117. Springer, Cham (2018). https://doi.org/10.1007/978-3-319-97088-2_7
16. Nickerson, K.L., Chen, Y., Wang, F., Hu, T.: Measuring evolvability and accessibility using the Hyperlink-Induced Topic Search algorithm. In: Proceedings of the 27th Genetic and Evolutionary Computation Conference (GECCO), pp. 1175–1182 (2018)
17. Brameier, M.F., Banzhaf, W.: Linear Genetic Programming. Springer, Boston (2007). https://doi.org/10.1007/978-0-387-31030-5
18. Barábasi, A.L.: Network Science. Cambridge University Press, Cambridge (2016)
19. Barrat, A., Barthélemy, M., Vespignani, A.: Dynamical Processes on Complex Networks. Cambridge University Press, Cambridge (2008)
20. Hu, T., Payne, J.L., Banzhaf, W., Moore, J.H.: Robustness, evolvability, and accessibility in linear genetic programming. In: Silva, S., Foster, J.A., Nicolau, M., Machado, P., Giacobini, M. (eds.) EuroGP 2011. LNCS, vol. 6621, pp. 13–24. Springer, Heidelberg (2011). https://doi.org/10.1007/978-3-642-20407-4_2
21. Hu, T., Banzhaf, W.: Quantitative analysis of evolvability using vertex centralities in phenotype network. In: Proceedings of the 25th Genetic and Evolutionary Computation Conference (GECCO), pp. 733–740 (2016)
22. Csardi, G., Nepusz, T.: The igraph software package for complex network research. InterJournal Complex Syst. **1695**, 1–9 (2006). http://igraph.org
23. Newman, M.E.J., Girvan, M.: Finding and evaluating community structure in networks. Phys. Rev. E **69**, 026113 (2004)
24. Masuda, N., Porter, M.A., Lambiotte, R.: Random walk and diffusion in networks. Phys. Rep. **716**, 1–58 (2017)

Improving Genetic Programming with Novel Exploration - Exploitation Control

Jonathan Kelly[✉], Erik Hemberg, and Una-May O'Reilly

MIT, Cambridge, MA 02139, USA
jgkelly@mit.edu, {hembergerik,unamay}@csail.mit.edu

Abstract. Low population diversity is recognized as a factor in premature convergence of evolutionary algorithms. We investigate program synthesis performance via grammatical evolution. We focus on novelty search – substituting the conventional search objective – based on synthesis quality, with a novelty objective. This prompts us to introduce a new selection method named knobelty. It parametrically balances exploration and exploitation by creating a mixed population of parents. One subset is chosen based on performance quality and the other subset is chosen based on diversity. Three versions of this method, two that adaptively tune balance during evolution solve program synthesis problems more accurately, faster and with less duplication than grammatical evolution with lexicase selection.

Keywords: Program synthesis · Novelty · Diversity

1 Introduction

Program synthesis is an open, relevant and challenging problem domain within genetic programming, (GP) [1]. Synthesis of introductory programming problems has been tackled using methods such as PUSHGP [2] and Grammar Guided Genetic Programming (G3P) [3]. These methods have, to date, not been able to solve the complete suite of program synthesis benchmark problems introduced in [1]. One possible explanation is convergence to local optima. Premature convergence is often correlated with low solution diversity. The population becomes concentrated within a small part of the solution space and crossover and mutation operators yield only sub-optimal solutions. In GP, for example, diversity has been narrowly examined in lexicase selection studies solving program synthesis [4]. In this paper, we present a broader study of diversity in GP with Grammatical Evolution (GE) and program synthesis.

Our central question is: **Can diversity be controlled within GP to improve synthesis performance and prevent premature convergence?**

An approach called novelty search, introduced in [5] provides us with partial inspiration. In novelty search, rather counterintuitively, selection is altered

© Springer Nature Switzerland AG 2019
L. Sekanina et al. (Eds.): EuroGP 2019, LNCS 11451, pp. 64–80, 2019.
https://doi.org/10.1007/978-3-030-16670-0_5

so its objective is not to select a solution of superior "quality", e.g. performance on synthesis, but of higher novelty, using a score calculated by measuring how different a solution is from others. Over the course of a search completely biased by novelty, untimely convergence is circumvented and solutions of better quality can be coincidentally identified as side effects, despite quality being ignored by selection, see [5–7]. Novelty search has a demonstrated track record in benchmark tunably deceptive problems and academic problems but there are differences between these domains and program synthesis. In GP, on robot controller, symbolic regression and genetic improvement, novelty search results are mixed [6–8].

Our study finds that some distance measures used in "pure" novelty search can struggle to find solutions that perform synthesis well. But, it reveals an intuitive similarity between lexicase selection and novelty search that considers output novelty. Time plots of novelty search point to a new hypothesis stating that a search relying upon a population which has members that are good at synthesis and others that are novel, will yield better solutions. Distinctly, this hypothesis does NOT propose a population composed of members that *individually* combine novelty and synthesis performance. Instead, it seeks a population of mixed composition.

To validate the hypothesis empirically, the study proposes a form of tunable selection that we call knobelty. The name knobelty, a porte-manteau of *novelty* and *knob*, conveys that the selection method has a parameterized threshold (vis. *knob*) that controls the likelihood of a novelty selection objective being used vs a performance-based one. In expectation, knobelty populates a population of parents, some fraction of which have been selected by a novelty measure, and the rest by synthesis performance. The study evaluates three versions of knobelty that differ by how the threshold between novelty and performance is controlled. The base case splits between novelty and performance selection using a static threshold, unchanged over a run. Two others vary how much novelty is selected for *during* evolution, i.e. dynamically. One decreases the threshold using a decreasing exponential function sensitive to time. The other is sensitive to duplication. Every generation, it re-adjusts the threshold according to the duplication that arises after selection, mutation and crossover. The study proceeds with a representative subset of program synthesis problems and reports the accuracy, speed, and efficiency of knobelty, concluding it is successful and worthy of more study.

The structure of the paper is as follows, Sect. 2 has background. Our methods are in Sect. 3. Experiments, results and discussion are in Sect. 4. Finally, Sect. 5 has conclusions and future work.

2   Background

The foundations of this study are grammar based GP – in the form of Grammatical Evolution, program synthesis and search convergence analysis through the lens of diversity and novelty search. We present background for each topic in this section.

2.1 Grammar Based Genetic Programming

Grammatical Evolution (GE) is a genetic programming algorithm, where a BNF-style, context free grammar is used in the genotype to phenotype mapping process [9]. A grammar provides flexibility because the solution language can be changed without changing the rest of the GP system. Different grammars, for the same language, can also be chosen to guide search bias. The genotype-phenotype mapping allows variation operators, crossover and mutation, to work on the genotype (an integer vector), or the derivation tree that is intermediate to the genotype and phenotype. Following natural evolution, selection is based on phenotype behavior, i.e. performance of program on required task. A drawback of GE is lack of locality [10].

A context free grammar (CFG) is a four-tuple $G = \langle N, \Sigma, R, S \rangle$, where: *(1)* N is a finite non-empty set of non-terminal symbols. *(2)* Σ is a finite non-empty set of terminal symbols and $N \cap \Sigma = \emptyset$, the empty set. *(3)* R is a finite set of production rules of the form $R : N \mapsto V^* : A \mapsto \alpha$ or (A, α) where $A \in N$ and $\alpha \in V^*$. V^* is the set of all strings constructed from $N \cup \Sigma$ and $R \subseteq N \times V^*$, $R \neq \emptyset$. *(4)* S is the start symbol, $S \in N$ [11].

GE uses a sequence of (many-to-one) mappings to transition from a genotype to its fitness:

(1) **Genotype:** An integer sequence
(2) **Derivation Tree:** Each integer in the genotype controls production rule selection in the grammar. This generates a rule production sequence which can be represented as a derivation tree
   Program/Phenotype: The leaves (terminals) of the derivation tree, which together form the sentence. For example, in the program synthesis domain this is the generated program.
(3) **Fitness:** For each test case, a value is assigned measuring the distance between the desired outputs and program outputs when executed with it. For a solution, fitness summarizes this value for all test cases. It can be a scalar statistic, e.g. sum, or a bit vector for each test case's success test.

2.2 Program Synthesis

In GP, program synthesis is formulated as an optimization problem: find a program q from a domain Q that minimizes combined error on a set of input-output cases $[X, Y]^N, x \in X, y \in Y$. Typically an indicator function measures error on a single case: $\mathbb{1} \colon q(x) \neq y$.

Initial forays into program synthesis considered specific programming techniques, such as recursion, lambda abstractions and reflection, or languages such as C or C++, or problems such as caching, exact integer and distributed algorithms, [12–18]. Other approaches consider implicit fitness sharing, co-solvability, trace convergence analysis, pattern-guided program synthesis, and behavioral archives of subprograms [19].

Subsequently, a general program synthesis benchmark suite of 29 problems systematically selected from sources of introductory computer science programming problems became available to the GP community [1]. Multiple studies, many with PUSHGP, have drawn on this suite, e.g. a study of different selection operators, in particular one named lexicase selection [4,20]. Diversity was measured in this work as a means of explaining why lexicase selection works. The goal of lexicase selection can be summarized as promoting into the next generation, parents that collectively solve different test cases. In this work, our baseline algorithm, GE_Perf and the performance selection modules of knobelty algorithms use lexicase selection. Referencing the same benchmarks G3P attempts to present a general grammar suitable for arbitrary program synthesis problems [21]. Another G3P effort analyses test set generalization [3].

2.3   Search Convergence

Solution discovery can be hindered by convergence to local optima. Diversity and behavioral novelty are two methods to address this.

Diversity. Early diversity related work in genetic algorithms includes "crowding" [22,23] where a solution is compared bitwise to a randomly drawn subpopulation and replaces the most similar member among the subpopulation. Later studies enforce some sort of solution niching. Separation of the population by age is done with ALPS [24]. Spatial separation of solutions is used in coevolutionary learning [25]. Behavioral information distance sustains diversity on the Tartarus problem [26]. In GP, diversity measures and diversity's correlation with fitness were studied as early as [27], and diversity has continually been demonstrated to play an important role in premature convergence [28]. Surprisingly, [29] showed that even variable-length GP trees can still converge genotypically.

Interestingly, bloat confounds tree distance. When trees get bigger, but do not change behaviorally because of bloat, distance becomes nonsensical, see [30]. This indicates bloat must be under control if we use diversity to guide the search. Overall, past efforts show that it is rare to use diversity to guide search. In addition, there is also limited work on diversity and program synthesis.

Novelty Search. Novelty search [5] is one approach to overcome convergence and lack of solution diversity. In pure novelty search, there is absolutely no selection pressure based on performance. The method uses a distance measure (defined over 2 solutions), a novelty measure defined for a solution (summarizing many distance measurements to others), a management policy for the memory holding solutions that the search has to date generated (from which distance/novelty will be calculated) and a selection objective maximizing novelty. Novelty search stresses selecting for novelty in a *behavioral* domain. The most intuitive representation of behavior in GP has been explored – program outputs. Since programs and derivation trees express behavior through statically analyzable semantic information, alternate representations remain to be explored. This study, for example, uses distance measures over GE programs and derivation trees. It also implements an updated memory management policy that uses an unbounded cache and sampling for novelty approximation.

GP explorations with novelty search [8] are comparable on different axes: problem domains and new methods. Problem domains include robot controller navigation, symbolic regression, classification among others [6–8,31,32]. New methods include crowding selection methods and a weighted combined fitness and novelty score [33]. Similarly, [7] combine diversity and performance into one objective convergence in a robot controller problem.

The background implies that there are research opportunities at the intersection of GP using grammars, program synthesis and novelty search. The next section introduces the methods we develop for this investigation.

3   Method

We proceed in three phases. In the first phase we evaluate novelty during GE program synthesis evolution. Methodologically this requires a distance measure (see Sect. 3.1), a novelty measure and a statistical summarization of novelty for a population (see Sect. 3.2).

3.1   Our Distance Measures

A distance measure is defined between two solutions $i, j, d(i, j) \in \mathbb{R}$. It describes how far apart solutions are, in some basis. We measure distance using each of the representations in GE that map from genotype to phenotype, plus outputs:

(1) Genotype: We measure with Hamming distance.
(2) Derivation Tree (DT): Distance is a common and different nodes count. Ignoring structure, we collapse the tree structure into node counts. We measure distance as the Euclidean norm of the nodes common to both trees plus how many nodes are different.
(3) Phenotype: We measure with (Levenshtein[1]) string edit distance and divide all measurements by the size of the largest phenotype.
(4) Output distance: We record the success of each test case in a binary vector cell. Output distance is the Hamming distance between the two vectors.

3.2   How We Measure Novelty

A novelty measure of a GE solution j is based on its pairwise distances to a set of programs C. We average these pairwise distances to obtain a scalar novelty value.

Conceptually C must function as a memory structure. One option is to make it an archive that is finite and selectively updated depending on some entry and retirement policy. A policy typically has multiple threshold parameters which makes it complex to manage. Alternatively, C can be an "infinite" cache but the

---

[1] https://en.wikipedia.org/wiki/String_metric.

Algorithm 1. $A(i, d, C, N)$ **Novelty approximation**
Parameters: I Individual, d distance measure, $\mathbf{C}$ cache, N sample size

1: $n \leftarrow 0$	▷ Initial novelty
2: **for** $i \in [1, \ldots, N]$ **do**	▷ Draw individuals to compare against
3: $\quad O \leftarrow \sim \mathcal{U}(\mathbf{C})$	▷ Uniform sampling with replacement
4: $\quad n \leftarrow n + d(I, O)$	▷ Get distance
5: **return** n/N	▷ Return average distance

cost of computing distance from j to every solution in a cache is computationally expensive.

We propose to simplify C and streamline its computation. We record every unique encountered individual in C. To approximate the novelty of j, we draw N samples (with replacement) from C and average $d(k, j)$ over them. To choose N, we scan a range of sample sizes and choose a value that is stable under many draws. Whereas the extreme case of a bottomless cache scales $O(P^2 T^2)$, (T is the number of generations and P is the population size), sampling reduces the complexity to $O(NPT)$. Algorithm 1 shows our approximate novelty calculation.

3.3 Knobelty Selection

We observe that convergence must be influenced by dual forces: the diversity of solutions in the population and the performance or quality of solutions in the population. The former fosters explorative search and the latter exploitive search; at issue is how to juggle these. They are not always conflicting so a multi-objective framing is inappropriate. A weighted score balancing each of them could be used as fitness but this would not explicitly yield either good performers or highly diverse solutions but solutions in between. We propose, therefore, to control this balance between exploration and exploration by creating a mixed parent pool. One subset will be selected based on novelty and the other based on performance. A parameterized threshold (knob) $\kappa \in [0, 1]$ choosing between novelty and performance selection will determine the subsets' sizes in expectation. We call this Knobelty Selection. Our hypothesis is that the decrease in fitness selection pressure and increase in novelty selection pressure will prevent convergence to local optima without severely degrading the efficiency for finding a global optima. Algorithm 4 supports three control methods for κ:

Static. Keep κ constant. (line 3)
Gen_Adapt. Change κ every generation, $\kappa(t) = 2^{-\lambda t}$, $t =$ generation, using an exponential schedule to initially boost novelty and then afterwards allow the population to slowly converge.
Dup_Adapt. Change κ according to duplication in the population. We initialize κ and, realizing crossover and mutation can disrupt the balance between novelty and performance selection, we check the duplication ratio of the population and adjust κ to $\kappa(t) = 1 - |\{x | x, y \in P, x \neq y\}|/|P|$.

Algorithm 2. $S_\nu(P, C)$ **Novelty Based Selection**
Parameters: P Population, C cache
Local: ω tournament size, d distance measure

1: $\tau \leftarrow S_t(\omega, P)$	▷ Randomly choose competitors for a tournament
2: **for** $i \in \tau$ **do**	▷ Calculate novelty of each competitor
3: $i_\nu \leftarrow A(i, d, C, N)$	▷ **Approximate novelty, see Alg. 1**
4: **return** max(τ_ν)	▷ Pick most novel competitor

Algorithm 3. $S(P, C, \kappa)$ **Knobelty Selection**
Parameters: P population, C cache, κ novelty probability
Global: S_π Performance selection function, S_ν Novelty selection function

1: $P' \leftarrow \emptyset$	▷ New population
2: **for** $i \in [1, \ldots, |P|]$ **do**	▷ Select an Individual
3: $k \leftarrow \sim \mathcal{U}([0, 1])$	▷ Uniform random value
4: **if** $k < \kappa$ **then**	▷ Get selection measurement
5: $P' \leftarrow P' \cup S_\pi(P, C)$	▷ **Performance based lexicase selection**
6: **else**	
7: $P' \leftarrow P' \cup S_\nu(P, C)$	▷ **Novelty based selection Alg. 2**
8: **return** P'	▷ Return new population

See Algorithms 2, 3, and 4 for more details. With these distance measures, novelty definitions and `knobelty` selection we proceed to experimental evaluation.

4   Experiments

This section presents experimental setup, and results and discussion.

4.1   Setup

In order to focus on the potential of explicit diversity control, we selected a small subset of problems from the general programming synthesis benchmark suite [1]. As our intent was not to match the previous standards set by related work done by PUSHGP [20] or G3P [3], but rather to explore the effect of explicit diversity control, we decided to use a fitness evaluation budget of 3.0×10^4, a budget that is less than one sixth of the budget used by PUSHGP, and one tenth the budget used in G3P. In doing so, we restrict the number of problems that our system is able to solve, but we significantly increase our investigative agility. To choose which problems to use in our experiments, we first ran a series of tests with the decreased fitness evaluations on many of the benchmark problems, and chose three that provided a range of difficulty. These tests were done with GE and lexicase selection. We selected one easy – Median (MED), another moderately difficult – Smallest (SML), and a third hard – String Lengths Backwards (SLB). Per convention, the I/O dataset is split in two, one used during evolution, the

Algorithm 4. Grammatical Evolution with Knobelty Selection
Parameters: KC: knob control method ∈ Static, Dup_Adapt,Gen_Adapt

1:	$P \leftarrow \iota()$	▷ Initialize population
2:	$C \leftarrow \emptyset$	▷ Initialize cache
3:	**if** KC = Static **then**	▷ **Static** κ
4:	$\kappa \leftarrow k$	▷ Set κ to a static value
5:	$f(g(P))$	▷ Map and evaluate population
6:	$C \leftarrow C \cup P$	▷ Add population to cache
7:	**for** $t \in [1, \ldots, T]$ **do**	▷ Iterate over generations
8:	**if** KC = Gen_Adapt **then**	▷ **Generation based** κ **update**
9:	$\kappa \leftarrow 2^{-\lambda t}$	▷ Update κ based
10:	**if** KC = Dup_Adapt **then**	▷ **Duplication sensitive** κ **adaptation**
11:	$\kappa \leftarrow \Delta(\text{inefficiency}(P), \text{inefficiency}(P_{t-1}))$	▷ Update κ, see Sec 3.3
12:	$P' \leftarrow S(P, C, \kappa)$	▷ **Knobelty Selection, Alg. 3**
13:	$P' \leftarrow \chi(P')$	▷ Crossover individuals
14:	$P' \leftarrow \mu(P')$	▷ Mutate individuals
15:	$f(g(P'))$	▷ Map and evaluate population
16:	$C \leftarrow C \cup P'$	▷ Add population to cache
17:	$P \leftarrow P'$	▷ Replace population
18:	**return** $max(C)$	▷ **Return best performing solution**

training set, and one for out of sample testing, the testing set. For each of our selected problems, the training set consists of 100 test cases and the testing set consists of 1000 test cases.

We report results on 100 runs. We report program synthesis performance in terms of how many runs out of 100 resulted in one or more programs that solved all the out of sample (test) cases. All other reported values are averages over 100 runs. We ran all experiments on a cloud VM with 24 cores, 24 GB of RAM, and 16 GB of disk.

To determine an efficient population to generation ratio, we swept the ratios while keeping fitness evaluations constant and found that a population size of 1500 with 20 generations produced better results on all three problems than other ratios. This contrasts significantly with the population to generation ratio that PUSHGP and G3P use, with our ratio of population size:generations $1500:20 = 75:1$ vs PUSHGP and G3P of $1000:300 = 3.3:1$. We believe that this is another example of diversity having an impact on performance. When choosing the original population the seeding operator is able to effectively space out individuals throughout the search space. Our results imply that this high initial diversity followed by a small number of generations to evolve is more effective than a smaller and thus less diverse initial population that has more generations to evolve.

Our implementation originates from the grammar based genetic programming repository PonyGE2 [34]. Building on PONYGE2, we added lexicase selection to create a conventional GE algorithm for program synthesis that uses lexicase selection. This algorithm, GE_Perf, uses performance based selection.

Table 1. Baseline performance for different GP variants on the MED, SLB and SML program synthesis problems. N.B. GE uses an order of magnitude lower fitness evaluation budget.

Heuristic	Test performance		
	MED	SLB	SML
GE_Perf	85	8	74
G3P [3]	79	68	94
PUSHGP [20]	55	94	100

We then designed and developed our various knobelty algorithm variants (see Algorithm 4 and Table 3)[2]. The set of parameters we used throughout all our experiments is listed in Table 2.

Table 2. Experimental settings

Parameter	Value
Codon size	100,000
Elite size	15 $(0.01P)$
Replacement	Generational
Initialisation	PI grow
Init genome length	200
Max genome length	500
Max init tree depth	10
Max tree depth	17
Max tree nodes	250
Max wraps	0
Crossover	Single point
Crossover probability	0.9
Mutate duplicates	False
Mutation	Int flip
Mutation probability	0.05
Novelty archive sample size (C)	100
Novelty tournament size (ω)	7

Table 1 presents the program synthesis performance of our baseline algorithm GE_Perf with those reported by PUSHGP and G3P, for each of the three test cases we chose. All three algorithms use lexicase selection. The "Perf" is implicit

---

[2] The code is available at https://github.com/flexgp/novelty-prog-sys.

in PUSHGP and G3P, as they are both entirely concerned with performance. We make it explicit in GE_Perf since we will later use GE for our knobelty algorithms.

Regarding the test performance of the baselines, bear in mind that GE_Perf uses between 1/6th and 1/10th fitness evaluations per run compared to PUSHGP and G3P. With this handicap, it performs moderately worse on SML, significantly worse in SLB, but is actually able to outperform the other approaches on MED. From this point forward, we will solely use the results from GE_Perf as our baseline.

Table 3. knobelty algorithm variants, see Algorithm 4

Abbreviation	Explanation
GenoNovelty	GE with Knobelty selection using genotype novelty approximation
DTreeNovelty	GE with Knobelty selection using derivation tree novelty approximation
PhenoNovelty	GE with Knobelty selection using phenotype novelty approximation
OutputsNovelty	GE witn Knobelty selection using standard outputs novelty approximation

All knobelty algorithm variants approximate novelty with Algorithm 1. One of its parameters is the distance measure it uses. We abbreviate the variants by this distance measure, see Table 3. The approximation's sampling size of 100 was experimentally set by a sweep that identified the lowest size that is stable over 1000 sample tests. The tournament size of knobelty selection tournaments is $\omega = 7$. For the exponentially decreasing novelty, we used $\lambda = $ Number of Generations/10 = 2. We did no experimentation with the range of λ.

Experimental Approach. We proceed in two steps.

1. We set $\kappa = 1.0$ and run OutputsNovelty to closely approximate the spirit of the original novelty [5]. We consider how this compares to GE_Perf, our baseline. Then, with $\kappa = 1.0$, we try DTreeNovelty, PhenoNovelty, GenoNovelty and compare to GE_Perf and OutputsNovelty.
2. We then use DTreeNovelty with Static knob control to conduct a sensitivity analysis of κ by sweeping it across a range of values for the MED problem. We look at program size, duplication, best performance and novelty. Then we run all three problems (MED, SLB and SML) using two novelty algorithms - DTreeNovelty, OutputsNovelty, with 3 knob controls – Static, Gen_Adapt and Dup_Adapt.

4.2 Results

Proceeding with step 1, Table 4 compares GE_Perf to PhenoNovelty, GenoNovelty, DTreeNovelty abd OutputsNovelty, run with a κ of 1.0, i.e. "pure" novelty. From these results, we can draw three insights. The first is that using pure novelty search with DTreeNovelty, PhenoNovelty, and GenoNovelty is unsuccessful. This observation is arguably to be expected. Genotypes in GE do not express behavior. Derivation trees and the programs defined by their leaves express behavior, but there is a many to one mapping between program and output behavior, so the search space of derivation trees and program overwhelms search based on novelty. The second observation is that OutputsNovelty, in contrast, does reasonably well, beating the baseline on two of the three problems. This can be explained by the observation that, while OutputsNovelty is not directly related to fitness, taking the Hamming Distance between two solu-

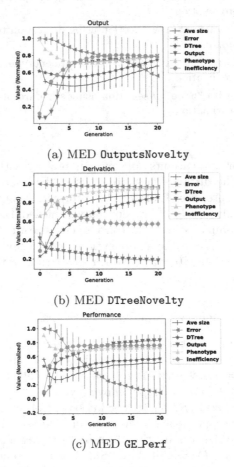

(a) MED OutputsNovelty

(b) MED DTreeNovelty

(c) MED GE_Perf

Fig. 1. Measurements per generation for MED. Y-axis is normalized measurement value and x-axis shows generation. Average and standard deviation over 100 runs.

tions' binary test case error vectors expresses their relative performance. Novelty search is favoring different but mutually compatible solutions that could be combined successfully with crossover. In fact, Fig. 1, which plots the average novelty at each generation for OutputsNovelty (Fig. 1a), DTreeNovelty (Fig. 1b), and GE_Perf (Fig. 1c), shows an inverse correlation between the output novelty and error trajectories. Since lower values of error are better, we observe that increasing OutputsNovelty also relates to better fitness. This finding is strengthened because for two out of three test cases, OutputsNovelty leads to better performance than GE_Perf.

Table 4 and Fig. 1 present *inefficiency*. Inefficiency is the ratio of the number of duplicate solutions to the product of number of fitness evaluations and generations. The third insight from Table 4 is that the inefficiency of GE_Perf is significantly higher then that of PhenoNovelty and DTreeNovelty. This means that GE_Perf can't generate novel solutions, while searching based on novelty can. While pure DTreeNovelty or PhenoNovelty fail, their ability to effectively explore the search space motivates the exploration of a combination of GE_Perf and novelty. Since DTreeNovelty is the algorithm that is the least inefficient, and thus explores the most solutions, and OutputsNovelty performs well on two of the three problems, we decide to move forward with them in our knobelty experiments.

Table 4. (Pure) Novelty experiments. We set $\kappa = 1.0$ and run OutputsNovelty, DTreeNovelty, PhenoNovelty, GenoNovelty. We show GE_Perf, our baseline which uses performance based selection, for comparison, above each problem's results.

Problem	Distance – knobelty Alg	Fitness		Time	Ineff.	Novelty (Total)			Ave size
		Train	Test			Geno.	Pheno.	DT	
MED	GE_Perf	86	85	789.34	71	0.89	0.29	9.73	24.52
MED	genotype – GenoNovelty	0	0	848.71	74	1.00	0.27	7.54	13.02
MED	phenotype – PhenoNovelty	0	0	1392.13	26	0.99	0.53	16.12	46.88
MED	derivation – DTreeNovelty	0	0	785.11	19	0.99	0.47	25.15	62.47
MED	outputs – OutputsNovelty	5	5	600.95	64	0.98	0.28	8.78	22.58
SLB	GE_Perf	8	8	1446.47	67	0.95	0.22	7.55	22.80
SLB	genotype – GenoNovelty	1	1	2039.14	67	1.00	0.18	6.67	14.06
SLB	phenotype – PhenoNovelty	0	0	2335.94	36	0.99	0.39	11.85	37.28
SLB	derivation – DTreeNovelty	0	0	2890.80	21	0.99	0.34	19.57	51.34
SLB	outputs – OutputsNovelty	13	13	2120.16	53	0.98	0.23	7.89	23.78
SML	GE_Perf	74	74	1350.96	81	0.93	0.30	8.45	19.68
SML	genotype – GenoNovelty	0	0	918.89	73	1.00	0.27	7.53	13.16
SML	phenotype – PhenoNovelty	0	0	1019.02	26	0.99	0.53	16.08	47.23
SML	derivation – DTreeNovelty	0	0	841.54	19	0.99	0.47	24.91	63.60
SML	outputs – OutputsNovelty	97	97	421.48	56	0.96	0.30	10.02	28.56

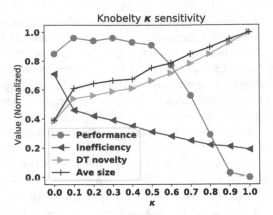

Fig. 2. Sensitivity analysis of κ constant for knobelty selection for the MED problem. Y-axis shows ratio value and x-axis shows κ. One line is performance, one line is D'Tree novelty, one line is inefficiency. The $\kappa = 0.2$ seems to display the best trade-off in performance, efficiency and novelty

We now proceed with step 2, where we start by analysing DTreeNovelty with Static knob control. We conduct a sensitivity analysis of κ by sweeping it across a range of values from 0.0 to 1.0 on 0.1 intervals for the MED problem. Figure 2 shows the results from these experiments. We can see that with low values of κ the performance improves compared to our baseline GE_Perf ($\kappa = 0.0$), confirming our hypothesis that pairing fitness selection with novelty selection can improve performance. The parameter setting of $\kappa = 0.2$ seems to display the best trade-off in performance, efficiency and novelty. As expected, average program size and novelty, specifically DTreeNovelty, increases as κ increases, and inefficiency decreases.

We select $\kappa = 0.2$ to go forward and we run all three problems (MED, SLB and SML) using the two novelty algorithms we picked from analysis of Table 4 – DTreeNovelty, OutputsNovelty. We ran the experiments with all three ways to set the parameter κ that controls the balance between using novelty based and performance based selection – Static, Gen_Adapt and Dup_Adapt. Table 5 shows the results obtained from these experiments. We see from these results that in all cases, using knobelty selection based on either DTreeNovelty or OutputsNovelty does better than our GE_Perf. We also see that despite the fact that OutputsNovelty did significantly better when run with a κ of 1.0 than DTreeNovelty (Table 4), when κ is less than 1.0, i.e. when each uses parents selected with a mixture of novelty and performance, they produce similar results. We also see that three of the algorithm plus κ control combinations yield perfect solutions 100% of the time. The DTreeNovelty algorithm for Dup_Adapt solves SML and MED perfectly, and the OutputsNovelty algorithm for Gen_Adapt solves SML perfectly. Additionally, on SLB, the number of test cases solved by just GE_Perf is increased by 100% by many of the algorithm and control combinations. These results on three problems strongly support our hypothesis that explicit diversity control can play a role in improving performance. They suggest going forward with more problems, and doing a more comprehensive comparison that includes other methods.

Table 5. Two `knobelty` algorithms, one basing novelty on derivation trees and the other on outputs are compared on the three problems for the three different ways to control the `knobelty` knob.

Algorithm	Control of κ	Test Performance		
		MED	SLB	SML
GE_Perf	-	85	8	74
DTreeNovelty	Static, $\kappa = 0.2$	+9	+10	+24
OutputsNovelty	Static, $\kappa = 0.2$	+12	+7	+23
DTreeNovelty	Dup_Adapt	+15	0	+26
OutputsNovelty	Dup_Adapt	+12	+13	+22
DTreeNovelty	Gen_Adapt	+11	+10	+24
OutputsNovelty	Gen_Adapt	+12	+9	+26

In the next section we conclude and present possible directions for future work.

5   Conclusions and Future Work

The contributions of this paper are:

1. We introduce a computationally tractable approximation of novelty for GP. It samples the cache rather than exhaustively referencing every item in it. This dispenses with a complex cache management policy.
2. We introduce novelty measures on genotype, derivation tree and program representation domains for GE.
3. Using these measures, we explore GE with a conventional performance objective, pure novelty, and `knobelty` for program synthesis.
4. We find evidence that `knobelty` can successfully balance a population's proportions of novel and high performing solutions, thus program synthesis can be improved in performance accuracy, speed and efficiency. Since these successful results are only based on three judiciously chosen problems, further investigation is merited.

There are a number of possible directions for future work. One is to mutate duplicate solutions, to increase the search space visited and drive the inefficiency to zero. Another is to evaluate the `knobelty` operators across more problems, with an increased number of fitness evaluations. These results would then be comparable with the results of PUSHGP and G3P. A third is to try `knobelty` with tree based operators and see how the results compare. In these tree based experiments, the effect that novelty has on bloat would be especially interesting to monitor, as tree based operators are known to have problems with bloat. Finally it would be interesting to investigate other representations of program behavior, including program traces, and other static program and derivation tree representations.

Acknowledgements. This material is based upon work supported by DARPA. The views and conclusions contained herein are those of the authors and should not be interpreted as necessarily representing the official policies or endorsements either expressed or implied of Applied Communication Services, or the US Government.

References

1. Helmuth, T., Spector, L.: General program synthesis benchmark suite. In: Proceedings of the 2015 Annual Conference on Genetic and Evolutionary Computation, pp. 1039–1046. ACM (2015)
2. Spector, L.: Autoconstructive evolution: push, pushGP, and pushpop. In: Proceedings of the Genetic and Evolutionary Computation Conference (GECCO-2001), vol. 137 (2001)
3. Forstenlechner, S., Fagan, D., Nicolau, M., O'Neill, M.: Towards understanding and refining the general program synthesis benchmark suite with genetic programming. In: 2018 IEEE Congress on Evolutionary Computation (CEC), pp. 1–6. IEEE (2018)
4. Helmuth, T., McPhee, N.F., Spector, L.: Lexicase selection for program synthesis: a diversity analysis. In: Riolo, R., Worzel, B., Kotanchek, M., Kordon, A. (eds.) Genetic Programming Theory and Practice XIII. GEC, pp. 151–167. Springer, Cham (2016). https://doi.org/10.1007/978-3-319-34223-8_9
5. Lehman, J., Stanley, K.O.: Exploiting open-endedness to solve problems through the search for novelty. In: ALIFE, pp. 329–336 (2008)
6. López-López, V.R., Trujillo, L., Legrand, P.: Novelty search for software improvement of a slam system. In: Proceedings of the Genetic and Evolutionary Computation Conference Companion, pp. 1598–1605. ACM (2018)
7. Doucette, J., Heywood, M.I.: Novelty-based fitness: an evaluation under the Santa Fe Trail. In: Esparcia-Alcázar, A.I., Ekárt, A., Silva, S., Dignum, S., Uyar, A.Ş. (eds.) EuroGP 2010. LNCS, vol. 6021, pp. 50–61. Springer, Heidelberg (2010). https://doi.org/10.1007/978-3-642-12148-7_5
8. Naredo, E.: Genetic programming based on novelty search. Ph.D. thesis, ITT, Instituto tecnologico de Tijuana (2016)
9. Ryan, C., O'Neill, M., Collins, J.J.: Introduction to 20 Years of Grammatical Evolution. In: Ryan, C., O'Neill, M., Collins, J.J. (eds.) Handbook of Grammatical Evolution, pp. 1–21. Springer, Cham (2018). https://doi.org/10.1007/978-3-319-78717-6_1
10. Thorhauer, A., Rothlauf, F.: On the locality of standard search operators in grammatical evolution. In: Bartz-Beielstein, T., Branke, J., Filipič, B., Smith, J. (eds.) PPSN 2014. LNCS, vol. 8672, pp. 465–475. Springer, Cham (2014). https://doi.org/10.1007/978-3-319-10762-2_46
11. Booth, T.L.: Sequential machines and automata theory (1967)
12. O'Neill, M., Ryan, C.: Evolving multi-line compilable C programs. In: Poli, R., Nordin, P., Langdon, W.B., Fogarty, T.C. (eds.) EuroGP 1999. LNCS, vol. 1598, pp. 83–92. Springer, Heidelberg (1999). https://doi.org/10.1007/3-540-48885-5_7

13. Lucas, S.M.: Exploiting reflection in object oriented genetic programming. In: Keijzer, M., O'Reilly, U.-M., Lucas, S., Costa, E., Soule, T. (eds.) EuroGP 2004. LNCS, vol. 3003, pp. 369–378. Springer, Heidelberg (2004). https://doi.org/10.1007/978-3-540-24650-3_35

14. Agapitos, A., Lucas, S.M.: Learning recursive functions with object oriented genetic programming. In: Collet, P., Tomassini, M., Ebner, M., Gustafson, S., Ekárt, A. (eds.) EuroGP 2006. LNCS, vol. 3905, pp. 166–177. Springer, Heidelberg (2006). https://doi.org/10.1007/11729976_15

15. Yu, T., Clack, C.: Recursion, lambda-abstractions and genetic programming. In: Poli, R., Langdon, W.B., Schoenauer, M., Fogarty, T., Banzhaf, W. (eds.) Late Breaking Papers at EuroGP 1998: The First European Workshop on Genetic Programming, pp. 26–30. CSRP-98-10, The University of Birmingham, UK, Paris, 14–15 April 1998

16. Wan, M., Weise, T., Tang, K.: Novel loop structures and the evolution of mathematical algorithms. In: Silva, S., Foster, J.A., Nicolau, M., Machado, P., Giacobini, M. (eds.) EuroGP 2011. LNCS, vol. 6621, pp. 49–60. Springer, Heidelberg (2011). https://doi.org/10.1007/978-3-642-20407-4_5

17. Weise, T., Tang, K.: Evolving distributed algorithms with genetic programming. IEEE Trans. Evol. Comput. **16**(2), 242–265 (2012). https://doi.org/10.1109/TEVC.2011.2112666

18. Weise, T., Wan, M., Tang, K., Yao, X.: Evolving exact integer algorithms with genetic programming. In: 2014 IEEE Congress on Evolutionary Computation (CEC), pp. 1816–1823, July 2014

19. Krawiec, K.: Behavioral Program Synthesis with Genetic Programming. SCI, vol. 618, pp. 1–19. Springer, Cham (2016). https://doi.org/10.1007/978-3-319-27565-9

20. Helmuth, T., McPhee, N.F., Spector, L.: Program synthesis using uniform mutation by addition and deletion. In: Proceedings of the Genetic and Evolutionary Computation Conference, pp. 1127–1134. ACM (2018)

21. Forstenlechner, S., Fagan, D., Nicolau, M., O'Neill, M.: A grammar design pattern for arbitrary program synthesis problems in genetic programming. In: McDermott, J., Castelli, M., Sekanina, L., Haasdijk, E., García-Sánchez, P. (eds.) EuroGP 2017. LNCS, vol. 10196, pp. 262–277. Springer, Cham (2017). https://doi.org/10.1007/978-3-319-55696-3_17

22. De Jong, K.A.: Analysis of the behavior of a class of genetic adaptive systems (1975)

23. Mengshoel, O.J., Goldberg, D.E.: The crowding approach to niching in genetic algorithms. Evol. Comput. **16**(3), 315–354 (2008)

24. Hornby, G.S.: ALPS: the age-layered population structure for reducing the problem of premature convergence. In: Proceedings of the 8th Annual Conference on Genetic and Evolutionary Computation, pp. 815–822. ACM (2006)

25. Mitchell, M., Thomure, M.D., Williams, N.L.: The role of space in the success of coevolutionary learning. In: Artificial Life X: Proceedings of the Tenth International Conference on the Simulation and Synthesis of Living Systems, pp. 118–124. MIT Press, Cambridge (2006)

26. Gomez, F.J.: Sustaining diversity using behavioral information distance. In: Proceedings of the 11th Annual Conference on Genetic and Evolutionary Computation, GECCO 2009, pp. 113–120. ACM, New York (2009). https://doi.org/10.1145/1569901.1569918

27. Burke, E.K., Gustafson, S., Kendall, G.: Diversity in genetic programming: an analysis of measures and correlation with fitness. IEEE Trans. Evol. Comput. **8**(1), 47–62 (2004)

28. Sudholt, D.: The benefits of population diversity in evolutionary algorithms: a survey of rigorous runtime analyses. arXiv preprint arXiv:1801.10087 (2018)
29. Burks, A.R., Punch, W.F.: An analysis of the genetic marker diversity algorithm for genetic programming. Genet. Program. Evolvable Mach. **18**(2), 213–245 (2017)
30. Affenzeller, M., Winkler, S.M., Burlacu, B., Kronberger, G., Kommenda, M., Wagner, S.: Dynamic observation of genotypic and phenotypic diversity for different symbolic regression GP variants. In: Proceedings of the Genetic and Evolutionary Computation Conference Companion, pp. 1553–1558. ACM (2017)
31. Shahrzad, H., Fink, D., Miikkulainen, R.: Enhanced optimization with composite objectives and novelty selection. arXiv preprint arXiv:1803.03744 (2018)
32. Goldsby, H.J., Cheng, B.H.C.: Automatically discovering properties that specify the latent behavior of UML models. In: Petriu, D.C., Rouquette, N., Haugen, Ø. (eds.) MODELS 2010. LNCS, vol. 6394, pp. 316–330. Springer, Heidelberg (2010). https://doi.org/10.1007/978-3-642-16145-2_22
33. Cuccu, G., Gomez, F.: When novelty is not enough. In: Di Chio, C., et al. (eds.) EvoApplications 2011. LNCS, vol. 6624, pp. 234–243. Springer, Heidelberg (2011). https://doi.org/10.1007/978-3-642-20525-5_24
34. Fenton, M., McDermott, J., Fagan, D., Forstenlechner, S., O'Neill, M., Hemberg, E.: PonyGE2: grammatical evolution in python. CoRR abs/1703.08535 (2017). http://arxiv.org/abs/1703.08535

Towards a Scalable EA-Based Optimization of Digital Circuits

Jitka Kocnova[✉] and Zdenek Vasicek[iD]

Faculty of Information Technology, IT4Innovations Centre of Excellence,
Brno University of Technology, Brno, Czech Republic
{ikocnova,vasicek}@fit.vutbr.cz

Abstract. Scalability of fitness evaluation was the main bottleneck preventing adopting the evolution in the task of logic circuits synthesis since early nineties. Recently, various formal approaches have been introduced to this field to overcome this issue. This made it possible to optimise complex circuits consisting of hundreds of inputs and thousands of gates. Unfortunately, we are facing to the another problem – scalability of representation. The efficiency of the evolutionary optimization applied at the global level deteriorates with the increasing complexity. In this paper, we propose to apply the concept of local resynthesis. Resynthesis is an iterative process based on extraction of smaller sub-circuits from a complex circuit that are optimized locally and implanted back to the original circuit. When applied appropriately, this approach can mitigate the problem of scalability of representation. Our evaluation on a set of non-trivial real-world benchmark problems shows that the proposed method provides better results compared to global evolutionary optimization. In more than 60% cases, substantially higher number of redundant gates was removed while keeping the computational effort at the same level.

Keywords: Cartesian Genetic Programming · Resynthesis · Logic optimization

1 Introduction

Logic synthesis, as understood by the hardware community, is a process that transforms a high-level description into a gate-level or transistor-level implementation. Due to the complexity of the problem, the synthesis process is typically broken into a sequence of steps. Among others, logic optimization represents an important part of the whole process. The goal of the logic optimization is to transform a suboptimal solution into an optimal gate-level implementation with respect to given synthesis goals. Due to the scalability issues, the problem is typically represented using a suitable internal representation. Current state-of-the-art logic synthesis tools, such as ABC, represent circuits using a directed acyclic graph composed of two-input AND nodes connected by direct or negated edges denoted as and-inverter graph (AIG). The optimization of AIGs is based

© Springer Nature Switzerland AG 2019
L. Sekanina et al. (Eds.): EuroGP 2019, LNCS 11451, pp. 81–97, 2019.
https://doi.org/10.1007/978-3-030-16670-0_6

on *rewriting*, a greedy algorithm which minimizes size of AIG by iteratively selecting subgraphs rooted at a node and replacing them with smaller precomputed subgraphs, while preserving the functionality of the root node [1]. AIG rewriting is local, however, the scope of changes becomes global by application of rewriting many times. In addition to that, *resubstitution* and *refactoring* can be employed. Resubstitution expresses the function of a node using other nodes present in the AIG [2]. Refactoring iteratively selects large cones of logic rooted at a node and tries to replace them with a more efficient implementation [1]. Refactoring can be seen as a variant of rewriting. The main difference is that rewriting selects subgraphs containing few leaves because the number of leaves determines the number of variables of a Boolean function whose optimal implementation is sought.

The AIG representation is simple and scalable, and leads to simple algorithms but it suffers from an inherent bias in representation. While eight of ten possible two-input logic gates may be represented by means of a single AIG node, XOR and XNOR gate require three AIG nodes each. The efficiency of synthesis is then limited as it mostly fully relies on transformations that disallow an increase the number of AIG nodes. It has been shown that there exists a huge class of real-world circuits for which the synthesis fails and provides very poor results [3–5]. In some cases, the area of the synthesized circuits is of orders of magnitude higher than the known optimum. If a large design is broken down to multiple smaller circuits and such a failure occurs during resynthesis, we obtain an unacceptably large circuit.

Various evolutionary approaches working directly at the level of gates were successfully applied to address this problem [3,6]. Vasicek demonstrated that the evolutionary synthesis using Cartesian Genetic Programming (CGP) conducted directly at the level of common gates is able to provide significantly better results compared to the state-of-the-art synthesis operating on AIGs [6]. On average, the method enabled a 34% reduction in gate count on an extensive set of benchmark circuits when executed for 15 min. It was observed, however, that the efficiency of the evolutionary approach deteriorates with an increasing number of gates. Substantially more generations were required to reduce circuits consisting of more than ten thousands gates. While [6] focuses strictly on the improvement of the scalability of the evaluation, Sekanina et al. employed a divide and conquer strategy to address the problem of scalability of representation [3]. The authors were able to obtain better results than other locally operating methods reported in the literature, however, the performance of this method was significantly worse than the evolutionary global optimization proposed in [6].

In order to improve the results of EA-based synthesis, we propose to combine the EA-based approach with refactoring while following the principle of local resynthesis applied in common logic synthesis tools. Firstly, a logic circuit is optimized by means of a common synthesis approach. Then, the optimized circuit is mapped to standard gates and optimized using the proposed method that extracts a relatively small sub-circuits that are subsequently optimized by Cartesian Genetic Programming (CGP). The original sub-circuit is then replaced by its optimized variant provided that there is an improvement at the global level and the whole process is repeated. Our approach is based on iterative optimization of large portions of the original circuit. Compared to rewriting, we do not impose any limitation on the number of leaves because the larger subgraphs offer more opportunities for potential area improvement.

2   Background

This section presents relevant background on conventional as well as EA-based optimization of logic circuits and introduces the notation used in the rest of the paper.

2.1   Boolean Networks

Every circuit can be represented using a Boolean network. A *Boolean network* is a directed acyclic graph (DAG) with nodes represented by Boolean functions [2]. The sources of the graph are the primary inputs (PIs) of the network and the sinks are the primary outputs (POs). The output of a node may be an input to other nodes called *fanouts*. The inputs of a node are called *fanins*. An edge connects two nodes that are in fanin/fanout relationship. Considering this notion, And-Inverter Graph is a Boolean network composed of two-input ANDs and inverters. The network primary inputs are signals that are driven by the environment, there is no node driving these signals in the network. Similarly, the primary outputs are signals that drive the environment and are needed by inner network nodes as well. The size of the network is the number of the nodes (primary inputs and outputs are not considered).

2.2   Limiting the Scope of Boolean Networks

Network scoping represents a key operation to ensure a good scalability of synthesis tools when working with large Boolean networks. In addition, it forms an integral part of rewriting as well as refactoring. Two approaches have been proposed to limit the scope of logic synthesis to work only on a small portion of a Boolean network – *windowing* and *cut computation* [2].

The windowing algorithm determining the window for a given node takes a node and two integers defining the number of logic levels on the fanin/fanout sides of the node to be included in the window. Two sets are produced as the result of windowing – leaf set and root set. The window of a Boolean network is the subset of nodes of the network containing nodes from root set together with all nodes on paths between the leaf set and the root set. The nodes in the leaf set are not included in the window. The main problem of this algorithm is that it is hard to predict how many logic levels have to be traversed to get a window of the desired size and required number of leaves.

A complementary approach based on computing so called k-feasible cuts is usually preferred to avoid determining the required number of logic levels. A cut of a node, called root node, is a set of nodes of the network, called leaves, such that each path from PI to the root node passes through at least one leaf. A cut is k-feasible if the number of nodes (i.e. cut size) in the cut does not exceed k. The volume of a cut is the total number of nodes encountered on all paths between the root node and the cut leaves. An example of two different 3-feasible cuts is shown in Fig. 1. To maximize the cut volume, a reconvergence-driven heuristic is applied in practice. The problem is that the cut computed using a naive bread-first-search algorithm may include only few nodes and leads to tree-like logic structures (see Fig. 1a showing a cut determined by the naive approach and Fig. 1b showing the output of reconvergence-driven heuristic). Such a structure does not lead to any don't cares in the local scope of the node and attempting optimization using such a cut would be wasted time. A simple and efficient cut computation algorithm producing a cut close to a given size while heuristically maximizing the cut volume and the number of reconvergent paths subsumed in the cut has been introduced in [2]. As our work is based on the reconvergence-driven cuts, we briefly discuss this algorithm. The algorithm starts with a set of leaves consisting of a single root node. This set is incrementally expanded by adding one node in each step of a recursive procedure. If the set consists of only PIs, the procedure quits. Otherwise, a non-PI node that minimizes a cost function is chosen from the set of leaves. The chosen node is removed from the leaf set and all its fanins are included instead of it. This causes expansion of the cut. If the cut-size limit is exceeded, the procedure quits and returns the cut before expansion. The cost function returns the number of new nodes that should be added to the leaf set instead of the removed node.

The k-feasible cuts are important not only for the gate-level logic synthesis but also for FPGA-based synthesis as a k-feasible cut can be implemented as a k-input LUT. For resubstitution and FPGA-based mapping, so called maximum

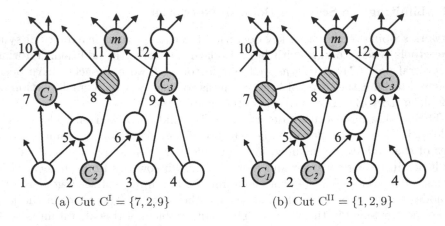

(a) Cut $C^I = \{7, 2, 9\}$ (b) Cut $C^{II} = \{1, 2, 9\}$

Fig. 1. Example of two possible 3-feasible cuts for root node m and given Boolean network. The cut C^{II} is preferred as its volume is four (root node m and contained nodes 5, 7, and 9). There is only one contained node (node 8) in the case of C^I.

fanout free cone (a subnetwork where no node in the cone is connected to a node not in the cone) is requested. It means that the cut-based scoping must always produce a single-output sub-circuits. Otherwise it would be impossible to replace the whole sub-circuit by a precomputed optimal implementation/a single LUT. Typically, 4-feasible and 5-feasible cuts are used for rewriting-based logic synthesis [2,7]. Small k is used not only to make the cut enumeration possible but also to manage memory requirements to store the precomputed optimal implementations of all k-input Boolean functions. For FPGA-based mapping, 5-input and 6-input LUTs are used. Apart from the rewriting, the reconvergence-driven cuts have been applied to refactoring and resubstitution [2]. Typically, k is between 5 and 12 for refactoring depending on the computation effort allowed [2].

2.3 Evolutionary Synthesis of Logic Circuits

Evolutionary algorithms (EAs) have been used to synthesize logic circuits since late nineties [8,9]. Miller et al., the author of Cartesian Genetic Programming (CGP) [10], is considered as a pioneer in the field of logic synthesis of gate-level circuits. He utilized his own variant of genetic programming to synthesize compact implementations of multipliers described by means of a behavioral specification [11]. Despite of many advantages of this unconventional technique, only small problem instances were typically addressed. To tackle the limited scalability, various decomposition strategies have been proposed. A good survey of the existing techniques is provided, for example, in [12]. The projection-based decomposition approaches such as [13] or [12] helped to increase the complexity of problem instances that can be solved by EAs. Despite of that, the gap between the complexity of problems addressed by EAs and in industry continued to widen as the advancements in technology developed. In 2011, the scalability of CGP has been significantly improved by introducing a SAT-based CGP. The SAT-based CGP replaces determining of Hamming distance done by exhaustive simulation with a modern SAT solver [14]. It exploits the fact that the candidate solutions must be functionally equivalent with their parent in logic optimization in order to be further accepted. In addition to that, it exploits the knowledge of differences between parental and candidate circuits. The efficiency of SAT-based method was further improved by combining a SAT solver with an adaptive high-performance circuit simulator used to quickly identify the potential functional non-equivalence. The most advanced SAT-based CGP employs a simulator that is driven by counterexamples produced by the SAT solver [6]. Neither the original nor the latter approach rely on a decomposition. The gate-level circuits are optimized directly.

Since its introduction, CGP remains the most powerful evolutionary technique in the domain of logic synthesis and optimization [9]. In this area, a linear form of CGP is preferred today. CGP models a candidate circuit having n_i PIs and n_o POs as a linear 1D array of n_n configurable nodes. Each node has n_a inputs and corresponds with a single gate with up to n_a inputs. The inputs can be connected either to the output of a node placed in the previous L columns or directly to PIs. This avoids a feedback. The function of a node can be chosen from a set of n_f functions. Depending on the function of a node, some of its inputs may become redundant. In addition to that, the fixed number of nodes n_n does not mean that all the nodes contribute to the POs. These key features allow redundancy and flexibility of CGP.

The candidate circuits are encoded as follows. Each PI as well as each node has associated an unique index. Each node is encoded using $n_a + 1$ integers $(x_1, \cdots, x_{n_a}, f)$ where the first n_a integers denote the indices of its fanins and the last integer determines the function of that node. Every candidate circuit is encoded using $n_n(n_a + 1) + n_o$ integers where the last n_o integers specify the indices corresponding with each PO.

CGP is a population oriented approach which operates with $1 + \lambda$ candidate solutions. The initial population is seeded by the original circuit ought to be optimized. Every new population consists of the best circuit chosen from the previous population and its λ offspring created using a mutation operator that randomly modifies up to h integers. Considering the CGP encoding, a single mutation causes either reconnection of a gate, reconnection of primary outputs or change in function of a gate. The selection of the individuals is typically based on a cost function (e.g. the number of active nodes). In the case that there are more individuals with the same score, the individual that has not served as a parent will be selected as the new parent. This procedure is typically repeated for a predefined number of iterations.

3   The Proposed Method

Let $\mathcal{C}$ be a combinational circuit described at the level of common gates represented by a Boolean network N consisting of $|N|$ nodes. Each node corresponds with a single gate in $\mathcal{C}$. The pseudo-code of the proposed optimization procedure based on evolutionary resynthesis is shown in Algorithm 1.

Firstly a node which may potentially lead to the best improvement of N is determined. Since the identification of this node itself is a nontrivial problem, some heuristic needs to be implemented. The size of transitive fan-in cone, level of the node or a more complex information can be used to determine the most suitable candidate. Then, a working area (window) is extracted from the Boolean network. This procedure starts with computation of the reconvergence-driven cut C as described in Sect. 2.2. From the practical reasons, is also beneficial to limit the size of C to be able to enumerate a large number of sub-circuits in a reasonable time. Hence, we can define four parameters: c_{min} and c_{max} restricting the volume of C ($c_{min} < c_{max}$), and k_{min} and k_{max} ($k_{min} \leq |C| \leq k_{max}$) limiting the size of cut (feasibility).

This step is followed by expansion of the cut C into a window W, i.e. expansion of the set of leaf nodes to a set of contained nodes. In addition to the nodes inside the cut, we should consider also all nodes that are not contained in the cut but have fanins inside the cut. Our expansion is similar to that employed in the resubstitution [2] where transitive fanout of C is considered, however, we do not impose any limit on the number of included nodes or their maximum level. The process of cut identification and the subsequent expansion is illustrated in Fig. 2.

Algorithm 1. EA-BASED RESYNTHESIS

Input: A Boolean network N
Output: Optimized network N', $cost(N') \leq cost(N)$
1 $N' \leftarrow N$
2 **while** *terminated condition not satisfied* **do**
3 　　$m \leftarrow$ identify the best candidate root node $m \in N'$
4 　　$C \leftarrow$ ReconvergenceDrivenCut(m)
5 　　$W \leftarrow$ ExpandCutToWindow(m, C)
6 　　**if** W *is not a suitable candidate* **then**
7 　　　　$\lfloor$ continue
8 　　$W' \leftarrow$ OptimizeNetworkUsingEA(W)
9 　　**if** $cost((N' \setminus W) \cup W') < cost(N')$ **then**
10 　　　$\lfloor$ $N' \leftarrow (N' \setminus W) \cup W'$

11 **return** N'

During the expansion, three set of nodes are created: the set of internal nodes I, the set of leaves L and the set of root nodes R. L contains nodes that will serve as PIs of the temporary network used in the subsequent optimization. Similarly R contains nodes whose outputs have to be connected to POs. Note that R contains not only the root node m but also other nodes whose fanouts are outside of the window (see Fig. 2). It holds that $C \subseteq L$ since the expansion may cause that some leaves of C become a fanout of a node inside the window. Two situations can happen for a leaf node. If all fanins are inside the window, the leaf can be simply removed from L. Otherwise, all fanins of the original leaf node need to be added to L (the case of C_1 in Fig. 2). This procedure has to be repeated iteratively to ensure that there are no leaves having a fanin already included the window.

Resynthesis is then applied to the window. Each window that is not suitable for the subsequent optimization is skipped. The motivation is to eliminate execution of a relatively time-consuming resynthesis for the windows that are unlikely to lead to any improvement. The identification of the suitable windows can be based on the size of W (small windows are filtered out) or a combination of size of C and size of W (thin windows are filtered out). In addition to that, we can use the information about the difference among level of the root node and leaves of C.

The resynthesis is performed by means of the CGP. At the beginning, each node in the window is assigned an unique index and chromosome corresponding with the nodes in the window is created. This chromosome is then used to seed the initial population. The evolutionary optimization is executed for a limited number of iterations. The number of iterations should be determined heuristically. The more iterations are allowed, the higher improvement can be achieved. On the other hand, many iterations on a small window wastes time.

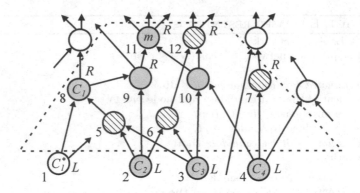

Fig. 2. Example of the window created using the proposed algorithm. The set of contained nodes of a 4-feasible cut $C = \{C_1, C_2, C_3, C_4\}$ rooted in node m is highlighted using the filled nodes. The hatched nodes are added to the window during the expansion of the cut. As a consequence of that, leave C_1 is replaced by C_1^*. The root and leaves of the window are denoted as R and L, respectively. The nodes in the window have assigned an index used to uniquely identify each node in the CGP. One of the many possibilities how to encode the window using CGP is for example: (2,3,AND) (2,3,OR) (4,1,INV) (1,5,XOR) (8,2,AND) (3,4,NOR) (9,10,AND) (6,10,OR) (7,8,9,11,12).

Finally, the optimized logic network W' is evaluated w.r.t. N' and if it performs better, it replaces all non-leaf nodes included in W. The whole optimization algorithm is terminated when a predefined number of iterations or a given runtime is exhausted.

Table 1 compares the proposed method with various methods for optimization of logic circuits available in the literature. Compared to the conventional approaches, we consider windows consisting of substantially higher number of gates. In addition to that, we do not impose any limits on the number of window inputs and outputs. Compared to the evolutionary approach [3], substantially larger sub-circuits identified using a different scoping method (windowing based on reconvergence driven cuts) are considered during resynthesis.

4   Experimental Evaluation

4.1   Experimental Setup

The proposed method was implemented in C++ as a part of Yosys open synthesis suite [17]. The advantage of this tool, among others, is that it allows us to directly manipulate with Verilog files and that it integrates ABC [18], a state-of-the-art academic tool for hardware synthesis and verification.

To evaluate the proposed approach, we used 28 highly optimized real-world circuits and optimized them by means of the proposed as well as current state-of-the-art approach. Nineteen Verilog netlists are taken from IWLS'05 Open

Table 1. Comparison of the optimization approaches according to the applied constraints. Parameters k_{min} and k_{max} determine the minimum allowed and maximum acceptable number of inputs of the accepted windows. Parameters c_{min} and c_{max} represent the restrictions applied to the volume of the windows.

Approach	k_{min}	k_{max}	c_{min}	c_{max}	Scoping method
Rewriting [2, 15]	–	4	–	–	Cut computation
Redundancy removal [2]	6	12	–	–	Windowing
Conventional resynthesis [16]	–	–	–	3360 (30%)	Windowing (various)
Evolutionary resynthesis [3]	1	10	8	50	Radius-based windowing
Proposed approach	–	–	10	10^4	Cut-based windowing

Cores benchmarks, the remaining nine netlists represent various arithmetic circuits[1]. The circuits were optimized by ABC (several iterations of ABC command 'resyn') and mapped to gates using a library of common 2-input gates including XORs/XNORs gates (ABC command 'map'). After mapping, optimization by the proposed and global method was executed and final number of mapped gates in circuits was examined. All of the optimized circuits were formally verified w.r.t their original form (ABC command 'cec').

The goal of this paper is to evaluate performance of the proposed method w.r.t. the state-of-the-art EA-based method (denoted as global) applied to the whole Boolean network and to compare both methods to the best result produced by the ABC. Both methods operate at the level of optimized and mapped Boolean networks to avoid the bias of AIG representation. The procedure OptimizeNetworkUsingEA is based on the CGP implemented as described in Sect. 2.3 with the following parameters: $n_a = 2$, $\lambda = 1$, $h = 2$, $n_n = |W|$. A single call of this procedure is executed for the global method (the procedure takes the whole Boolean network and returns its optimized version). On contrary, several calls of this procedure are executed in the proposed method. The termination conditions are designed as follows. The global method terminates when n_{iters} iterations are exhausted. One iteration corresponds with evaluation of a single candidate solution. In the case of the proposed method a simple divide-and-conquer strategy is employed. The proposed method is allowed to create n_{cuts} cuts. For each cut, the OptimizeNetworkUsingEA is allowed to perform n_{iters}/n_{cuts} iterations. This strategy is relatively naive because it supposes that the computation effort does not depend on the size of the window but it helps to fairly evaluate the impact of the proposed method. It ensures that exactly the same number of generations are evaluated in both cases. In this paper, we use $n_{iters} = 10^{10}$ iterations. Only windows whose volume is larger than 10 and less than 10^4 nodes are accepted, i.e. $c_{min} = 10$, $c_{max} = 10^4$. The root node m is chosen randomly in this study. This strategy simplifies the problem but it may lead to degradation of the performance especially if many unacceptable windows are produced. If this happens

---

[1] All the benchmarks are taken from https://lsi.epfl.ch/MIG.

in 10% cases, for example, the total number of effective generations is in fact reduced to 90%. The only criterion in the fitness function considered in this paper is the area on a chip expressed as the number of gates. For each method and each benchmark, five independent runs were executed to obtain statistically reasonable results.

4.2   Experimental Results

The overall results are summarized in Table 2. The first three columns contain information related to the benchmarks (name, number of PIs and POs). The next two columns show parameters of the mapped circuits and those numbers serve as a baseline for our comparison – the number of gates and logic depth is provided. Then, the achieved results expressed as the relative reduction with respect to the baseline are reported for the proposed and global method. For each method, we report the average and the best obtained improvement. The numbers are calculated across all independent runs.

The best results are very close to the average ones which suggests that the both EA-based methods are stable although they are in principle non-deterministic. According to the number of highlighted cases showing the better results, the proposed method performs substantially better considering the average as well as the best results. It won in 22 out of 28 cases. There are even cases, when the global method provided none or nearly none improvement (see e.g. benchmarks DSP, des_perf, ethernet, systemcaes and so on). The average reduction on the IWLS'05 benchmarks is slightly better in favor of the global method, but it is affected mostly by five cases (mem_ctrl, pci_spoci_ctrl, spi, systemcdes, and tv80), where the global method provided substantially better results. Looking at the arithmetic circuits, the global method was able to slightly improve only two circuits. In other cases, the reduction is negligible. We analyzed the five cases where the global method outperformed the proposed one and concluded that the global method works well especially for small instances (less than 10^4 gates) that have a reasonable depth (10 to 25 levels). The global optimization of circuits with large depth is unsatisfactory. A substantial improvement is achieved on the arithmetic circuits. The number of gates is reduced by nearly 15% on average. The highest reduction, 30.1%, is recorded for hamming benchmark. The detailed analysis revealed that this was possible due to better handling of XORs/XNORs compared to ABC and also by a relatively huge redundancy of the original circuit optimized by ABC. The relative number of AND/OR/NAND/NOR gates remained nearly the same (around 74%). The number of XORs/XNORs increased from 10% to 15%.

A more detailed analysis is shown in Table 3 where we reported the computational effort required to reduce the benchmark circuits by 1%, 5% and 10%. The table shows the mean number of generations that have to be evaluated to obtain a circuit whose number of gates is reduced by a given level. The empty cells mean that none of the evolutionary runs produced circuit satisfying the required condition. This can happen either because of the insufficient number of generations or because it is in principle impossible to obtain such a circuit (we are already at

Table 2. Comparison of the proposed and global method against ABC. The columns 'Impr. proposed' and 'Impr. global' report the relative improvement in the number of gates compared to the optimized circuits obtained using ABC. Column 'ABC' contains parameters of the optimized circuits after mapping ('gates' is the number of gates, D is logic depth).

Benchmark	PIs	POs	ABC		Impr. proposed		Impr. global [6]	
			Gates	D	Avg	Best	Avg	Best
DSP	4223	3792	43491	45	**3.6%**	**3.6%**	0.0%	0.0%
ac97_ctrl	2255	2136	11433	10	**2.9%**	**2.9%**	1.4%	1.4%
aes_core	789	532	21128	20	**2.9%**	**2.9%**	0.6%	1.7%
des_area	368	70	5199	25	**6.0%**	**6.1%**	2.1%	2.3%
des_perf	9042	1654	78972	16	**1.8%**	**1.8%**	0.0%	0.1%
ethernet	10672	10452	60413	23	**0.5%**	**0.5%**	0.0%	0.0%
i2c	147	127	1161	12	9.2%	9.2%	**10.0%**	**10.7%**
mem_ctrl	1198	959	10459	24	7.0%	7.0%	**24.8%**	**25.4%**
pci_bridge32	3519	3136	19020	21	**3.5%**	**3.5%**	0.5%	0.6%
pci_spoci_ctrl	85	60	1136	15	18.3%	18.5%	**34.8%**	**35.7%**
sasc	133	123	746	8	**6.2%**	**6.2%**	2.4%	2.8%
simple_spi	148	132	822	11	**5.5%**	**5.7%**	4.4%	4.6%
spi	274	237	3825	26	5.6%	5.6%	**13.5%**	**20.2%**
ss_pcm	106	90	437	7	**5.7%**	**6.7%**	2.3%	2.3%
systemcaes	930	671	11352	27	**11.9%**	**12.3%**	0.0%	0.0%
systemcdes	314	126	2601	25	4.8%	5.0%	**9.1%**	**9.9%**
tv80	373	360	8738	39	6.6%	6.9%	**11.1%**	**11.3%**
usb_funct	1860	1692	15405	23	**5.8%**	**5.9%**	2.6%	2.6%
usb_phy	113	73	452	9	**13.9%**	**14.0%**	12.2%	12.2%
Average (IWLS'05 benchmarks)			15620	20	6.4%	6.5%	**7.0%**	**7.6%**
mult32	64	64	8225	42	**16.5%**	**16.6%**	0.0%	0.0%
sqrt32	32	16	1462	307	**22.3%**	**24.3%**	3.0%	3.0%
diffeq1	354	193	20719	218	**11.5%**	**11.5%**	0.0%	0.0%
div16	32	32	5847	152	**15.7%**	**15.8%**	0.0%	0.0%
hamming	200	7	2724	80	**28.6%**	**30.1%**	14.6%	14.6%
MAC32	96	65	7793	55	**7.7%**	**7.8%**	0.0%	0.0%
revx	20	25	8131	171	**14.5%**	**14.5%**	0.0%	0.1%
mult64	128	128	21992	190	**7.4%**	**7.4%**	0.3%	0.5%
max	512	130	3719	117	**5.3%**	**5.3%**	0.7%	0.8%
Average (arithmetic benchmarks)			8956	148	**14.4%**	**14.8%**	2.1%	2.1%

Table 3. The average number of CGP generations needed to achieve 1%, 5%, and 10% reduction

Benchmark	1% improvement		5% improvement		10% improvement	
	Global	Proposed	Global	Proposed	Global	Proposed
DSP	$>10^{10}$	$8 \cdot 10^8$	–	–	–	–
ac97_ctrl	$45 \cdot 10^7$	$7 \cdot 10^8$	–	–	–	–
aes_core	$>10^{10}$	$1 \cdot 10^9$	–	–	–	–
des_area	$4 \cdot 10^7$	$98 \cdot 10^7$	$>10^{10}$	$11 \cdot 10^8$	–	–
des_perf	$>10^{10}$	$3 \cdot 10^9$	–	–	–	–
i2c	$5 \cdot 10^5$	$28 \cdot 10^7$	$35 \cdot 10^5$	$5 \cdot 10^8$	$7 \cdot 10^9$	$>10^{10}$
mem_ctrl	$5 \cdot 10^5$	$27 \cdot 10^7$	$5 \cdot 10^5$	$45 \cdot 10^8$	$5 \cdot 10^5$	$>10^{10}$
pci_bridge32	$>10^{10}$	$78 \cdot 10^7$	–	–	–	–
pci_spoci_ctrl	$5 \cdot 10^5$	10^7	$5 \cdot 10^5$	$14 \cdot 10^7$	10^6	$42 \cdot 10^7$
sasc	$21 \cdot 10^6$	$15 \cdot 10^5$	$>10^{10}$	$43 \cdot 10^6$	–	–
simple_spi	$5 \cdot 10^6$	$86 \cdot 10^6$	$>10^{10}$	$72 \cdot 10^7$	–	–
spi	$5 \cdot 10^6$	$416 \cdot 10^6$	$65 \cdot 10^6$	$3 \cdot 10^9$	$72 \cdot 10^6$	$>10^{10}$
ss_pcm	$4 \cdot 10^6$	10^8	$>10^{10}$	$2 \cdot 10^8$	–	–
systemcaes	$>10^{10}$	$12 \cdot 10^7$	$>10^{10}$	$17 \cdot 10^8$	$>10^{10}$	$7 \cdot 10^9$
systemcdes	$65 \cdot 10^5$	$17 \cdot 10^7$	$55 \cdot 10^6$	$>10^{10}$	$74 \cdot 10^7$	$>10^{10}$
tv80	$5 \cdot 10^5$	$231 \cdot 10^6$	$26 \cdot 10^6$	$3 \cdot 10^9$	$18 \cdot 10^7$	$>10^{10}$
usb_funct	$94 \cdot 10^6$	$575 \cdot 10^6$	$>10^{10}$	$65 \cdot 10^8$	–	–
usb_phy	$5 \cdot 10^5$	$12 \cdot 10^6$	$25 \cdot 10^5$	$19 \cdot 10^6$	$55 \cdot 10^7$	$17 \cdot 10^7$
Average	$2.8 \cdot 10^9$	$5.2 \cdot 10^8$	$4.6 \cdot 10^9$	$2.4 \cdot 10^9$	$3 \cdot 10^9$	$7.2 \cdot 10^9$
mult32	$>10^{10}$	$72 \cdot 10^6$	$>10^{10}$	$48 \cdot 10^7$	$>10^{10}$	$2 \cdot 10^9$
sqrt32	$5 \cdot 10^5$	$19 \cdot 10^6$	$37 \cdot 10^5$	$11 \cdot 10^7$	$>10^{10}$	$39 \cdot 10^7$
diffeq1	$>10^{10}$	$2 \cdot 10^8$	$>10^{10}$	$16 \cdot 10^8$	$>10^{10}$	$67 \cdot 10^8$
div16	$>10^{10}$	$13 \cdot 10^7$	$>10^{10}$	$5 \cdot 10^8$	$>10^{10}$	$24 \cdot 10^8$
hamming	$5 \cdot 10^5$	$17 \cdot 10^6$	$5 \cdot 10^5$	$12 \cdot 10^7$	$2 \cdot 10^6$	$5 \cdot 10^8$
MAC32	$>10^{10}$	$6 \cdot 10^7$	$>10^{10}$	$96 \cdot 10^7$	–	–
revx	$>10^{10}$	$36 \cdot 10^7$	$>10^{10}$	$94 \cdot 10^7$	$>10^{10}$	$5 \cdot 10^9$
mult64	$>10^{10}$	$39 \cdot 10^7$	$>10^{10}$	$73 \cdot 10^8$	–	–
max	$>10^{10}$	$91 \cdot 10^6$	$>10^{10}$	$96 \cdot 10^7$	–	–
Average	$7.7 \cdot 10^9$	$2 \cdot 10^8$	$7.9 \cdot 10^9$	$1.5 \cdot 10^9$	$8.3 \cdot 10^9$	$2.8 \cdot 10^9$

the optimum or close to the optimum). Looking at the first two columns showing the computation effort required for reduction by 1%, we can easily identify that the global method converges faster compared to the proposed method. On the other hand, it has tendency to stuck at a local optima which is evident especially on more complex benchmarks (arithmetic circuits having large logic depth and complex circuits consisting of tens thousands of gates). Nearly none improve-

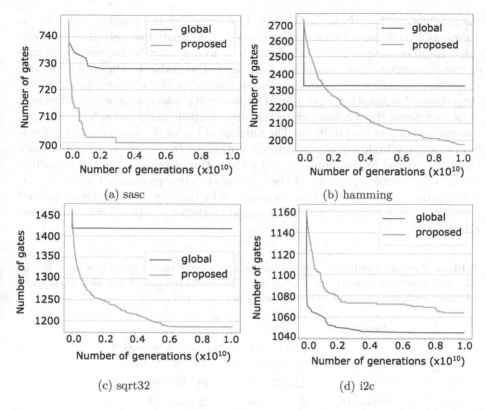

(a) sasc

(b) hamming

(c) sqrt32

(d) i2c

Fig. 3. Typical convergence curves for four chosen benchmark circuits. The lower value (number of gates) the better result.

ment was achieved for arithmetic circuits. The only exception is the benchmark circuit 'hamming'. The proposed method converges in some cases slowly but it provides better results when we enable to run it longer. See for example benchmarks 'des_area', 'simple_spi', 'ss_pcm', 'usb_funct', or 'hamming'. In these cases the proposed method requires more generations to reduce the circuits by 1%, but substantially less generations are needed on average to achieve 5% reduction. The effect of slow convergence is especially noticeable on 'hamming' circuit, where approximately 250 times more generations were needed to reduce the original circuit by 5% and 10% percent. Despite of that, the proposed method was able to reach 30.1% reduction while the global method got stuck at 14.6%. The typical convergence curves for four benchmark circuits are shown in Fig. 3. The first three plots show how the global methods usually got stuck at local optima. The last plot depicts the situation where the global method performs better compared to the proposed one.

We assume that the slow convergence is caused by the fact that each subcircuit produced by the proposed windowing algorithm is optimized for a fixed number of generations independently on its parameters (the number of gates, the number of PIs or POs, and so on). This simplifies the problem but leads to a potential inefficiency. Many generations can be wasted to optimize small circuits. In order to investigate this fact, we analyzed what is the average volume of the sub-circuit. The results are summarized in Table 4. The table contains the average number of leaves, roots and volume of the windows produced by the proposed windowing algorithm. Despite using a simple strategy for selecting a root node, the window parameters are relatively good and sub-circuits of a reasonable volume are produced. The number of leaves $|L|$ determining the number of primary inputs of the sub-circuit optimized by evolution is substantially higher compared to the numbers used in rewriting. Compared to the rewriting, a relatively complex portions of the original circuits are chosen for subsequent optimization. This could explain the reason, why the proposed EA-based method is able to achieve such reduction compared to the conventional state-of-the-art synthesis.

Table 4. Average number of leaves, roots and volume of the windows produced by the proposed windowing algorithm. The averages are reported for all windows (first three columns) and those leading to a reduction (last three columns).

Benchmark	Windows					
	All created			Causing reduction		
	$|L|$	$|R|$	Volume	$|L|$	$|R|$	Volume
DSP	32	26	53	46	38	86
mem_ctrl	27	25	38	28	26	44
pci_spoci_ctrl	14	13	21	18	19	32
systemcaes	22	15	35	14	13	26
systemcdes	27	26	51	38	39	78
mult32	20	16	34	26	21	52
sqrt32	33	29	62	20	17	37
diffeq1	30	27	53	28	26	55
div16	32	28	50	25	24	44
hamming	30	26	44	26	24	45

We analyzed all the evolutionary runs across all benchmarks circuits and determined the maximum number of generations that caused a reduction of a sub-circuit. For each run of CGP we recorded the last generation that caused a change in the number of gates together with the volume of the optimized sub-circuit. The obtained numbers are plotted as a function of sub-circuit volume using a boxplot in Fig. 4. As expected, the more nodes are there in the sub-circuit the more CGP generations are typically used to optimize it. We can also see that the dependence between these two values is exponential – this is illustrated also by the blue curve representing polynomial interpolation of the median value. As the volume of the window increases, the number of occurrences

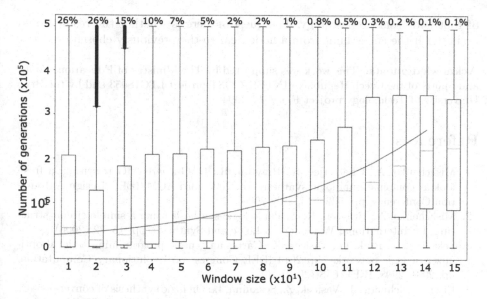

Fig. 4. Boxplots showing the number of generations that caused removal of a gate. The numbers above each boxplot show the number of occurrences of the window of a certain volume.

of such cases decreases (see the numbers above each boxplot showing how many times we seen a window having volume between X and X + 10). Usually, small windows are produced. Windows up to 45 nodes were produced in more than 77% cases. Due to this fact, the interpolation is limited to 150 nodes because there is insufficient number of results for the bigger windows.

5 Conclusion

Compared to the conventional logic synthesis, state-of-the-art EA-based optimization is able to produce substantially better results but at the cost of a higher run time. Unfortunately, the run time increases with the increasing complexity of the Boolean networks. This paper addresses this problem by combining the EA-based optimization with windowing that allows to work on a smaller portions of the original Boolean network. Even though we used a very simple strategy of root node selection which may degrade the capabilities of the resynthesis, the proposed method is able to outperform the original EA-based optimization applied to the whole Boolean networks. The number of nodes w.r.t the original method was improved by 9.2% on average. Even though only area was analyzed in this study, the depth of the optimized circuits is comparable with the original circuits.

In our future work, we would like to implement an adaptive strategy that modifies the maximum number of generations according to the size of the optimized logic circuit. In addition to that, we would like to focus on improvement of root node selection strategy. The question here is whether the result would be better if the cut is built from a node near to the previously chosen one.

Acknowledgments. This work was supported by The Ministry of Education, Youth and Sports of the Czech Republic – INTER-COST project LTC18053 and by the Brno University of Technology project FIT-S-17-3994.

References

1. Mishchenko, A., Chatterjee, S., Brayton, R.: DAG-aware AIG rewriting: a fresh look at combinational logic synthesis. In: 2006 43rd ACM/IEEE Design Automation Conference, pp. 532–535, July 2006
2. Mishchenko, A., Brayton, R.: Scalable logic synthesis using a simple circuit structure. In: International Workshop on Logic and Synthesis, pp. 15–22 (2006)
3. Sekanina, L., Ptak, O., Vasicek, Z.: Cartesian genetic programming as local optimizer of logic networks. In: 2014 IEEE Congress on Evolutionary Computation, pp. 2901–2908. IEEE CIS (2014)
4. Fiser, P., Schmidt, J., Vasicek, Z., Sekanina, L.: On logic synthesis of conventionally hard to synthesize circuits using genetic programming. In: 13th IEEE Symposium on Design and Diagnostics of Electronic Circuits and Systems, pp. 346–351 (2010)
5. Fiser, P., Schmidt, J.: Small but nasty logic synthesis examples. In: Proceedings of the 8th International Workshop on Boolean Problems, pp. 183–190 (2008)
6. Vasicek, Z.: Cartesian GP in optimization of combinational circuits with hundreds of inputs and thousands of gates. In: Machado, P., et al. (eds.) EuroGP 2015. LNCS, vol. 9025, pp. 139–150. Springer, Cham (2015). https://doi.org/10.1007/978-3-319-16501-1_12
7. Li, N., Dubrova, E.: AIG rewriting using 5-input cuts. In: Proceedings of the 29th International Conference on Computer Design, pp. 429–430. IEEE CS (2011)
8. Lohn, J.D., Hornby, G.S.: Evolvable hardware: using evolutionary computation to design and optimize hardware systems. IEEE Comput. Intell. Mag. **1**(1), 19–27 (2006)
9. Miller, J.F., Thomson, P.: Cartesian genetic programming. In: Poli, R., Banzhaf, W., Langdon, W.B., Miller, J., Nordin, P., Fogarty, T.C. (eds.) EuroGP 2000. LNCS, vol. 1802, pp. 121–132. Springer, Heidelberg (2000). https://doi.org/10.1007/978-3-540-46239-2_9
10. Miller, J.F.: Cartesian Genetic Programming. Springer, Heidelberg (2011). https://doi.org/10.1007/978-3-642-17310-3
11. Vassilev, V., Job, D., Miller, J.F.: Towards the automatic design of more efficient digital circuits. In: Lohn, J., Stoica, A., Keymeulen, D., Colombano, S. (eds.) Proceedings of the 2nd NASA/DoD Workshop on Evolvable Hardware, pp. 151–160. IEEE Computer Society, Los Alamitos (2000)
12. Tao, Y., Zhang, L., Zhang, Y.: A projection-based decomposition for the scalability of evolvable hardware. Soft Comput. **20**(6), 2205–2218 (2016)
13. Stomeo, E., Kalganova, T., Lambert, C.: Generalized disjunction decomposition for the evolution of programmable logic array structures. In: First NASA/ESA Conference on Adaptive Hardware and Systems (AHS 2006), pp. 179–185 (2006)
14. Vasicek, Z., Sekanina, L.: Formal verification of candidate solutions for post-synthesis evolutionary optimization in evolvable hardware. Genet. Program. Evolvable Mach. **12**(3), 305–327 (2011)

15. Fiser, P., Halecek, I., Schmidt, J.: Are XORs in logic synthesis really necessary? In: IEEE 20th International Symposium on Design and Diagnostics of Electronic Circuits and Systems (DDECS), pp. 138–143 (2017)
16. Fiser, P., Schmidt, J.: It is better to run iterative resynthesis on parts of the circuit. In: Proceedings of the 19th International Workshop on Logic and Synthesis, pp. 17–24. University of California Irvine (2010)
17. Wolf, C., Glaser, J., Kepler, J.: Yosys-a free Verilog synthesis suite. In: Proceedings of the 21st Austrian Workshop on Microelectronics (Austrochip) (2013)
18. Brayton, R., Mishchenko, A.: ABC: an academic industrial-strength verification tool. In: Touili, T., Cook, B., Jackson, P. (eds.) CAV 2010. LNCS, vol. 6174, pp. 24–40. Springer, Heidelberg (2010). https://doi.org/10.1007/978-3-642-14295-6_5

Cartesian Genetic Programming as an Optimizer of Programs Evolved with Geometric Semantic Genetic Programming

Ondrej Koncal and Lukas Sekanina[✉]

Faculty of Information Technology, IT4Innovations Centre of Excellence,
Brno University of Technology, Božetěchova 2, 612 66 Brno, Czech Republic
koncalo@gmail.com, sekanina@fit.vutbr.cz

Abstract. In Geometric Semantic Genetic Programming (GSGP), genetic operators directly work at the level of semantics rather than syntax. It provides many advantages, including much higher quality of resulting individuals (in terms of error) in comparison with a common genetic programming. However, GSGP produces extremely huge solutions that could be difficult to apply in systems with limited resources such as embedded systems. We propose Subtree Cartesian Genetic Programming (SCGP) – a method capable of reducing the number of nodes in the trees generated by GSGP. SCGP executes a common Cartesian Genetic Programming (CGP) on all elementary subtrees created by GSGP and on various compositions of these optimized subtrees in order to create one compact representation of the original program. SCGP does not guarantee the (exact) semantic equivalence between the CGP individuals and the GSGP subtrees, but the user can define conditions when a particular CGP individual is acceptable. We evaluated SCGP on four common symbolic regression benchmark problems and the obtained node reduction is from 92.4% to 99.9%.

Keywords: Cartesian Genetic Programming ·
Geometric Semantic Genetic Programming · Symbolic regression

1 Introduction

Geometric Semantic Genetic Programming (GSGP) is a recent branch of genetic programming (GP) in which specific genetic operators, the so-called *geometric semantic genetic operators*, directly work at the level of semantics rather than syntax [1]. In this context, the *semantics* is defined as the vector of outputs of a program on the different training data. GSGP is successful because geometric semantic operators induce a unimodal fitness landscape which is known to be relatively easy for search-based optimization algorithms. On many various symbolic regression and classification problems it has been shown that GSGP provides statistically better results than a common genetic programming and other

© Springer Nature Switzerland AG 2019
L. Sekanina et al. (Eds.): EuroGP 2019, LNCS 11451, pp. 98–113, 2019.
https://doi.org/10.1007/978-3-030-16670-0_7

machine learning methods in terms of the error score [2,3]. However, since the genetic operators used in GSGP produce offspring that are larger than their parents, the evolved programs unprecedentedly grow in their size. Some approaches addressing this issue have been developed (see Sect. 2.1), but the problem is still considered as unsolved.

This paper presents a new method capable of reducing the size of a program evolved with GSGP. This work is motivated by the fact that there could be a high-quality program created by GSGP for a given application, but the program is unfeasible for implementation on a platform with limited resources (e.g., in an embedded system with a small memory). In our approach, we consider this program (i.e. the result of GSGP) as a golden (reference) solution in terms of functionality and try to minimize its size. The optimized program should then be implemented in the target embedded system. In our preliminary experiments, we employed GP to reduce the number of nodes in several programs evolved with GSGP and keep the error at the same level. Because we optimized the entire programs without any decomposition and the programs were too complex, no reduction in the number of nodes has been achieved at all.

We introduce *Subtree Cartesian Genetic Programming* (SCGP) as an efficient optimizer of the size of programs evolved with GSGP. The *reference solution* (i.e. the program evolved with GSGP) is converted to the representation used in Cartesian Genetic Programming (CGP). In order to avoid scalability problems of the aforementioned preliminary approach, SCGP executes, in the first step, a series of CGP runs with the aim to minimize the number of nodes in all subtrees belonging to the CGP representation of the reference solution. These optimized subtrees are then paired and again optimized by CGP. Several iterations of the pairing strategy then lead to a single optimized program. The optimization process is thus decomposed into a number of CGP runs solving low-complexity optimization problems. CGP is used because of its well-known capabilities to optimize the phenotype size (as exemplary demonstrated, for example, for digital circuits in [4]). The proposed method is evaluated using four symbolic regression problems (%F, LD50, PPB and P3D, see Sect. 4.1) for which a significant reduction is reported in the number of nodes in the evolved trees.

2   Relevant Work

2.1   Geometric Semantic Genetic Programming

GP operators traditionally work in the syntactic space and manipulate the syntax of parents. The parents can also be modified based on their semantics which is defined as a vector of outputs of a program on the different training data [1].

GSGP creates new candidate solutions using *geometric semantic operators* working at the level of semantics. *Geometric semantic crossover* (GSC) and *geometric semantic mutation* (GSM) usually work as follows

$$\text{GSC: } T_{XO} = (T_1 \cdot T_R) + ((1 - T_R) \cdot T_2) \tag{1}$$

$$\text{GSM: } T_M = T + ms \cdot (T_{R1} - T_{R2}) \tag{2}$$

where $T, T_1, T_2 : \mathbb{R}^n \to \mathbb{R}$ are parents, $T_R, T_{R1}, T_{R2} : \mathbb{R}^n \to \mathbb{R}$ are random real functions with output values in interval $\langle 0, 1 \rangle$ and ms is a *mutation step*. Generating the output values in interval $\langle 0, 1 \rangle$ is ensured by the sigmoid function

$$T_R = (1 + e^{-T_{rand}})^{-1}, \tag{3}$$

where T_{rand} is a random tree with no constraints on the output values.

By applying these operators, one can effectively create a unimodal error surface for problems such as symbolic regression. The search process conducted in such a search space is then more efficient than in the case of a common GP. However, geometric semantic operators, by construction, always produce offspring that are larger than their parents, causing a fast growth in the size of the individuals. The growth is linear for GSM and exponential for GSC [1].

In order to reduce the size of candidate solutions, various approaches have been developed. One branch of the methods is based on simplifying the off-spring during the evolution, for example, using a computer algebra system [1] or developing specific genetic operators such as subtree GSC and subtree GSM [5]. Recently, an on-the-fly simplification of trees was proposed capitalizing the fact that the individuals are always linear combinations of trees and repeated structures can be aggregated [3]. The original huge individuals created by GSGP were significantly reduced to several thousands of nodes.

Another approach relies on an efficient GSGP implementation, in which pointers to existing structures (trees) are recorded rather than all the new trees [6]. The method only stores the initial population and a set of randomly generated trees. A new record is created after performing GSC or GSM in form of (*crossover*, $\&T_1, \&T_2, \&T_R$) or (*mutation*, $\&T, \&T_{R1}, \&T_{R2}, ms$). The semantics of the individuals is also stored and used to compute the semantics of the off-spring, again by means of the pointers to stored semantics records. This method enables to reduce the evaluation time as pre-calculated partial results are always available in the memory.

2.2 Cartesian Genetic Programming

In CGP [7], a candidate individual is modeled as a two-dimensional grid of nodes, where the type of nodes depends on a particular application. Each individual utilizes n_i primary inputs and n_o primary outputs. A unique address is assigned to all primary inputs and to the outputs of all nodes to define an addressing system enabling connections to be specified. As no feedback connections are allowed in the basic version of CGP, only directed acyclic graphs can be created. Each candidate individual is represented using $r \times c \times (n_a + 1) + n_o$ integers, where $r \times c$ is the grid size and n_a is the maximum arity of node functions. In this representation, the $n_a + 1$ integers specify one programmable node in such a way that n_a integers specify source addresses for its inputs and one integer determines its function. Finally, the l-back parameter defines how many columns of nodes in front of i-th column can be used as the data source for the i-th column.

New candidate individuals are created by mutation of selected genes (integers) of the chromosome. It is important to ensure that all randomly created gene values are within a valid interval (i.e. a valid candidate individual is always produced). Crossover is not normally used in CGP. CGP employs a simple search algorithm denoted $(1 + \lambda)$ which operates with a set of $1 + \lambda$ candidate individuals [7]. The initial population is created either randomly or heuristically, for example, an existing program can be employed. A new population is constructed by applying the mutation operator on the parent individual to generate λ offspring individuals. These offspring are then evaluated and the best performing individual is taken as a new parent. These steps are repeated until the time available for the evolution is exhausted or a suitable solution is discovered.

CGP was used to evolve new implementations of digital circuits and to optimize existing circuits, for example, in terms of the number of gates [4]. This is a very similar task to our problem – the program size reduction in GSGP.

3   Subtree CGP

Subtree CGP is a method that analyzes a log file created by a single run of GSGP in the framework [6] and minimizes the number of nodes in the best evolved individual. The result of GSGP will be called the *Golden Tree* (GT) in the paper. It has to be emphasized that GT is, in fact, distributed in the log file because the framework [6] only stores: (i) the trees used in the initial population, (ii) randomly created trees needed for geometric semantic operators and (iii) particular records about geometric semantic operators. These entities are linked using pointers. In order to extract GT, the log file has to be parsed and analyzed from the initial to the last population.

SCGP operates in two steps: (i) SCGP converts GT to the CGP representation. (ii) SCGP repeatedly executes a CGP-based optimizer on all elementary subtrees of GT and on various compositions of these optimized subtrees in order to create one compact representation of GT. SCGP does not guarantee the (exact) semantic equivalence between the CGP individuals and subtrees of GT, but the user can define conditions when a particular result of CGP is acceptable. In order to simplify this initial study, we will only consider GSGP utilizing GSM.

3.1   Obtaining the CGP Representation

As our benchmark problems are symbolic regression tasks with d independent variables and one output variable y, the basic SCGP parameters are $n_i = d$, $n_o = 1$, $r = 1$, $n_a = 2$, $l - \text{back} = c$ and $c = u$, where u is the total number of nodes in GT. Each CGP node can operate either as a constant (in interval $\langle -10.0;$ $10.0\rangle$) or function (taken from a function set $\Gamma = \{+, -, *, /, (1 + e^{-x})^{-1}, e^x\}$, where division is protected, i.e. $x/0 = 1$). Despite the fact that this setup is application-specific and is typically given in the experimental part, it is provided here to simplify the following description of the method.

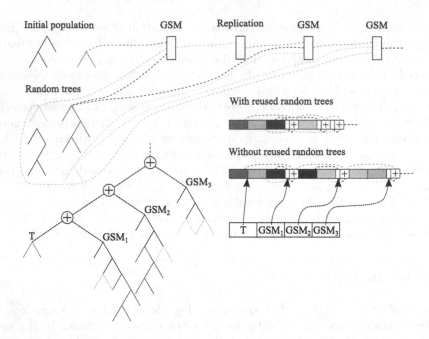

Fig. 1. GSGP to CGP conversion: original records created by GSGP (top); corresponding Golden Tree (bottom-left); CGP chromosome with and without reused subtrees (bottom-right).

Because GSGP operates with common syntax trees representing arithmetic expressions, their conversion to the CGP chromosome is straightforward; a tree is expressed as a string in the postfix notation and during the reading of this string from left to right, appropriate nodes (represented using three integers – two pointers to the inputs of the node and one node function) are created in the CGP chromosome. Note that the primary inputs representing the independent variables are internally handled as special nodes.

In order to build GT, SCGP parses the log file of a single GSGP run (generation by generation), but in such a way that only the nodes contributing to GT are identified and converted to the CGP representation. In particular,

- all subtrees representing the initial population of GSGP,
- all subtrees representing the randomly generated functions and
- arithmetic operators representing (GSM or simple replication) that are connecting these subtrees

are converted to the CGP representation using the procedure given above[1]. Figure 1 illustrates how the subtrees are converted to a CGP chromosome. The number of nodes in the CGP representation is equal to the number of nodes in GT if it were represented as a single tree. Please note that ms constants are represented by small white empty boxes in the CGP chromosome in Fig. 1.

---

[1] As all these trees are used in a single resulting solution, we will call them subtrees.

Some subtrees (e.g. random trees) can, however, be used multiple times in GT. In order to shorten the CGP chromosome, these multiple instances of a given subtree can be detected and only one instance of each subtree can be included into the CGP chromosome as seen in Fig. 1. For the benchmark problems used in this paper, this technique reduces the number of nodes 1.4–2.2×. It should be noted that this technique is NOT used in the current version of SCGP.

3.2   CGP-based Optimization of Subtrees

The proposed SCGP assumes that (i) only GSM was used in GSGP (i.e. no GSC) and (ii) the GSGP to CGP conversion does not apply any size reduction techniques, i.e. all (even multiple instances of) subtrees are preserved. These assumptions are important as they ensure that the CGP-based optimization can independently be performed on all the subtrees. Since the subtrees are connected via (associative and commutative) addition operators their optimization can be performed in an arbitrary order.

During the conversion process an array of indexes p is created, pointing to the subtrees that will further be optimized. These subtrees include a random tree from the initial population (T) and all subtrees whose root is the $\cdot$ operator from the $ms \cdot (T_{R1} - T_{R2})$ expressions created by GSM. Let s be the size of p. By means of p, all s subtrees can be identified in the CGP chromosome, their corresponding chromosomes extracted and used as initial seeds (denoted $\alpha[i]$) for the CGP-based subtree optimization.

A single CGP run is executed for each subtree in p. The objective of CGP is to minimize the number of nodes and keep the error at desired level. In order to accelerate the fitness (error) evaluation, responses of a particular subtree (that was identified in GT) are pre-calculated for the whole data set. In other words, the semantics $ST[i]$ is created for subtrees $\alpha[i]$, $i = 1, \ldots, s$ and data set D. Please note that these particular CGP runs operate in different search spaces because the chromosome sizes can, in principle, be different. The remaining CGP parameters (such as the population size, the number of generations, mutation rate etc.) are identical for all the CGP runs.

Thanks to the addition operators connecting all the subtrees, one can define an auxiliary vector $AUX[i]$ for each subtree $p[i]$

$$AUX[i](j) = y(j) - \sum_{\forall k: k \neq i} ST[k](j), \tag{4}$$

whose role is to define a contribution of the i-th subtree to the overall fitness, where $y(j)$ is a correct result for j-th fitness case. This vector is used as a golden solution (in terms of the error) during the subtree optimization conducted by CGP.

The fitness function reflecting the error of a candidate CGP individual g in the optimization of i-th subtree is defined as

$$f^{err}[i](g) = \frac{\sum_{j \in D} |AUX[i](j) - g(j)|}{|D|}, \tag{5}$$

where D is the training data set. The final fitness score is constructed hierarchically

$$f^{cgp}[i](g) = \begin{cases} \#\text{ of active nodes in g} & \text{if } f^{err}[i](g) \leq f^{err}[i](\alpha[i]) \\ \infty & \text{otherwise,} \end{cases} \tag{6}$$

where $\alpha[i]$ is the seed used in the initial population of CGP. The goal is to minimize the number of active nodes and, at the same time, keep the error at least identical with respect to the original subtree.

An individual satisfying the quality condition (i.e. its fitness f^{cgp} is no worse than the fitness of the seed individual) is called an *acceptable solution*. This quality condition can be re-defined according to user's requirements.

The best performing solution obtained in the CGP run is then used in the following steps of SCGP. As this solution can be semantically different w.r.t the seed, it is necessary to recalculate its $ST[i]$.

All remaining subtrees are optimized using the aforementioned procedure; however, updated versions of all $ST[i]$ vectors are always employed. All active nodes in these CGP chromosomes are marked in order to avoid useless evaluations of the inactive nodes in the next steps.

3.3 Subtree Pairing

After optimizing all elementary subtrees and with the aim of minimizing the total number of nodes of this set of subtrees (formally denoted Collection in the following pseudo-code), selected subtrees are paired and then minimized. The pairing procedure can be executed on arbitrarily chosen subtrees because all subtrees are connected using the associative and commutative addition operators (Fig. 1). As the following pseudo-code illustrates the pairing is performed in iterations until only one final tree (optimized GT) is obtained:

```
while (the number of subtrees > 1) do
    New_Collection = empty_set
    while (the number of subtrees in Collection > 1) do
        select two subtrees A and B according to Pairing Strategy
        create subtree C by joining A and B
        compute ST[C] = ST[A] + ST[B]
        run CGP (Sect. 3.2) to optimize the number of nodes in C
        remove A and B from Collection
        update ST[C]
        insert C to New_Collection
    end while
    Collection = Union (Collection, New_Collection)
end while
run CGP (Sect. 3.2) to optimize the final tree
update ST[0]
```

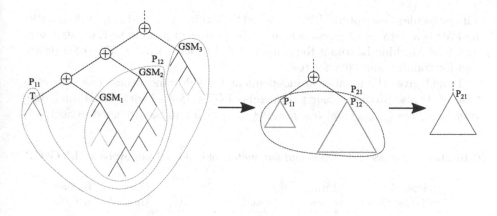

Fig. 2. Subtree pairing example

Figure 2 shows the pairing mechanism on a structure containing four subtrees. We propose three pairing strategies whose impact on the SCGP performance will be analyzed in Sect. 4.3:

- MAXDIFF – subtrees showing a maximum difference in their sizes are paired;
- MINDIFF – subtrees showing a minimum difference in their sizes are paired;
- RAND – randomly selected subtrees are paired;

The final (sub)tree is the result of SCGP. In total, SCGP executes $2s - 1$ elementary CGP runs, where s is the number of subtrees in GT.

4 Results

This section provides a basic experimental evaluation of the proposed method. It starts with a description of the data sets used in our experiments. It presents the results of the conventional GSGP (i.e. GTs for our data sets) that were obtained with the framework from [6]. SCGP capabilities to reduce the GT size are analyzed with respect to the number of generations allowed in a single CGP run and the pairing strategy used.

4.1 Data Sets

The proposed method is evaluated using four data sets that are commonly used in connection with the GSGP research. Three data sets are from the area of pharmacokinetics: (i) predicting the value of human oral bioavailability of a candidate new drug as a function of its molecular descriptors (*bioav*, denoted %F in the literature), (ii) predicting the value of the plasma protein binding level of a candidate new drug as a function of its molecular descriptors (LD50) and (iii) predicting the value of the toxicity of a candidate new drug as a function of its molecular descriptors (PPB; denoted %PPB in the literature) [8,9]. Finally, the P3D is a data set of physicochemical properties of protein tertiary structure from UCI Machine Learning Repository [10]. In each data set, 70% records are used for training and 30% for test.

Table 1 gives the number of independent variables and fitness cases for each data set. It also provides basic parameters of GTs evolved with GSGP including *training fitness* (f_{train}) and *test fitness* (f_{test}) as it will be discussed in Sect. 4.2.

Table 1. Parameters of data sets and parameters of Golden Trees evolved with GSGP.

	Independent variables (rows)	Fitness cases	f_{train} (best)	f_{test} (best)	Size (nodes)	Reduced size (nodes)
bioav	241	359	19.455	26.938	22,285	9,863
PPB	628	131	4.312	30.768	22,907	10,291
LD50	626	234	1336.960	1495.621	21,568	9,934
P3D	9	45,730	3.981	3.999	5,764	3,992

4.2   Obtaining Reference Solutions with GSGP

In order to evaluate the proposed method, we first establish reference solutions (i.e. Golden Trees) for each benchmark problem. The GSGP implementation from [6] is used with the parameters summarized in Table 2, function set $\{+, -, *, /\}$ and the fitness function defined as the mean absolute error between the output of a candidate individual and the golden output (y). GSGP only employs GSM in which the ms step is randomly generated from interval $\langle 0, 1 \rangle$. This parameter setting is almost identical with [2]; the main difference lies in allowing 10× more evaluations for GSGP to ensure that resulting GTs are of high quality and non-trivial in order to later illustrate the performance of SCGP.

Figure 3 shows box plots constructed from 40 independent runs for each data set. In particular, (i) the program size, (ii) the reduced program size and (iii) the training fitness are reported. Properties of GTs are summarized in Table 1.

As we used a state of the art GSGP implementation, a common setup of GSGP and the obtained results are consistent with [9] in terms of quality, we can conclude that relevant reference solutions (GTs) were generated with GSGP.

Table 2. Parameters of GSGP

Population size	2000	Mutation prob.	0.9
Generations	1000 (300 for *P3D*)	Max. tree depth	8
# of random trees	500	Single-node trees	Not used
Tree initialization	Ramped Half-and-Half	Tournament size	4

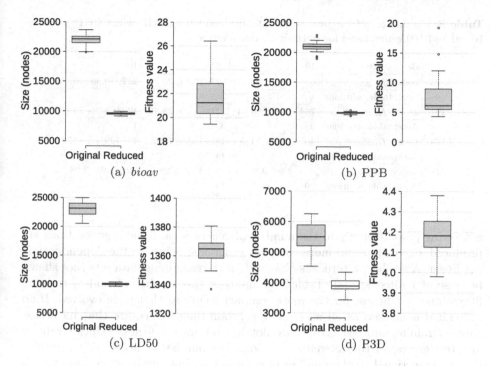

(a) *bioav*

(b) PPB

(c) LD50

(d) P3D

Fig. 3. Program size and training fitness values obtained from 40 independent GSGP runs.

4.3   Experiments with SCGP

From GTs reported in Table 1, all subtrees are extracted and used as the initial solutions α_i for the size optimization conducted by SCGP. Each subtree is optimized separately in a single CGP run in which the population size is 8 and the mutation probability is 0.08 (per gene). The impact of these parameters on the SCGP performance was not the subject of a detailed analysis because these values are typical for CGP. As the problem instances that are optimized by CGP are relatively simple, only a low number of generations seems to be sufficient in comparison with typical numbers from the CGP literature [7]. Our search for the most suitable number of generations was conducted with 10, 50 and 100 generations. The fitness computation exploits the pre-computed values $ST[i]$ as explained in Sect. 3.2. The impact of three pairing strategies (MAXDIFF, MIN-DIFF and RAND) on the SCGP performance is also investigated.

The Number of Generations. Figure 4 shows the impact of the number of generations (used by CGP when it optimizes one subtree) on the execution time and the size of resulting programs when applied on the *bioav* data set. The box plots are derived from 20 independent SCGP runs in which the RAND pairing strategy is employed. Because the population size is always 8, the execution time

Table 3. Parameters of the best solutions obtained with SCGP when CGP produces 10, 50 and 100 generations for optimizing one subtree.

	Generations	10	50	100
bioav	nodes; f_{train}; f_{test}	653; 19.46; 26.51	322; 19.45; 26.39	242; 19.40; 26.65
	Acceptable solutions	1	2	5
PPB	nodes; f_{train}; f_{test}	7643; 4.29; 30.04	2178; 4.29; 28.86	1723; 4.26; 26.80
	Acceptable solutions	3	3	1
LD50	nodes; f_{train}; f_{test}	46; 1334.8; 1489.1	19; 1333.0; 1492.1	17; 1327.2; 1473.8
	Acceptable solutions	11	13	9
P3D	nodes; f_{train}; f_{test}	939; 3.98; 3.99	281; 3.97; 3.98	202; 3.98; 3.98
	Acceptable solutions	20	18	18

is growing proportionally to the number of generations. If more generations are produced, one can obtain more compact programs; however, the dependency is not linear. A similar pattern was observed for the remaining data sets (not shown because of limited space). Table 3 summarizes the number of nodes, training fitness and test fitness for the most compact solutions that were evolved. If an individual is marked as an acceptable solution then we require that its fitness values (training and test errors) are not worse than the fitness values (training and test errors) of the reference solution. The number of acceptable solutions is not proportional to the number of generations, but the resulting program is always smaller if more generations are produced. As we are primarily interested in reducing the number of nodes, we will consider 100 generations as a reasonable setting in the final experiments.

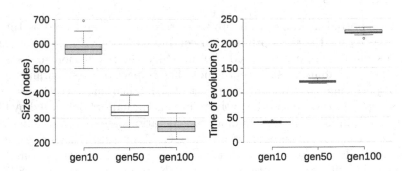

Fig. 4. The program size and the time of evolution obtained with SCGP on *bioav* when CGP produces 10, 50 and 100 generations for optimizing one subtree.

Subtree Pairing Strategies. Figure 5 shows the impact of three pairing strategies on the execution time and the size of programs resulting from SCGP when applied on particular data sets. The boxplots are derived from 20 independent runs of SCGP in which 50 generations are produced per CGP run. While MAXD-IFF is the best performing approach, RAND gives slightly worse results than

MAXDIFF and MINDIFF is clearly the worst one (as there are, in principle, more limited options to reduce the program size in comparison with the other approaches). Regarding the execution time, MAXDIFF is the most expensive approach except one case (P3D). Table 4 summarizes the number of nodes, the training fitness and test fitness for the smallest, but still acceptable programs. The number of acceptable solutions is very low in some cases, but recall that only 50 generations per CGP run are used in this set of experiments.

Table 4. Parameters of the best solutions obtained with SCGP utilizing different pairing strategies

	Pairing strategy	MAXDIFF	MINDIFF	RAND
bioav	nodes; f_{train}; f_{test}	288; 19.45; 25.38	383; 19.45; 26.69	322; 19.45; 26.39
	Acceptable solutions	3	1	2
PPB	nodes; f_{train}; f_{test}	1816; 4.29; 30.67	3755; 4.22; 28.94	2178; 4.29; 28.86
	Acceptable solutions	1	1	3
LD50	nodes; f_{train}; f_{test}	27; 1334.0; 1406.9	21; 1308.5; 1382.9	19; 1333.0; 1492.1
	Acceptable solutions	11	8	13
P3D	nodes; f_{train}; f_{test}	198; 3.98; 3.98	457; 3.98; 3.99	281; 3.97; 3.98
	Acceptable solutions	19	17	18

4.4 Final Results

For the final set of experiments we use the SCGP setup given in Sect. 4.3, but with the best performing number of generations (100) and pairing strategy (MAXDIFF) identified in the previous experiments. Figure 6 shows the program size and the execution time in form of box plots derived from 20 independent SCGP runs. The best obtained solutions are compared in Table 5 against GTs for all data sets. In the case of PPB, the setup used for SCGP did not provide any acceptable solution (a very compact solution with 1270 nodes was discovered, but its f_{test} is slightly higher than the golden tree exhibits). Hence, we took the best solution $PPB_{Tab.3}$ from Table 3 in order to report the best performing solutions for all data sets in one table.

4.5 Discussion

We have shown that if GSGP is followed by SCGP a significant reduction in the number of nodes can be obtained while the (error) fitness is not worsened.

In paper [3], where PPB was used as one of benchmark problems, the median size of the best individual obtained by a common GSGP (utilizing GSM and GSC) is 2.29e+64 nodes and the median size of the best individual evolved with their optimized GSGP-Red method is 12,185 nodes. Both numbers are much bigger with respect to our results. For the remaining benchmarks, no relevant results are available in the literature.

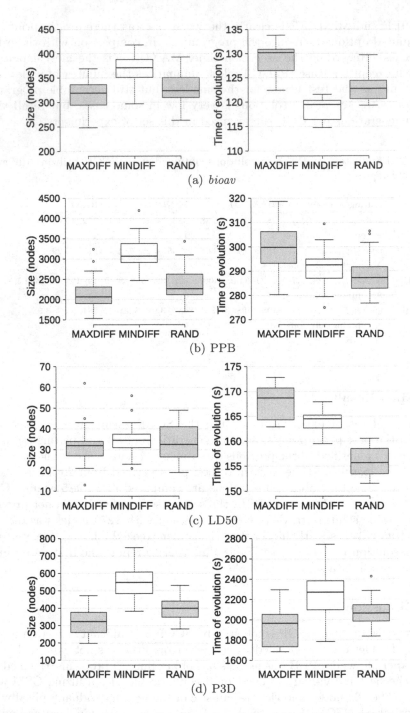

Fig. 5. The program size and the time of evolution obtained with SCGP for three subtree pairing strategies.

In order to further investigate how the pairing mechanism works, we report Table 6 which gives the number of subtrees after pairing. The initial number of subtrees is given by GT. One can observe that the number of subtrees is reduced to half in each iteration of SCGP. The number of nodes is significantly reduced in the first iterations of pairing; however, at some point it remains unchanged as seen in Fig. 7 for *bioav*, PPB and P3D. It means that a future improved version of our method could detect the iteration of pairing in which the number of nodes remains unchanged and skip the remaining pairing steps of the algorithm to save the computation time.

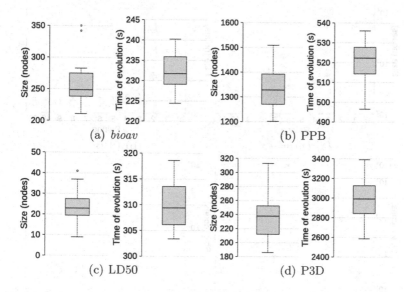

Fig. 6. Program sizes and time of evolution obtained with the final setup of SCGP.

Table 5. The number of nodes and the training and test (error) fitness for the best individuals obtained with GSGP and SCGP. As there is no acceptable solution created by SCGP for PPB, the best acceptable solution obtained in previous experiment ($PPB_{Tab.3}$) is reported.

	GSGP				SCGP			
	nodes	nodes (reduced)	f_{train}	f_{test}	nodes	f_{train}	f_{test}	accept.
bioav	22,285	9,863	19.455	26.938	224	19.39	26.69	4
PPB	22,907	10,291	4.312	30.768	1270	4.27	32.30	–
$PPB_{Tab.3}$					1723	4.26	26.80	1
LD50	21,568	9,934	1336.960	1495.621	13	1333.8	1417.6	12
P3D	5,764	3,992	3.981	3.999	186	3.98	3.98	16

Table 6. The number of subtrees after 1–10 iterations of pairing.

	Initial	1	2	3	4	5	6	7	8	9	10
bioav	911	456	228	114	57	29	15	8	4	2	1
PPB	892	446	223	112	56	28	14	7	4	2	1
LD50	921	461	231	116	58	29	15	8	4	2	1
P3D	259	130	65	33	17	9	5	3	2	1	–

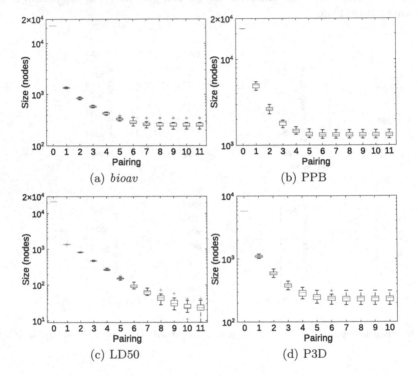

Fig. 7. The number of nodes in subtrees after pairing. Box plots constructed from 20 independent SCGP runs.

5   Conclusions

We proposed a CGP-based method capable of reducing the number of nodes in programs generated by GSGP. The obtained node reduction is 98.9% for *bioav*, 92.4% for PPB, 99.9% for LD50 and 96.7% for P3D. One SCGP run required hundreds of seconds for data sets containing hundreds of records (*bioav*, PPB, LD50) and thousands of seconds for P3D in which the data set contains ten thousands of records. The method can directly be used in GSGP utilizing a local search [11].

Our future work will be devoted to improving the key steps of SCGP (pairing strategies, fitness function for subtrees and termination conditions) and involving GSC into the process.

Acknowledgments. This work was supported by the Ministry of Education, Youth and Sports, under the INTER-COST project LTC 18053. The authors would like to thank Dr. Mauro Castelli for his support regarding the C++ framework for GSGP.

References

1. Moraglio, A., Krawiec, K., Johnson, C.G.: Geometric semantic genetic programming. In: Coello, C.A.C., Cutello, V., Deb, K., Forrest, S., Nicosia, G., Pavone, M. (eds.) PPSN 2012. LNCS, vol. 7491, pp. 21–31. Springer, Heidelberg (2012). https://doi.org/10.1007/978-3-642-32937-1_3
2. Vanneschi, L., Silva, S., Castelli, M., Manzoni, L.: Geometric semantic genetic programming for real life applications. In: Riolo, R., Moore, J.H., Kotanchek, M. (eds.) Genetic Programming Theory and Practice XI. GEC, pp. 191–209. Springer, New York (2014). https://doi.org/10.1007/978-1-4939-0375-7_11
3. Martins, J.F.B.S., Oliveira, L.O.V.B., Miranda, L.F., Casadei, F., Pappa, G.L.: Solving the exponential growth of symbolic regression trees in geometric semantic genetic programming. In: Proceedings of the Genetic and Evolutionary Computation Conference, GECCO 2018, 15–19 July 2018, Kyoto, Japan, pp. 1151–1158. ACM (2018)
4. Miller, J.F., Job, D., Vassilev, V.K.: Principles in the evolutionary design of digital circuits - part I. Genet. Program. Evolvable Mach. **1**(1), 8–35 (2000)
5. Nguyen, Q.U., Pham, T.A., Nguyen, X.H., McDermott, J.: Subtree semantic geometric crossover for genetic programming. Genet. Program. Evolvable Mach. **17**(1), 25–53 (2016)
6. Castelli, M., Silva, S., Vanneschi, L.: A C++ framework for geometric semantic genetic programming. Genet. Program. Evolvable Mach. **16**(1), 73–81 (2015). https://doi.org/10.1007/s10710-014-9218-0
7. Miller, J.F.: Cartesian Genetic Programming. Springer, Berlin (2011). https://doi.org/10.1007/978-3-642-17310-3
8. Archetti, F., Lanzeni, S., Messina, E., Vanneschi, L.: Genetic programming for computational pharmacokinetics in drug discovery and development. Genet. Program. Evolvable Mach. **8**(4), 413–432 (2007). https://doi.org/10.1007/s10710-007-9040-z
9. Vanneschi, L.: An introduction to geometric semantic genetic programming. In: Schütze, O., Trujillo, L., Legrand, P., Maldonado, Y. (eds.) NEO 2015. SCI, vol. 663, pp. 3–42. Springer, Cham (2017). https://doi.org/10.1007/978-3-319-44003-3_1
10. Dua, D., Karra Taniskidou, E.: UCI machine learning repository (2017). http://archive.ics.uci.edu/ml
11. Castelli, M., Trujillo, L., Vanneschi, L., Silva, S., Z.-Flores, E., Legrand, P.: Geometric semantic genetic programming with local search. In: Proceedings of the Genetic and Evolutionary Computation Conference, GECCO 2015, 11–15 July 2015, Madrid, Spain, pp. 999–1006. ACM (2015)

Can Genetic Programming Do Manifold Learning Too?

Andrew Lensen$^{(\boxtimes)}$ ⓘ, Bing Xue, and Mengjie Zhang

School of Engineering and Computer Science, Victoria University of Wellington,
PO Box 600, Wellington 6140, New Zealand
{andrew.lensen,bing.xue,mengjie.zhang}@ecs.vuw.ac.nz

Abstract. Exploratory data analysis is a fundamental aspect of knowledge discovery that aims to find the main characteristics of a dataset. Dimensionality reduction, such as manifold learning, is often used to reduce the number of features in a dataset to a manageable level for human interpretation. Despite this, most manifold learning techniques do not explain anything about the original features nor the true characteristics of a dataset. In this paper, we propose a genetic programming approach to manifold learning called GP-MaL which evolves functional **mappings** from a high-dimensional space to a lower dimensional space through the use of interpretable trees. We show that GP-MaL is competitive with existing manifold learning algorithms, while producing models that can be interpreted and re-used on unseen data. A number of promising future directions of research are found in the process.

Keywords: Manifold learning · Genetic programming ·
Dimensionality reduction · Feature construction

1 Introduction

Manifold learning has risen to prominence in recent years due to significant improvement in autoencoders and the widespread use of the t-Distributed Stochastic Neighbour Embedding (t-SNE) visualisation algorithm [11]. Manifold learning is the main area in the non-linear dimensionality reduction literature, and consists of algorithms which seek to discover an embedded (non-linear) manifold within a high-dimensional space so that the manifold can be represented in a much lower-dimensional space. Hence, they aim to perform dimensionality reduction while preserving as much of the *structure* of the high-dimensional space.

Within manifold learning, there are two broad categories of algorithms: those that produce a mapping between the high and low-dimensional spaces ("mapping methods"), and those which provide only the found embedding[1] ("embedding

---

[1] An embedding here refers to the low-dimensional representation of the structure present in a dataset.

© Springer Nature Switzerland AG 2019
L. Sekanina et al. (Eds.): EuroGP 2019, LNCS 11451, pp. 114–130, 2019.
https://doi.org/10.1007/978-3-030-16670-0_8

methods"). Mapping methods are particularly attractive, as they allow future examples to be processed without re-running the algorithm, and they have the potential to be *interpretable*, which is often desirable in machine learning tasks.

Genetic Programming (GP) is well known for producing *functions* which map inputs (the domain) to outputs (the codomain) using tree-based GP [15]. GP appears to have several promising characteristics for solving this problem:

- It is a global learner, and so should be less prone to producing partial-manifolds (i.e. local minima) unlike many existing methods which use gradient descent or other approaches;
- As it uses a population-based search method, it does not require a differentiable fitness function (unlike auto-encoders, t-SNE, etc.) and so could be used with a range of optimisation criteria; and
- It is intrinsically suited to producing interpretable mappings, as tree-based GP in particular can be understood by evaluating the tree from bottom to top. A wide range of tools are also available for producing interpretable GP models, including automatic program simplification and parsimony pressure.

Despite these traits, we are not aware of any work that uses GP to learn a manifold by mapping an input dataset to a set of lower-dimensional outputs.

1.1 Goals

In this work, we propose the first approach to using GP to perform Manifold Learning (**GP-MaL**). In particular, we will:

1. Propose a multi-tree GP representation and function and terminal sets for performing manifold learning;
2. Construct an appropriate fitness function to evaluate how effectively a GP individual preserves the structure of the high-dimensional space;
3. Evaluate how GP-MaL fares compared to existing manifold learning algorithms on a variety of classification tasks; and
4. Investigate the viability of GP-MaL for producing interpretable mappings of manifolds.

2   Background

2.1   Dimensionality Reduction

Broadly speaking, dimensionality reduction (DR) is the task of reducing an existing feature space into a lower-dimensional one, which can be better understood and processed more efficiently and effectively [9]. Two main approaches in DR are feature selection (FS) and feature extraction/construction [9]. While FS approaches—which select a small subset of the original features—are sufficient when a dataset has significant intrinsic redundancy/irrelevancy, there is a limit to how much the dimensionality can be reduced by FS alone. For example, if we want to reduce the dimensionality to two or three features, using FS alone is likely to poorly retain the structure of the dataset. In such a scenario, feature extraction/construction methods are able to reduce dimensionality more effectively by *combining* aspects of the original features in some manner.

One of the most well-known FC methods is Principle Component Analysis (PCA) [5]. PCA produces *components* (constructed features) which are linear combinations of the original features, such that each successive component has the largest variance possible while being orthogonal to the preceding components. Variance is a fundamental measurement of the amount of *information* in a feature, and so PCA is optimal for performing linear dimensionality reduction under this framework. However, linear combinations are not sufficient when data has a complex underlying structure; linear methods tend to focus on maintaining global structure while struggling to maintain local neighbourhood structure in the constructed feature space [11]. Thus, there is a clear need for nonlinear dimensionality reduction, of which a major research area is manifold learning.

2.2   Manifold Learning

Manifold learning algorithms are based on the assumption that the majority of real-world datasets have an intrinsic redundancy in how they represent information they contain through their features. A manifold is the inherent underlying structure which contains the information held within that dataset, and often this manifold can be represented using a smaller number of features than that of the original feature set [1]. Thus, manifold learning algorithms attempt to learn/extract this manifold into a lower-dimensional space. PCA, for example, can be seen as a linear manifold learning algorithm; of course, most real-world manifolds are strongly non-linear [1].

Multidimensional Scaling (MDS) [6] was one of the first approaches to manifold learning proposed, and attempts to maintain between-instance distances as well as possible from the high to the low dimensional space. Metric MDS often uses a loss function called *stress*, which is then minimised using a majorizing function from convex analysis. Another well-known, more recent method is Locally-Linear Embedding (LLE) [17], which describes each instance as a linear combination of its neighbours[2], and then seeks to maintain this combination in the low-dimensional space using eigenvector-based optimisation. MDS performs a non-parametric transformation of the original feature space, and so is not interpretable with respect to the original features; LLE is also difficult to interpret given it is based on preserving neighbourhoods.

t-Distributed Stochastic Neighbor Embedding (t-SNE) [11] is considered by many to be the state-of-the-art method for performing visualisation (i.e. 2D/3D manifold learning); it models the original feature space as a joint probability distribution in terms of how close an instances' neighbours are and then attempts to produce the same joint distribution in the low-dimensional space by using Kullback-Leibler divergence to measure the similarity of the two distributions.

---

[2] Here, neighbours refer to the closest instances to a point by (Euclidean) distance.

However, t-SNE was developed purely for visualisation (2/3D dimensionality reduction) and so it is not specifically designed as a general manifold learning algorithm [11]. It is also similar to MDS in that it produces an embedding with no mapping back to the original features. Finally, autoencoders are often regarded to do a type of manifold learning [1], but again they tend to be quite opaque in the meaning of their learnt representation, while requiring significantly more computational resources than the classical manifold learning methods.

2.3   Related Work

Evolutionary Computation (EC) has seen very recent use in evolving autoencoders for image classification tasks using Genetic Algorithms [19], GP [16], and Particle Swarm Optimisation [18]. Historically, auto-encoders have had to be manually designed or require significant domain knowledge to get good results, and so automatic evolution of auto-encoder structure is a clear improvement. However, these methods are still a somewhat indirect use of EC for representation/manifold learning, as they do not allow an EC method to directly learn the underlying structure as our proposed GP approach may.

GP has also been used for visualisation (i.e. the 2D form of manifold learning) in a supervised learning context using a multi-objective fitness function to optimise both classification performance and clustering-based class separability measures [2]. Recently, a GP method was proposed to evolve features for feature selection algorithm testing [8], which also used a multi-tree representation, but used a specialised fitness function to encourage redundant feature creation based on mutual information (MI). The use of GP for visualisation of solutions for production scheduling problems has also been recently investigated [13]. GP has also been applied to other tangential unsupervised learning tasks for feature creation, such as clustering [7], as well as extensive use in supervised learning domains [12,20]. Clearly, GP has shown significant potential as a feature construction method, and so it is hoped that it can be extended to directly perform manifold learning as well.

3   GP for Manifold Learning (GP-MaL)

The proposed method, GP-MaL, will be introduced in three stages. Firstly, the design of the terminal and function set is discussed. Then, a fitness function appropriate for manifold learning is formulated and explained. Finally, a method to improve the computational efficiency of GP-MaL (while maintaining good performance) is developed.

3.1   GP Representation

In this work, we utilise a multi-tree GP representation, whereby each tree represents a single dimension in the output (low-dimensional) space. While multi-tree

GP is known to scale poorly as the number of trees (t) increases, manifold learning usually assumes a low output dimensionality (e.g. $t < 10$). The terminal set consists of the d scaled real-valued input features, as well as random constants drawn from $U[-1, +1]$ to allow for variable sub-tree weighting. The output of each tree is not scaled or normalised in any way as this may introduce bias to the evolved trees or affect tree interpretability.

The function set (Table 1) chosen is inspired by existing feature construction and manifold learning literature. It includes the standard "+" and "×" arithmetic operators to allow simple combinations of features/sub-trees, as well as a "5+" operator which sums over five inputs[3] to encourage the use of many input features on large datasets. Subtraction and division were not included as they are the complements of addition and subtraction and so are redundant in the "way" in which they combine sub-trees. To encourage the learning of non-linear manifolds, two common non-linear activation functions from auto-encoders were added: the sigmoid and rectified linear unit (ReLU) operators. The function set also includes two conditional (non-differentiable!) operators, "max" and "min", which may allow GP to produce more advanced functions. Finally, the "if" function is also included, which takes three inputs a, b, c and outputs b if $a > 0$ or c otherwise, to allow for more flexible conditions to be learnt.

Table 1. Summary of the function set used in GP-MaL.

Category	Arithmetic			Non-Linear		Conditional		
Function	+	×	5+	Sigmoid	ReLU	Max	Min	If
No. of Inputs	2	2	5	1	1	2	2	3
No. of Outputs	1	1	1	1	1	1	1	1

Mutation is performed by selecting a random tree in a GP individual, and then selecting a random sub-tree to mutate within that tree, as standard. Crossover is performed in a similar way, by selecting a random tree from each candidate individual, and then performing standard crossover.

3.2   Fitness Function

A common optimisation strategy among manifold learning algorithms is to encourage preserving the high-dimensional neighbourhood around each instance in the low-dimensional space. For example, MDS attempts to maintain distances between points, whereas t-SNE uses a probabilistic approach to model how related different points are, and attempts to produce an embedding with a similar joint probability distribution. We refrain from using a distance-based approach

---

[3] Five inputs were found to be a good balance between encouraging wider trees and minimising computing resources required.

due to the associated issues with the curse of dimensionality [4], and instead try to preserve the *ordering* of neighbours from the high to low dimensions.

Consider an instance I which has ordered neighbours $N = \{N_1, N_2, ..., N_{n-1}\}$ for n instances neighbours in the high-dimensional space, and neighbours N' in the low-dimensional space. If we were to perfectly retain all structure in the dataset, then the ordering of N' must be identical to that of N, i.e. $N = N'$. In other words, the quality of the low-dimensional space can be measured by how similar N' is to N. In this work, we propose measuring similarity by how far each instances' neighbours deviate in their ordering in the low-dimensional space compared to the high-dimensional space. For example, if $N = \{N_1, N_2, N_3\}$ and $N' = \{N_2, N_3, N_1\}$, the neighbours deviate by 2, 1, and 1 positions respectively. Clearly, the larger the deviation, the more inaccurately the orderings have been retained. Let $Pos(a, X)$ give the index of a in the ordering of X. We propose the following similarity measure:

$$Similarity(N, N') = \sum_{a \in N} Agreement(|Pos(a, N) - Pos(a, N')|) \qquad (1)$$

where *Agreement* is a function that gives **higher** values for **smaller** deviations. GP-MaL uses an *Agreement* function based on a Gaussian weighting to allow for small deviations without significant penalty, while still penalising large deviations harshly. In this work, a Gaussian with a μ of 0 and $\theta = 20$ is used. θ controls how harshly deviations are punished – in preliminary testing we found a high θ gave best results as it created a smoother fitness landscape. The weighting for a given deviation dev is $1 - prob(-dev, +dev)$, i.e. the area of the Gaussian not in this range. In this way, when there is no deviation, the weighting is 1 (perfect), whereas when it is maximally deviated the weighting tends to 0.

The complete fitness function is the normalised similarity across all instances in the dataset (X):

$$Fitness = \frac{1}{n^2} \sum_{I \in X} Similarity(N_I, N_I') \qquad (2)$$

Fitness is in the range $[0, 1]$ and should be maximised.

3.3   Tackling the Computational Complexity

Unfortunately, computing the above fitness requires ordering every instance's neighbours by their distances in the low-dimensional space, at a cost of $O(n \log(n))$ using a comparison sort. This gives a net complexity of $O(n^2 \log(n))$ for each individual in the population. This scales poorly with the number of instances in the dataset. Consider a given neighbour N_b, which comes after N_a and before N_c. Even if we do not optimise the deviation of N_b, it seems likely that it will still be near N_a and N_c in the low-dimensional ordering, as it is likely to have similar feature values to N_a and N_c and hence will have a similar output from the evolved function. Based on this observation, we can omit some neighbours from our similarity function in order to reduce the computational complexity. Clearly, removing any neighbours will slightly reduce the accuracy of the fitness function, but this is made up by the significantly decreased computational cost (similar to surrogate model approaches). An example of this can be seen in Fig. 1, where the number of edges are decreased significantly by only considering the two nearest neighbours. Despite this, the global structure of the graph is still preserved well, with E and G only having two edges to the rest of the (distant) nodes, and C, D, and F sharing many edges as they are in close proximity.

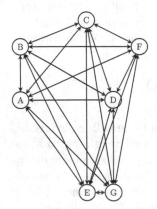
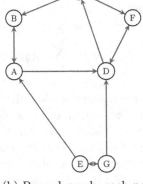

(a) Complete graph: 42 directed edges.   (b) Pruned graph: each node is connected to its two nearest neighbours. 14 edges.

Fig. 1. Pruning of a graph to reduce computational complexity.

When considering which neighbours to omit, it is more important to consider the closer neighbours' deviations, in order to preserve local structure, as this is most likely to preserve useful information about relationships in the data. However, it is still important to consider more distant neighbours, so that the global structure is also preserved. Based on this, we propose choosing neighbours more infrequently the further down the nearest-neighbour list they are. One approach is to choose the first k neighbours, followed by k of the next $2k$ neighbours (evenly spaced), then k of the next $4k$, etc. This gives η neighbours according to the following equation:

$$\eta = k \log_2\left(\frac{n}{k} + 1\right) \tag{3}$$

thus η is proportional to $\log(n)$ ($k \ll n$). The complexity per GP individual is then $O(\eta \log(\eta)) = O(\log(n) \log(\log(n)))$, which gives a sublinear complexity. In preliminary testing, we found using $k = 10$ to give only minor differences in learning performance which was significantly outweighed by the ability to train for many more generations in the same computational time. We use this approach in all experiments in this paper. While k could be perhaps be decreased further, it would not reduce computational time significantly, as tree evaluation is now the main cost of the evolutionary process.

4   Experiment Design

To evaluate the quality of our proposed GP-MaL algorithm, we focus mainly on the attainable accuracy on classification datasets using the evolved low-dimensional datasets. High classification accuracy generally requires as much of the structure of the dataset to be retained as possible in order to find the best decision boundaries between classes, and so is a useful proxy for measuring the amount of retained structure. We refrain from using the fitness function (or similar optimisation criteria) to measure the manifold "quality" so as not to introduce bias towards any specific manifold learning method. The scikit-learn [14] implementation of the Random Forest (RF) classification algorithm (with 100 trees) is used as it is a widely used algorithm with high classification accuracy, is stable across a range of datasets, and has reasonably low computational cost [21]. While other algorithms could also be compared, we found the results to be generally consistent across algorithms, and so do not include these for brevity. 10-fold cross-validation is used to evaluate every generated low-dimensional dataset, and 40 evolved datasets (40 GP runs) are used for each tested dataset in order to account for evolutionary stochasticity.

The characteristics of the ten datasets we used for our experiments are shown in Table 2. A range of datasets from varying domains were chosen with different numbers of features, instances, and classes.

Table 2. Classification datasets used for experiments. Most datasets are sourced from the UCI repository [3] which contains original accreditations.

Dataset	Instances	Features	Classes	Dataset	Instances	Features	Classes
Wine	178	13	3	COIL20	1440	1024	20
Movement Libras	360	90	15	Madelon	2600	500	10
Dermatology	358	34	6	Yale	165	1024	15
Ionosphere	351	34	2	MFAT	2000	649	10
Image Segmentation	2310	19	7	MNIST 2-class	2000	784	2

We compare the proposed GP-MaL method to a number of baseline manifold learning methods: PCA (as a linear baseline), MDS (which uses a similar optimisation criteria), LLE (a popular MaL method) and t-SNE (state of the art for 2D/3D manifold learning). Scikit-learn [14] was used for all the baseline methods, except for t-SNE, with default settings. For t-SNE, we used van der Maaten's more efficient Barnes-Hut implementation [10]. For each method and dataset, we produce transformed datasets for two, three, five, and the cube root (CR) of the number of original features. Two/three features are useful for visualisation but are unlikely to be sufficient to preserve all structure, whereas the

Table 3. GP parameter settings.

Parameter	Setting	Parameter	Setting
Generations	1000	Population size	1024
Mutation	20%	Crossover	80%
Elitism	Top 10	Selection type	Tournament
Min. tree depth	2	Max. tree depth	8
Tournament size	7	Pop. initialisation	Half-and-half

cube root approach was found in preliminary testing to be the point at which all tested methods could capture maximal structure from the datasets. Five features are used as a "middle-ground". As all of these implementations have stochastic components, we also ran each 40 times for each dataset.

We use standard GP parameter settings, as per Table 3. One notable setting is that we use 1000 generations; as we are interested primarily in exploring the *potential* of GP for this task, we are not particularly concerned with optimising the number of generations for best efficiency; this will be explored in future work.

5   Results and Analysis

The full set of results for each method and dataset are shown in Table 4. For each baseline method on each dataset, we label the result with a "+" if the baseline was significantly **better** than GP-MaL, a "−" if it was significantly **worse**. If neither of these notations appear, there was no significant difference in the results. We used one-tailed Mann-Whitney U tests with a 95% confidence interval to compute significance. For convenience, a summary of these results are provided in Table 5 by totalling the number of "wins" (significantly better), "losses" (significantly worse) and "draws" (no significant difference) the proposed GP-MaL method has compared with each baseline. We compare GP-MaL's performance to PCA and MDS, and LLE and t-SNE in the following subsections, as these pairs of methods exhibit similar patterns.

5.1   GP-MaL Compared to PCA & MDS

GP-MaL has a clear advantage over PCA when the most significant amount of feature reduction—to two or three features—is required. Given that PCA is a linear manifold learning method, it is not surprising that GP-MaL is able to preserve more structure in 2 or 3 dimensions by performing more complex, non-linear reductions. At higher dimensions, the gap narrows somewhat, as at 5 or CR features there are enough available output dimensions in order to make linear combinations able to model the underlying structure of the data more accurately. PCA weights every input feature in each component it creates, which means the way in which it models this structure is rather opaque when there are many input features. The MDS results have a similar pattern to the PCA ones, except that MDS and GP-MaL are quite even on 3 and 5 features. It is interesting to note that MDS uses a similar optimisation criterion to GP-MaL, but struggles significantly more at 2 features.

5.2 GP-MaL Compared to LLE & T-SNE

Overall, GP-MaL is the most consistent of all the methods across the different numbers of features produced. LLE wins on one more dataset than GP-MaL for 2 features, but otherwise GP-MaL has a clear advantage with 7 wins on 3/5/CR features. The performance of LLE fluctuates quite widely across the datasets, and generally loses to PCA as the number of features is increased.

While GP-MaL is clearly worse than t-SNE on the 2 and 3D results, it outperforms t-SNE on 5 or CR features. On the Ionosphere and COIL20 datasets, t-SNE's performance actually decreases as the number of output features are increased, which means it is much more sensitive to the number of components that the user chooses than GP-MaL; GP-MaL almost strictly improves as more output features are produced, which is what we generally expect from dimensionality reduction techniques.

Table 4. Experiment results. GPM refers to the proposed GP-MaL method. The number after each method specifies the dimensionality of the low-dimensional manifold; "cr" means the cube root approach determined the dimensionality.

Method	Wine	Move.	Derm.	Iono.	Image.	COIL20	Mad.	Yale	MFAT	MNIST
GPM2	0.955	0.485	0.914	0.826	0.797	0.628	0.605	0.382	0.639	0.909
PCA2	0.764−	0.405−	0.769−	0.776−	0.675−	0.647+	0.572−	0.244−	0.643	0.906−
MDS2	0.711−	0.476−	0.723−	0.837+	0.716−	0.732+	0.574−	0.339−	0.687+	0.909
LLE2	0.659−	0.499+	0.803−	0.833	0.809+	0.850+	0.601−	0.120−	0.843+	0.980+
tSNE2	0.718−	0.782+	0.852−	0.890+	0.921+	0.948+	0.712+	0.455+	0.935+	0.986+
GPM3	0.964	0.579	0.924	0.872	0.892	0.773	0.688	0.472	0.765	0.925
PCA3	0.793−	0.608+	0.780−	0.877	0.805−	0.823+	0.681−	0.374−	0.749−	0.932+
MDS3	0.726−	0.594+	0.774−	0.910+	0.883−	0.849+	0.677−	0.404−	0.830+	0.932+
LLE3	0.667−	0.513−	0.824−	0.847−	0.831−	0.923+	0.648−	0.297−	0.847+	0.984+
tSNE3	0.712−	0.768+	0.847−	0.756−	0.924+	0.952+	0.731+	0.394−	0.935+	0.987+
GPM5	0.960	0.673	0.951	0.915	0.958	0.847	0.864	0.553	0.888	0.940
PCA5	0.913−	0.705+	0.899−	0.923+	0.911−	0.887+	0.881+	0.531−	0.885	0.945+
MDS5	0.732−	0.719+	0.817−	0.928+	0.901−	0.886+	0.685−	0.564+	0.881−	0.948+
LLE5	0.683−	0.684+	0.825−	0.817−	0.837−	0.930+	0.665−	0.456−	0.870−	0.985+
tSNE5	0.718−	0.747+	0.835−	0.714−	0.930−	0.878+	0.763−	0.532−	0.939+	0.987+
GPMcr	0.962	0.681	0.941	0.899	0.891	0.913	0.863	0.661	0.935	0.952
PCAcr	0.789−	0.704+	0.852−	0.879−	0.804−	0.950+	0.857−	0.648−	0.939+	0.957+
MDScr	0.725−	0.722+	0.792−	0.920+	0.884−	0.911	0.670−	0.651	0.889−	0.957+
LLEcr	0.669−	0.685	0.814−	0.803−	0.828−	0.924+	0.679−	0.577−	0.912+	0.984+
tSNEcr	0.710−	0.759+	0.853−	0.713−	0.925+	0.730−	0.765−	0.650−	0.944+	0.987+

In a number of cases, t-SNE does actually outperform GP-MaL while using fewer features—however, consider that t-SNE (and LLE) are embedding method which do not have to manipulate the original feature space to produce the output

Table 5. Summary of experiment results. The number of "wins", "losses", and "draws" are shown for GP-MaL compared to each baseline.

Baseline	Wins	Losses	Draws	Baseline	Wins	Losses	Draws
PCA2	8	1	1	PCA3	6	3	1
MDS2	6	3	1	MDS3	5	5	0
LLE2	4	5	1	LLE3	7	3	0
tSNE2	2	8	0	tSNE3	4	6	0
PCA5	4	5	1	PCAc	6	4	0
MDS5	5	5	0	MDSc	5	3	2
LLE5	7	3	0	LLEc	7	2	1
tSNE5	6	4	0	tSNEc	6	4	0

feature space (i.e. it is not a functional mapping). It is clearly more difficult to evolve such a mapping, but also has significant benefits in that GP-MaL's output dimensions can be interpreted in terms of how they combine the original features, which is often as important as visualisation alone in exploratory data analysis. This behaviour will be explored further in Sect. 6.

5.3 Summary

GP-MaL shows promising performance for an initial attempt at directly using GP for manifold learning, winning against all baselines on at least two of the four configurations tested. While GP-MaL faltered somewhat on some datasets such as MNIST, it achieved much better performance on other lower-dimensional datasets such as Wine and Dermatology. This suggests that with further improvements to its learning capacity, GP-MaL may have the potential to outperform existing methods on these higher-dimensional datasets too.

Another important consideration is the interpretability of the models produced by each baseline. t-SNE, LLE, MDS, and PCA (to a lesser extent) are almost black-boxes as they give little information about how the manifolds in the data are represented in terms of the original features. Interpretability is an increasing concern in data mining, and feature reduction is often touted as a way to improve it; we will examine in the following section if GP-MaL can be interpreted any more easily than these existing methods.

6   Further Analysis

6.1   GP-MaL for Data Visualisation

A common use of manifold learning techniques such as t-SNE and PCA is for visualisation of datasets in two- or three-dimensions. Figures 2 and 3 plot the two-dimensional outputs of each manifold learning method for two datasets that GP-MaL performed best on (Dermatology) and worst on (COIL20) respectively.

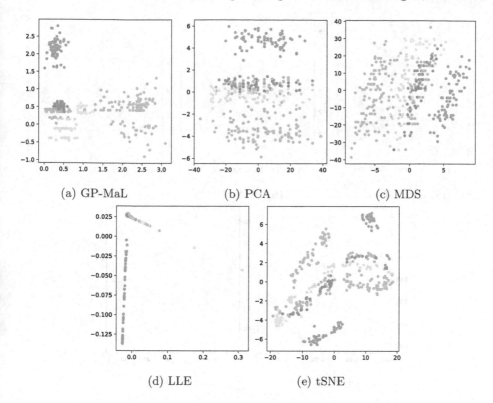

Fig. 2. The two created features on Dermatology dataset, coloured by class label. (Color figure online)

To show the potential of each method, we used the result that had the highest classification accuracy for plotting.

On the Dermatology dataset, GP-MaL clearly separates each class better than the baseline manifold methods. PCA, MDS, and t-SNE struggle to seperate the purple, green, and pink classes apart, whereas GP-MaL is able to separate them while keeping them reasonably close to signify their similarities. t-SNE splits both the purple and blue classes into two disjoint groups with other classes appearing in the middle of the split. LLE clearly struggles to give a good visualisation at all—it is only able to split the blue class from the others.

On the COIL20 dataset, LLE is able to separate the classes somewhat more effectively along one dimension, but still fails to produce a reasonable visualisation. t-SNE clearly does very well, but does continue to separate some classes into disjoint clusters (all of the green ones). It is not clear which of GP-MaL, PCA, and MDS produces the best result; MDS tends to incorrectly separate some classes like t-SNE, whereas GP-MaL and PCA have poorer separation of different classes overall. Unlike the other methods, the two dimensions produced by GP-MaL can be interpreted in terms of how they use the original features— this will be explored further in the next subsection.

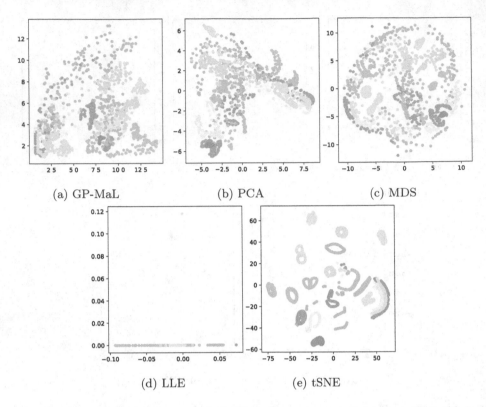

(a) GP-MaL   (b) PCA   (c) MDS

(d) LLE   (e) tSNE

Fig. 3. The two created features on COIL20 dataset, coloured by class label. (Color figure online)

6.2   Tree Interpretability

Part of an individual evolved on the MFAT dataset is shown in Fig. 4. While the left tree is quite large (containing 73 features), the right tree is very simple: it adds two features together, and outputs the sum. This gives a strong indication that these features are very important for modelling the underlying structure of the data. While the left tree is harder to analyse, it is useful to note that four of the children of the root node are again very simple: three features (two of which are transformed non-linearly) and a constant weighting. This again suggests that these features particularly model the instances in the MFAT dataset, with $X292$ appearing in both trees. $X292$ and $X294$ are the first and third Karhunen-Loève coefficients extracted from the original images; these coefficients are extracted in a similar way to PCA, so it makes sense that GP would recognise these to be very useful features: being the first and third coefficients, they represent a significant amount of the variance present in this dataset.

Figure 5 shows another example GP individual evolved on the 500-dimensional Madelon dataset, with five trees used. Of the five trees, four are simple enough to be human-interpretable, with the fifth being somewhat larger, but still

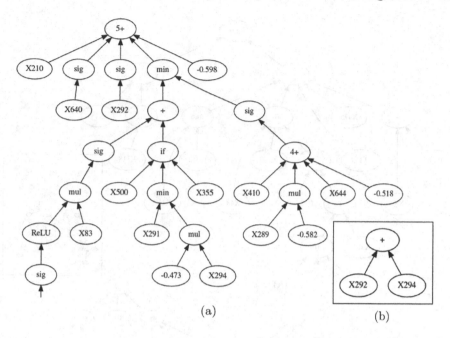

(a)   (b)

Fig. 4. An example of two simplified trees (features) evolved on the MFAT dataset, giving 65% classification accuracy. Only the top of the left tree is shown.

interpretable at the root. Trees 5b–e each combine between three and five features in ways that make sense, but which would not be able to be represented by many existing manifold learning methods. For example, consider Tree 5c, which uses either the original value of $X475$, or a non-linear sigmoid transformation of $X475$ depending on the value of $X138$. This suggests that there is a particular feature interaction between $X138$ and $X475$ which may be important to the underlying structure of the dataset. In fact, $X475$ is the feature which has the second-highest information gain (IG) on this dataset[4]. Tree 5e is just a single selected feature— $X455$ clearly is important in the manifold of this dataset.

Although Tree 5a is clearly more complex, the top of the tree still provides an interesting picture of the most important aspects of the dataset. For example, if $X48$ is a very low value, then this is simply the output of the tree. Examining $X48$ more closely reveals that it is in the top 3% of features in terms of IG, and that at its smallest values it always predicts the positive class. Also of note is that $X475$ appears twice again in the top of this tree as well as in Tree 5c.

Summary: As the focus of this work was to show the *potential* of GP for direct manifold learning, no parsimony pressure (or other such methods) were applied to encourage simple trees. Nevertheless, aspects of the evolved individuals can

---

[4] Information gain (mutual information) is often used in feature selection for classification to measure the dependency between a feature and the class label.

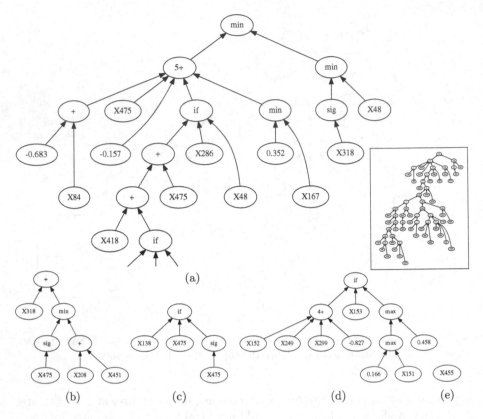

Fig. 5. An example of five simplified trees (features) evolved on the Madelon dataset, giving 87.8% classification accuracy. Tree (a) was truncated to save space, but the full tree in the box to the right to show that is reasonably small. The original dataset had 500 features.

be analysed with ease and provide insight into the structure of the datasets. This is a clear advantage over existing manifold learning techniques which are black (or very grey) boxes, and bodes well for future work. The use of GP also has the nice upside of allowing the evolved trees to be re-used on future examples without having to perform the whole manifold learning process again (as t-SNE or the other methods would require).

7   Conclusion

Manifold learning has become significantly more popular in recent years due to the emergence of autoencoders and visualisation techniques such as t-SNE. Despite this, the learnt manifolds tend to be completely or nearly uninterpretable with respect to the original feature space. With interpretability being a key goal in exploratory data mining, we proposed a new GP method called GP-MaL for directly performing manifold learning in this paper. Appropriate terminal and function sets were presented, along with a fitness function tailored to the task, and further techniques for reducing computational complexity. We showed that GP-MaL was competitive, and in some cases clearly better than, existing manifold learning methods, and was generally more stable across that datasets tested. GP-MaL was also shown to produce interpretable models that help the user to gain concrete insight into their dataset, unlike many existing manifold learning methods. The potential of GP-MaL for visualisation purposes was highlighted. Furthermore, the functional nature of the models produced by GP-MaL allows it to be applied to future data without re-training. As the first work using GP for directly performing manifold learning, these findings showcase the potential of GP to be applied further to this domain.

GP-MaL is quite flexible, and could easily be extended further with other function sets and fitness functions (that do not have to be differentiable!). We also hope to explore a multi-objective approach in the future that balances the often-conflicting objectives of maintaining global and local structure. It is also clear that techniques to encourage simpler/more concise trees such as parsimony pressure would further improve the usefulness of GP-MaL.

References

1. Bengio, Y., Courville, A.C., Vincent, P.: Representation learning: a review and new perspectives. IEEE Trans. Pattern Anal. Mach. Intell. **35**(8), 1798–1828 (2013)
2. Cano, A., Ventura, S., Cios, K.J.: Multi-objective genetic programming for feature extraction and data visualization. Soft Comput. **21**(8), 2069–2089 (2017)
3. Dheeru, D., Karra Taniskidou, E.: UCI machine learning repository (2017). http://archive.ics.uci.edu/ml
4. François, D., Wertz, V., Verleysen, M.: The concentration of fractional distances. IEEE Trans. Knowl. Data Eng. **19**(7), 873–886 (2007)
5. Jolliffe, I.T.: Principal component analysis. In: Lovric, M. (ed.) International Encyclopedia of Statistical Science, pp. 1094–1096. Springer, Berlin (2011). https://doi.org/10.1007/978-3-642-04898-2
6. Kruskal, J.B.: Multidimensional scaling by optimizing goodness of fit to a nonmetric hypothesis. Psychometrika **29**(1), 1–27 (1964)
7. Lensen, A., Xue, B., Zhang, M.: New representations in genetic programming for feature construction in k-means clustering. In: Shi, Y., et al. (eds.) SEAL 2017. LNCS, vol. 10593, pp. 543–555. Springer, Cham (2017). https://doi.org/10.1007/978-3-319-68759-9_44
8. Lensen, A., Xue, B., Zhang, M.: Automatically evolving difficult benchmark feature selection datasets with genetic programming. In: Proceedings of the Genetic and Evolutionary Computation Conference, GECCO, pp. 458–465. ACM (2018)
9. Liu, H., Motoda, H.: Feature Selection for Knowledge Discovery and Data Mining, vol. 454. Springer, Boston (2012). https://doi.org/10.1007/978-1-4615-5689-3
10. van der Maaten, L.: Accelerating t-SNE using tree-based algorithms. J. Mach. Learn. Res. **15**(1), 3221–3245 (2014)
11. van der Maaten, L., Hinton, G.E.: Visualizing high-dimensional data using t-SNE. J. Mach. Learn. Res. **9**, 2579–2605 (2008)

12. Neshatian, K., Zhang, M., Andreae, P.: A filter approach to multiple feature construction for symbolic learning classifiers using genetic programming. IEEE Trans. Evol. Comput. **16**(5), 645–661 (2012)
13. Nguyen, S., Zhang, M., Alahakoon, D., Tan, K.C.: Visualizing the evolution of computer programs for genetic programming [research frontier]. IEEE Comput. Intell. Mag. **13**(4), 77–94 (2018)
14. Pedregosa, F., et al.: Scikit-learn: machine learning in Python. J. Mach. Learn. Res. **12**, 2825–2830 (2011)
15. Poli, R., Langdon, W.B., McPhee, N.F.: A Field Guide to Genetic Programming. Lulu.com, Morrisville (2008)
16. Rodriguez-Coayahuitl, L., Morales-Reyes, A., Escalante, H.J.: Structurally layered representation learning: towards deep learning through genetic programming. In: Castelli, M., Sekanina, L., Zhang, M., Cagnoni, S., García-Sánchez, P. (eds.) EuroGP 2018. LNCS, vol. 10781, pp. 271–288. Springer, Cham (2018). https://doi.org/10.1007/978-3-319-77553-1_17
17. Roweis, S.T., Saul, L.K.: Nonlinear dimensionality reduction by locally linear embedding. Science **290**(5500), 2323–2326 (2000)
18. Sun, Y., Xue, B., Zhang, M., Yen, G.G.: A particle swarm optimization-based flexible convolutional auto-encoder for image classification. IEEE Trans. Neural Netw. Learn. Syst. (2018). https://doi.org/10.1109/TNNLS.2018.2881143
19. Sun, Y., Yen, G.G., Yi, Z.: Evolving unsupervised deep neural networks for learning meaningful representations. IEEE Trans. Evol. Comput. (2018). https://doi.org/10.1109/TEVC.2018.2808689
20. Tran, B., Xue, B., Zhang, M.: Genetic programming for feature construction and selection in classification on high-dimensional data. Memet. Comput. **8**(1), 3–15 (2016)
21. Zhang, C., Liu, C., Zhang, X., Almpanidis, G.: An up-to-date comparison of state-of-the-art classification algorithms. Expert Syst. Appl. **82**, 128–150 (2017)

Why Is Auto-Encoding Difficult for Genetic Programming?

James McDermott[✉]

National University of Ireland, Galway, Ireland
james.mcdermott@nuigalway.ie

Abstract. Unsupervised learning is an important component in many recent successes in machine learning. The autoencoder neural network is one of the most prominent approaches to unsupervised learning. Here, we use the genetic programming paradigm to create autoencoders and find that the task is difficult for genetic programming, even on small datasets which are easy for neural networks. We investigate which aspects of the autoencoding task are difficult for genetic programming.

Keywords: Unsupervised learning · Autoencoder ·
Linear genetic programming · Symbolic regression

1 Introduction

"If intelligence was a cake, unsupervised learning would be the cake, supervised learning would be the icing on the cake, and reinforcement learning would be the cherry on the cake." – Yann LeCun.[1]

There have been many recent successes in machine learning which take advantage in one way or another of unsupervised learning. Arguably the seminal paper of modern deep learning was that of Hinton et al. [1] in which multiple shallow networks were trained, one after the other, each in an unsupervised manner, and then joined together with the addition of a supervised final layer, to give a *deep belief net*.

The *autoencoder* is a neural network trained to reproduce its input at its output. No labels are involved, so it can be seen as unsupervised learning, or it is sometimes called "self-supervised" instead. The key idea is the *bottleneck* or *latent representation*, an intermediate layer of fewer neurons than the original data. By forcing the data through such a reduced layer, and then reconstructing it, we can hope to learn a new representation or *code* which loses very little information but has useful properties, in particular lower dimensionality. This may be possible as long as the training data has support only on a lower-dimensional manifold within its space. The mapping from input to code is termed the encoder and the mapping from code to output is termed the decoder. Both components are useful in applications: the encoder provides us with a new data representation

---

[1] https://www.facebook.com/yann.lecun/posts/10153426023477143.

© Springer Nature Switzerland AG 2019
L. Sekanina et al. (Eds.): EuroGP 2019, LNCS 11451, pp. 131–145, 2019.
https://doi.org/10.1007/978-3-030-16670-0_9

of reduced dimensionality (and sometimes, a desired distribution); the decoder can be used as a generative model. Successes involving autoencoder neural networks and other unsupervised models include the ability to generate fake images given textual captions [2] and high-quality de-noising of images [3].

In this context, it is interesting to see that unsupervised learning in the genetic programming (GP) paradigm is much less common than supervised learning. The symbolic regression (SR) problem, a supervised problem, is by far the best-known example of GP for ML. For the avoidance of doubt, by supervised and unsupervised learning here we are referring to the presence or absence of labels or label-like information for data points in a data set used in the GP fitness function. We are not seeing the evolutionary process itself as a supervised or unsupervised process. Neither are we referring to the neuro-evolution paradigm [4,5] in which evolutionary methods are used to create neural network architectures [6] or to train neural networks [7].

Instead, our focus is on replicating the autoencoder model using the genetic programming paradigm (functions in a population, created by flexible mutation and recombination, represented as combinations of functional arithmetic primitives) in place of the neural paradigm (adjustable weights). We will call this the genetic programming autoencoder (GPAE). We will investigate a simple multiple-output Linear GP (LGP) system as the GP representation, using a Step-Counting Hill-Climbing (SCHC) algorithm as the search method. The architecture of the system is shown in Fig. 1.

The motivations for producing such a model are the same as for autoencoders in the first place (representation learning, dimensionality reduction, ability to generate new samples). An evolutionary approach offers also the potential for more interpretable GP-style encoder and decoder functions. A further motivation for using GP for the task is that, since it is black-box, it would be possible to optimise multiple objectives (e.g. both the reconstruction error and a desired distribution at the hidden layer, as in the variational autoencoder [8]), which is not possible in neural networks trained by back-propagation, other than by combining them in a weighted sum. Moreover, non-differentiable objectives could be used easily, again in contrast to neural network models. Non-differentiable objectives may arise, e.g., with unusual desired hidden layer distributions.

However, the motivations for the research are broader. Autoencoding is an interesting and non-trivial task which can be framed as a type of multiple-output regression (with constraints). If GP symbolic regression is a powerful, general approach to regression as has been repeatedly observed [9], then autoencoding ought to be within its scope. But in fact, our results seem to show that it doesn't work well. Therefore we investigate several gradually easier versions of the task, to understand which aspects of the task cause the difficulty. In this way we may hope to learn something about GP's limitations.

Reader's Guide. In the next section, Sect. 2, we present some related work in more detail. The proposed GPAE model is presented in Sect. 3. Experiments and results are in Sect. 4, and we conclude with Sect. 5.

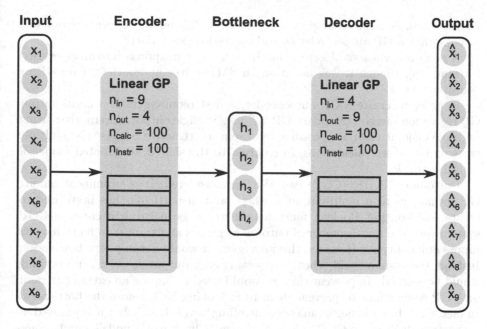

Fig. 1. GPAE architecture. It is composed of two LGP programs, an encoder and a decoder. The size of the encoder input matches that of the decoder output and is set by the number of variables in the dataset (all regarded as "independent" variables). The encoder output becomes the decoder input; it is referred to as the hidden or bottleneck layer with size n_{hidden}. For example, in one set of our experiments, the number of variables is 9 and $n_{hidden} = 4$ as shown. Each of the two programs has its own set of hyperparameters: input size n_{in}, output size n_{out}, number of calculation registers n_{calc} and number of instructions n_{instr} (to be discussed in detail in Sect. 3.1).

2   Related Work

Rumelhart, Hinton and Williams introduced the autoencoder model in the 1980s [10], and Hinton et al. brought it to attention again in 2006 as a general method of dimensionality reduction [11] and as a tool in training deep networks [1]. As already stated, many recent successes in machine learning have taken advantage of autoencoders and their variants, including variational [8], denoising [3], and adversarial autoencoders [12], and generative adversarial networks [13].

It is clear that one of the main advantages of the neural network model is that its distributed, differentiable representation is amenable to fast training by gradient descent. Several modern GP regression approaches have attempted to introduce some or all of this advantage to GP, e.g. [14,15]. However, other researchers have shown that black-box methods – not taking advantage of the gradient – are sufficient to train neural networks also [7]. A recent "bake-off" has found that the advantage of neural networks for classical regression tasks (as opposed to tasks of regression or classification on datasets such as images and audio) is not as large as might be expected [9]. Therefore, it seems reasonable to hope that a GP autoencoder should succeed on such data.

A program synthesis algorithm has been used as an approach to unsupervised learning [16], though it was based on an SMT solver approach, not on GP or neural networks.

In order to create a GP autoencoder, a first requirement is a multi-output GP regression method. Standard GP [17] uses a single output. Alternatives using multiple disjoint trees are possible, but seem unattractive because the multiple trees do not share calculations, in contrast to the shared, distributed model of neural networks.

Cartesian GP [18] (CGP) can also be considered. It is prominent among GP techniques as a multi-output method, and in particular has been used to create multi-output Boolean functions. It would be natural to create a CGP system with the same number of outputs as inputs and train it to reproduce its inputs at its outputs. However, this arrangement would not impose a bottleneck. Instead, the system could "cheat" by setting each output $\hat{x}_i = 1x_i$, and nothing would be learned. To prevent this, we would need to impose an extra constraint on CGP connections to prevent them from looking back before the bottleneck. In fact, CGP has a hyperparameter controlling "levels-back", but it is phrased as a constant: how many CGP *columns* ("layers") back can a node's inputs come from. That would not be sufficient for our autoencoder application since unless it is set to 1, it would not clearly divide the program into two parts with a single, clearly-defined bottleneck. In any case, it allows a node to take input from the original input nodes regardless of its value.

Other GP systems which could naturally be used as multi-output systems include the DAG-oriented systems proposed by Jackson [19] (known as single-node GP or SNGP) and by Schmidt and Lipson [20]. Neuro-evolution of augmenting topologies (NEAT) [5] could certainly be used to create an autoencoder, but is best seen as a neural network system rather than GP, since it uses adjustment of weights, and so does not quite fit with our goals.

Linear GP as studied by, e.g., Brameier and Banzhaf [21] is another natural option as it can easily provide multiple outputs: rather than a single output register, we can designate as many as we need. However, as with CGP, there is no natural bottleneck. An LGP program could be considered as an autoencoder as long as a constraint was added to the transmission of information from the first half of the program to the second half. This is even less natural in LGP than in CGP, since LGP uses registers to store and transmit information.

Perhaps the easiest solution to this problem, and the one adopted in this paper, is to model the autoencoder as a *pair* of multi-input, multi-output programs. Given this decision, any of the above approaches (CGP, LGP, DAG-oriented) would be suitable as the representation for the two programs in the pair. We could even choose two representations, e.g. CGP for the encoder and LGP for the decoder. However, for simplicity, we have chosen to use LGP for both.

The next issue is how to train the programs. One possibility is a coevolution-ary system of two symbiotic populations, i.e. one population of encoders and one population of decoders. Cooperative coevolutionary systems have been studied in many previous works with interesting results, e.g. [22]. The disadvantage to such a system in our context is that a good encoder and a good decoder aren't necessarily good together, and that the algorithm becomes complex. The alter-native we have chosen instead is to regard each *individual* as a pair of programs, with genetic operators whose outputs are pairs rather than single programs. The advantage is that good encoder material and good decoder material stay together. The required assumption is then that small changes in the behaviour of one can improve the performance of the pair even without a simultaneous change in the behaviour of the other. As we will see, small incremental improvements are indeed possible throughout evolution.

The details of our proposed system are described in the following section. The main open question in this context is simply whether a GP autoencoder can work.

3   Proposed Model

The essence of the autoencoder is a vector of inputs, a vector of outputs, and somewhere in between, a bottleneck which can be seen as a new, learned repre-sentation of reduced dimensionality.

We have chosen to implement this model as a *pair* of multi-input, multi-output Linear GP (LGP) programs, one for the encoder and one for the decoder. The only communication between them is via the outputs of the encoder, which become the inputs to the decoder. This framework ensures that there is a true bottleneck (i.e. no "cheating" – see above) and good re-use of code (we can use a single LGP implementation, taking a pair of instances as an autoencoder).

In order to train the autoencoder, we see each individual as a *pair* of pro-grams. Initialisation and mutation are defined on pairs of programs and work by delegating to natural operators designed on programs as described below.

For the GP population model, we have chosen a hill-climbing GP approach. That is, there is a single GP individual, not a full population, and mutation is the only variation operator. The motivation is simplicity (few hyper-parameters).

3.1   Linear GP for Multi-output Regression

As the main component of our GPAE we propose to use a simple LGP repre-sentation [21] suitable for generic multi-output regression. We will not use loops or conditionals as considered in some LGP systems.

Each program is a list of instructions. A program runs in a simple environ-ment consisting of memory registers. Each instruction carries out a calculation with 1 or 2 inputs, which are values previously stored in registers, and writes the result to a register.

The registers are partitioned into three types. (1) The input registers are used to hold initial values of the variables. The number of these n_{in} is given by the number of variables in the dataset. There is no restriction against over-writing these initial values during execution. (2) Calculation registers are used as temporary storage. The number of them is given by n_{calc}. At initialisation, numerical constants ranging evenly from 0 to 1 inclusive are supplied as initial values of the calculation registers. Again, there is no restriction on over-writing these values during execution. (3) Finally, the outputs are read, after execution, from the output registers. The number of them n_{out} is given by the number of outputs required in the multi-output problem (we are in the context of multi-output regression rather than autoencoding).

Initialisation and mutation operators are defined in a natural way for this representation. Initialisation creates a list of instructions of length n_{instr}. For each instruction it chooses two registers uniformly from the list of all registers as the operands, then one operator uniformly from the list of possible operators, and one register uniformly from all registers as the location to write the result. Mutation uniformly chooses an instruction in the program, then uniformly chooses one of the four components (result location, operator, operand 1, or operand 2) for mutation. The chosen component is replaced by a uniformly-chosen component of the same type (i.e. a uniformly-chosen operator or a uniformly-chosen register). Crossover is also easily defined, but is not used in this work since we use a hill-climbing method (see below).

We will refer to n_{calc} and n_{instr} as "capacity" parameters. Larger values give the program more capacity to fit target data.

The arithmetic operators used are $+$, $-$, $\times$, the analytic quotient $AQ(a, b) = a/\sqrt{1.0 + b^2}$ [23], a protected logarithm $\text{plog}(x) = \log(1 + |x|)$, and sin. The AQ and plog operators avoid numerical errors and avoid the large discontinuities characteristic of some other protected operators. Both AQ and sin are recommended by Nicolau and Agapitos [24].

3.2   From Programs to Pairs of Programs

In order to train an autoencoder, we use a pair of programs each of the form described above. That is, we treat an autoencoder as an encoder-decoder pair: $\hat{x} = D(E(x))$. The number of inputs to the encoder E is given by the dataset; the number of outputs is a hyperparameter n_{hidden}, the size of the "hidden layer". We borrow this term from the neural networks field since it is familiar, but strictly speaking our representation does not use layers. An alternative name is the *latent representation*. The output of the encoder is taken as the input of the decoder. n_{hidden} then also gives the number of inputs to the decoder D, and its number of outputs is again given by the size of the dataset.

Operators on *pairs* of programs work in a natural way. Initialisation delegates to initialisation on programs; mutation chooses one of the pair randomly, then delegates to mutation on that.

3.3 Step-Counting Hill-Climbing

The step-counting hill-climbing (SCHC) method [25] is a hill-climbing method which is not much more complex than simple hill-climbing, with just a few extra lines of code and one extra hyper-parameter, but offers a remarkable performance improvement relative to simple hill-climbing.

In particular, it has some ability to escape local optima. It does so by maintaining a threshold fitness value, equal to or usually somewhat *worse* than the fitness of the current individual. A new individual formed by mutation from the current individual is *accepted* (i.e. it replaces the current individual) if its fitness is better than the threshold. The threshold is updated to the current individual's fitness value at regular intervals, and so, it changes monotonically. This simple scheme allows for some disimproving moves. It is similar in this respect to simulated annealing (SA) and to late-acceptance hill-climbing (LAHC) [26], but is claimed to be robust to re-scaling of the objective function (which is not true of SA) and to be simpler to configure than either. It has been previously used in (tree-structured) GP [27], finding results as good as standard GP with a population but in a much simpler configuration. Moreover, even simple hill-climbing has been shown to do well in genetic programming [28,29].

The objective function for the GPAE is the mean squared error (MSE) of the reconstruction $\hat{x}$ against the training point x, averaged across training points.

4 Experiments and Results

The first goal of our experiments is to attempt the autoencoder task. In our pilot experiments, we observed that the task was very difficult for the Linear GP system, across several datasets and many configurations. Rather than reporting a comprehensive experiment on this, we have chosen to report this failure for a single dataset, and a few configurations, and then to expend our experimental effort in a different direction. We attempt several cut-down versions of the task, in order to understand just which aspects of the task make it difficult.

4.1 Datasets

Three datasets are used throughout these experiments: the Glass dataset, Australian Credit Approval (ACA) dataset, and PenDigits dataset, all sourced from the UCI repository [30]. All are provided as supervised datasets, but have previously been used as test datasets for unsupervised algorithms (discarding the labels), and are used in this way here. In fact, they have been shown to be relatively easy for autoencoder neural networks [31], even relatively low-capacity ones and without large computational requirements. In our experiments, the labels are discarded, a train-test split is performed, and each column is normalised by the max-abs method, using the training set only to choose the normalisation parameters. The Glass dataset has 9 variables, and we set $n_{\text{hidden}} = 4$; ACA has 14 and we set $n_{\text{hidden}} = 4$; PenDigits has 16 and we set $n_{\text{hidden}} = 5$.

These choices are for compatibility with previous work [31]. For the avoidance of doubt in replication, datasets and code are provided[2].

4.2   Experimental Design

We have used the following parameters:

- $n = 100,000$ (fitness evaluation budget)
- $L = 10, 100,$ or 1000 (SCHC fitness history length)
- $n_{instr} = 5, 50,$ or 500 (number of instructions per program)
- $n_{calc} = 5, 50,$ or 500 (number of calculation registers)
- $n_{hidden} = 4$ (hidden "layer" or latent representation size, for GPAE).

For n_{instr} and n_{calc}, not all combinations make sense: there is no point in having $n_{instr} < n_{calc}$ since the extra calculation registers can never be used. Other than this restriction, we try all combinations of n_{instr}, n_{calc} and L. We use 16 repetitions of each setup for the main autoencoder tasks, and 8 repetitions per setup for the regression tasks.

4.3   Results: Autoencoding Experiments

Results for autoencoding experiments on the Glass dataset are shown in Fig. 2.

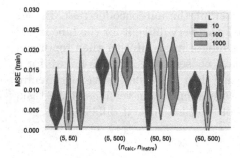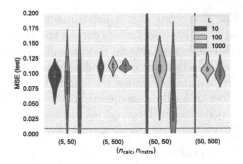

Fig. 2. Auto-encoding the Glass dataset: distribution of final train error (left) and test (right), analysed by hyperparameters. Lower is better. PCA baselines (see text) shown as horizontal lines.

We compare against a strong baseline: a principal components analysis (PCA) reconstruction of the data fitted on the training data with the number of components given by n_{hidden}. These results show that the GPAE does not succeed in matching the PCA baseline, and so cannot be deemed to have succeeded. This is true across different settings of hyperparameters.

---

[2] https://github.com/jmmcd/GPAE.

We proceed to view fitness per iteration of the SCHC algorithm in Fig. 3, and these results show the algorithm is learning, but not enough, and that adding more iterations would not likely help, at least with the current SCHC algorithm and hyperparameters. Both Figs. 2 and 3 demonstrate over-fitting.

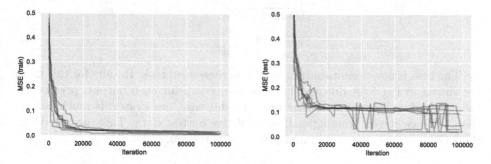

Fig. 3. Auto-encoding the Glass dataset with $n_{calc} = n_{instr} = 50$: train error (left) and test (right) on individual runs, per iteration.

Next, we attempt autoencoding on two other datasets, ACA and PenDigits. To save space we focus only on a single set of hyperparameter settings. The results are shown in Fig. 4. Again, we compare with a PCA baseline. Results confirm that autoencoding these datasets is also difficult for our GPAE.

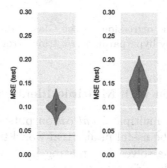

Fig. 4. Auto-encoding the datasets ACA (left) and PenDigits (right) with $n_{calc} = n_{instr} = 50$: distribution of final test error. PCA baselines (see text) are shown as horizontal lines.

4.4 Results: Regression with Multiple Distinct Outputs

Since we regard an autoencoder as a compound function $D(E(x))$, i.e. decode the encoding of a point x, we may ask whether the difficult aspect of the problem is in D or E. In our autoencoder with the Glass dataset, E is a function of $n_{in} = 9$ variables with $n_{hidden} = 4$ outputs. We next attempt a task analogous to that of learning E alone: learn a multi-output function of x. Using the same Glass dataset, we define 4 synthetic functions of the variables x as follows:

$$x_0^2 \tag{1}$$

$$x_1 - x_2^3 \tag{2}$$

$$\sqrt{x_3 \cdot x_4 \cdot x_5} \tag{3}$$

$$\sin(x_6 + \frac{x_7}{1 + |x_8|}) \tag{4}$$

The task is now a multi-output symbolic regression task. Results for this task are shown in Fig. 5. A different baseline is needed: for each output, predicting the mean of its values on training data. In this case, our system is much better than the baseline on training, and competitive on test data. This suggests that the LGP system is capable of multi-output regression.

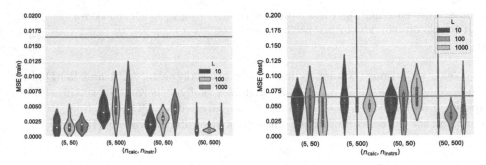

Fig. 5. Multi-output symbolic regression on the Glass dataset with multiple **distinct** synthetic targets, analysed by hyperparameters: train error (left) and test (right).

4.5   Results: Regression with Multiple Identical Outputs

We next attempt to learn multiple *identical* outputs. We make no claim that solving such a task would be useful in any real-world task, and indeed it is easy to see that the task can be solved by treating it as a single-output regression task and then hacking a "copy" instruction, e.g. using $r_i = 1 \times r_j$ where r_j is a register to be copied to r_i. Nevertheless, in order to confirm that our system can solve this problem, we define a new dataset in which the multiple outputs consist of the same output repeated four times. That output is $\sin(x_6 + \frac{x_7}{1+|x_8|})$, perhaps the most difficult of the outputs required in the previous task.

The task is again a multi-output symbolic regression task. Results for this task are shown in Fig. 6, and show that the LGP system does well on this problem.

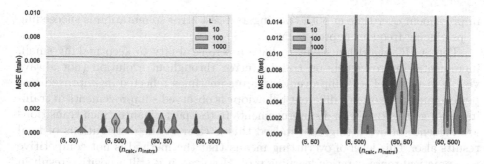

Fig. 6. Multi-output symbolic regression on the Glass dataset with multiple **identical** synthetic targets, analysed by hyperparameters: train error (left) and test (right).

4.6   Results: Regression with a Single Output

Finally, to demonstrate the LGP system's ability to work as a standard SR system, we next attempt to learn a *single* output. That output is again $\sin x_6 + \frac{x_7}{1+|x_8|}$. Results for this task are shown in Fig. 7, and are similar to before: the system far out-performs the baseline on training data with any hyperparameters, and appears better than the baseline on test data, though with a stronger effect of the hyperparameters this time.

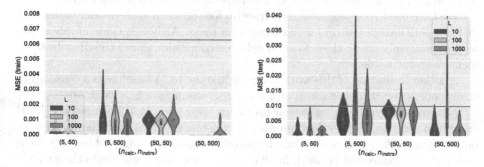

Fig. 7. Single-output symbolic regression on the Glass dataset with a synthetic target, analysed by hyperparameters: train error (left) and test (right).

4.7   Discussion and Limitations

We have passed from an autoencoding task, to several successively easier symbolic regression tasks, and observed some interesting results.

The autoencoding task appears to be difficult, but the main "ingredient" – multi-output symbolic regression – is less so. Perhaps this suggests that it is the solving of two tasks simultaneously and in particular the requirement for mutation to make an improvement in one component while the other is held constant which is difficult. On the other hand, the runs do show a dramatic improvement over time in both training and test error so mutation is succeeding in producing frequent improvements.

The SCHC algorithm seems to have a strong ability to keep making small, incremental improvements on training error throughout evolution (not always visible in plots). These improvements are sometimes reflected in improved test performance, but often a different behaviour is observed – improvements in training error go together with disimprovements in test performance. Such transitions are evidence of overfitting, and indeed this is clear in the distributions of final results also. This level of overfitting means that the method is not competitive as a practical representation learning tool. However, it is still a positive result in one sense: the ability to overfit the data is a prerequisite for the ability to learn it correctly.

Throughout the results, we have found some evidence (strong in some cases, and mixed in others) in favour of medium values for the LGP system's "capacity" hyperparameters $n_{calc} = n_{instr} = 50$, and a long SCHC history length $L = 1000$. Increased values for the system's capacity allow it more power to fit the data, assuming sufficient training power is given through the fitness evaluation budget and the search methods. Making the capacity hyperparameters too large would however make the system more prone to overfitting. Given the observed results, we cannot clearly say that the system needs either more or less capacity on autoencoding tasks, since although it has not succeeded in beating a baseline on training data (perhaps suggesting more capacity is needed), it has also failed to generalise (suggesting regularisation is needed, perhaps through reduced capacity).

We have used 16 repetitions of each setup for the main autoencoder tasks, and just 8 repetitions per setup for the regression tasks. Although these numbers are seen as relatively small, they do allow reliable conclusions since the distributions seem clear-cut as shown.

Another limitation of this work is that while the PCA baseline is a reasonably strong one, predicting the mean of y for a regression task is a weak one.

None of the synthetic regression problems used (Sect. 4.4) required the LGP system to synthesize real-valued numerical constants (e.g. as coefficients). It may be that this task, which is certainly required for autoencoding problems, is itself an important ingredient in the difficulty of autoencoding. This would then partly explain why our system does so much better on multi-output regression tasks. This suggests that a technique for directly optimising constants, e.g. [15], could be a useful addition to the LGP system – but how to incorporate this into the autoencoding objective function is not clear.

5   Conclusions and Future Work

In this paper we have asked whether the important unsupervised learning task of autoencoding can be approached using a GP representation. We have proposed a linear GP system for the task, and demonstrated that it is capable of learning multi-output regression, but found that it does not succeed on autoencoding. We have proceeded to test out the system on multi-output and single-output regression problems to demonstrate that the GPAE system is not hobbled by a weak regression component. We have also reported tentative results on hyper-parameters for the GPAE system.

Several other "decompositions" of the AE task could be proposed, in order to continue the investigation into which aspects of it are difficult. For example, given the PCA baseline, we can use the PCA's learned representation as a target for multi-output symbolic regression from the original dataset (i.e. emulate the PCA encoder) and separately use the original dataset as a target for multi-output symbolic regression from the PCA representation (i.e. emulate the PCA decoder). This work is under way. Multi-output regression tasks are of practical interest but are under-explored, relative to single-output regression.

As already discussed, some alternatives to LGP are possible for this system. In particular, a pair of Cartesian GP programs, with a mutation-only algorithm, are considered a natural next step. Separately, a *single* CGP program with number of rows equal to n_{hidden} and with levels-back equal to 1, and with an extra restriction that columns after a designated hidden layer are disallowed from using the input variables.

A population-based search algorithm would also be possible as an alternative to SCHC. A coevolutionary approach to autoencoding with GP – one population of encoders and one of decoders – would be very interesting to investigate in the future, in particular because the adversarial setting now popular in adversarial autoencoders and generative adversarial networks [12,13] has been claimed to have its origins in coevolution[3].

One way of thinking of unsupervised representation learning is as a problem of "invention". There is no fixed target for the hidden layer representation: the GPAE has to "invent" it, and clearly this is a task of a different nature from that of supervised learning. The fact that the GPAE succeeds at all – dramatic improvements on train and test error relative to the starting-point, even if it does not match the performance of a neural network – is an interesting result, and is, to our knowledge, a novelty for GP.

Our objective function for GPAE is purely based on reconstruction error. In this it is similar to a classic autoencoder. However, modern formulations including the variational autoencoder [8] include in the objective function a regularisation term rewarding the encoder for producing a standard Gaussian distribution at the hidden layer. This has a practical purpose in that it allows sampling at the hidden layer. In our context, adding such a term might also help to control encoder behaviour so that it requires less "invention", making the decoder's job easier.

Acknowledgements. This work was carried out while JMcD was at University College Dublin. Thanks to members of the University College Dublin Natural Computing Research and Applications group, in particular Takfarinas Saber and Stefano Mauceri, for useful discussions. Thanks to Van Loi Cao for data-processing code and for discussion. Thanks also to the anonymous reviewers.

---

[3] http://togelius.blogspot.com/2018/05/empiricism-and-limits-of-gradient.html.

References

1. Hinton, G.E., Osindero, S., Teh, Y.W.: A fast learning algorithm for deep belief nets. Neural Comput. **18**(7), 1527–1554 (2006)
2. Zhang, H., et al.: StackGAN: text to photo-realistic image synthesis with stacked generative adversarial networks. In: Proceedings of the IEEE International Conference on Computer Vision, pp. 5907–5915 (2017). https://arxiv.org/abs/1612.03242
3. Vincent, P., Larochelle, H., Bengio, Y., Manzagol, P.A.: Extracting and composing robust features with denoising autoencoders. In: Proceedings of the 25th International Conference on Machine Learning, pp. 1096–1103. ACM (2008)
4. Yao, X.: A review of evolutionary artificial neural networks. Int. J. Intell. Syst. **8**(4), 539–567 (1993)
5. Stanley, K.O., Miikkulainen, R.: Evolving neural networks through augmenting topologies. Evol. Comput. **10**(2), 99–127 (2002)
6. Zoph, B., Le, Q.V.: Neural architecture search with reinforcement learning. arXiv preprint arXiv:1611.01578 (2016)
7. Such, F.P., Madhavan, V., Conti, E., Lehman, J., Stanley, K.O., Clune, J.: Deep neuroevolution: genetic algorithms are a competitive alternative for training deep neural networks for reinforcement learning. arXiv preprint arXiv:1712.06567 (2017)
8. Kingma, D.P., Welling, M.: Auto-encoding variational Bayes. arXiv preprint arXiv:1312.6114 (2013)
9. Orzechowski, P., La Cava, W., Moore, J.H.: Where are we now? A large benchmark study of recent symbolic regression methods. In: Proceedings of GECCO (2018). arXiv preprint arXiv:1804.09331
10. Rumelhart, D.E., Hinton, G.E., Williams, R.J.: Learning internal representations by error propagation. California University San Diego La Jolla Institute for Cognitive Science, Technical report (1985)
11. Hinton, G.E., Salakhutdinov, R.R.: Reducing the dimensionality of data with neural networks. Science **313**(5786), 504–507 (2006)
12. Makhzani, A., Shlens, J., Jaitly, N., Goodfellow, I., Frey, B.: Adversarial autoencoders. arXiv preprint arXiv:1511.05644 (2015)
13. Goodfellow, I., et al.: Generative adversarial nets. In: Advances in Neural Information Processing Systems, pp. 2672–2680 (2014)
14. McConaghy, T.: FFX: fast, scalable, deterministic symbolic regression technology. In: Riolo, R., Vladislavleva, E., Moore, J. (eds.) Genetic Programming Theory and Practice IX, pp. 235–260. Springer, Heidelberg (2011). https://doi.org/10.1007/978-1-4614-1770-5_13
15. Trujillo, L., et al.: Local search is underused in genetic programming. In: Riolo, R., Worzel, B., Goldman, B., Tozier, B. (eds.) Genetic Programming Theory and Practice XIV. GEC, pp. 119–137. Springer, Cham (2018). https://doi.org/10.1007/978-3-319-97088-2_8
16. Ellis, K., Solar-Lezama, A., Tenenbaum, J.: Unsupervised learning by program synthesis. In: Advances in Neural Information Processing Systems, pp. 973–981 (2015)
17. Koza, J.R.: Genetic Programming: On the Programming of Computers by Means of Natural Selection. MIT Press, Cambridge (1992)
18. Miller, J.F., Thomson, P.: Cartesian genetic programming. In: Poli, R., Banzhaf, W., Langdon, W.B., Miller, J., Nordin, P., Fogarty, T.C. (eds.) EuroGP 2000. LNCS, vol. 1802, pp. 121–132. Springer, Heidelberg (2000). https://doi.org/10.1007/978-3-540-46239-2_9

19. Jackson, D.: A new, node-focused model for genetic programming. In: Moraglio, A., Silva, S., Krawiec, K., Machado, P., Cotta, C. (eds.) EuroGP 2012. LNCS, vol. 7244, pp. 49–60. Springer, Heidelberg (2012). https://doi.org/10.1007/978-3-642-29139-5_5

20. Schmidt, M., Lipson, H.: Distilling free-form natural laws from experimental data. Science **324**(5923), 81–85 (2009). http://www.sciencemag.org/content/324/5923/81.abstract

21. Brameier, M., Banzhaf, W.: Linear genetic programming. Springer, Heidelberg (2006). https://doi.org/10.1007/978-0-387-31030-5

22. Potter, M.A., Jong, K.A.D.: Cooperative coevolution: an architecture for evolving coadapted subcomponents. Evol. Comput. **8**(1), 1–29 (2000)

23. Ni, J., Drieberg, R.H., Rockett, P.I.: The use of an analytic quotient operator in genetic programming. Trans. Evol. Comput. **17**(1), 146–152 (2013)

24. Nicolau, M., Agapitos, A.: On the effect of function set to the generalisation of symbolic regression models. In: Proceedings of the Genetic and Evolutionary Computation Conference Companion, pp. 272–273. ACM (2018)

25. Bykov, Y., Petrovic, S.: A step counting hill climbing algorithm applied to university examination timetabling. J. Sched. **19**(4), 479–492 (2016)

26. Burke, E.K., Bykov, Y.: The late acceptance hill-climbing heuristic. Eur. J. Oper. Res. **258**(1), 70–78 (2017)

27. Cao, V.L., Nicolau, M., McDermott, J.: Late-acceptance and step-counting hill-climbing GP for anomaly detection. In: Proceedings of the Genetic and Evolutionary Computation Conference Companion, pp. 221–222. ACM (2017)

28. O'Reilly, U.-M., Oppacher, F.: Program search with a hierarchical variable length representation: genetic programming, simulated annealing and hill climbing. In: Davidor, Y., Schwefel, H.-P., Männer, R. (eds.) PPSN 1994. LNCS, vol. 866, pp. 397–406. Springer, Heidelberg (1994). https://doi.org/10.1007/3-540-58484-6_283. http://www.springer.de/cgi-bin/searchbook.pl?isbn=3-540-58484-6

29. Chellapilla, K.: Evolutionary programming with tree mutations: evolving computer programs without crossover. In: Genetic Programming, Stanford, CA, USA, pp. 431–438 (1997)

30. Dheeru, D., Karra Taniskidou, E.: UCI machine learning repository (2017). http://archive.ics.uci.edu/ml

31. Cao, V.L., Nicolau, M., McDermott, J.: Learning neural representations for network anomaly detection. IEEE Trans. Cybern. **99**, 1–14 (2018). Early access

Solution and Fitness Evolution (SAFE): Coevolving Solutions and Their Objective Functions

Moshe Sipper[1,2]([✉]), Jason H. Moore[1], and Ryan J. Urbanowicz[1]

[1] Institute for Biomedical Informatics, University of Pennsylvania,
Philadelphia, PA 19104-6021, USA
{sipper,jhmoore,ryanurb}@upenn.edu
[2] Department of Computer Science, Ben-Gurion University,
84105 Beer Sheva, Israel

Abstract. We recently highlighted a fundamental problem recognized to confound algorithmic optimization, namely, *conflating* the objective with the objective function. Even when the former is well defined, the latter may not be obvious, e.g., in learning a strategy to navigate a maze to find a goal (objective), an effective objective function to *evaluate* strategies may not be a simple function of the distance to the objective. We proposed to automate the means by which a good objective function may be discovered—a proposal reified herein. We present **S**olution **A**nd **F**itness **E**volution (**SAFE**), a *commensalistic* coevolutionary algorithm that maintains two coevolving populations: a population of candidate solutions and a population of candidate objective functions. As proof of principle of this concept, we show that SAFE successfully evolves not only solutions within a robotic maze domain, but also the objective functions needed to measure solution quality during evolution.

Keywords: Evolutionary computation · Coevolution ·
Novelty search · Objective function

1 Objective ≠ Objective Function

The goal of any evolutionary algorithm (EA) is to solve a problem, i.e., obtain a specified *objective*. This invariably entails defining an *objective function* (e.g., fitness function), which is the function we want to minimize or maximize. In a recent paper, we targeted a fundamental problem that one might face in practice. Specifically, while the objective may be known and well defined, the objective function may be deceptive, and one might easily conflate the objective with the objective function [1]. Consider the mazes in Fig. 1, wherein the challenge is to evolve a robotic controller (i.e., a model that determines movement given the robot's current state) such that the robot, when placed in the start position, is able to make its way to the goal. The controller defines the *behavior* (i.e., decides the direction of movement) of the robot when it is in a given *state* (position in

© Springer Nature Switzerland AG 2019
L. Sekanina et al. (Eds.): EuroGP 2019, LNCS 11451, pp. 146–161, 2019.
https://doi.org/10.1007/978-3-030-16670-0_10

the maze, distances to obstacles). The set of movement decisions over a fixed number of time steps defines the robot's *path*, and the robot's *endpoint* is its position at the final time step.

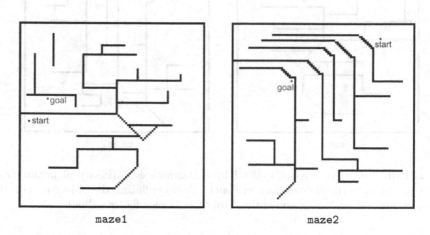

maze1   maze2

Fig. 1. In these maze problems a robot begins at the start square and must make its way to the goal square (objective).

It seems intuitive that the fitness f of a given robotic controller be defined as a function of the distance from the robot to the objective, as done, e.g., by [2]. However, reaching the objective may be difficult since the robot is faced with a deceptive landscape, where higher fitness (i.e., being reasonably close to the goal) may not imply that the robot is "almost there". It is quite easy for the robot to attain a fairly good fitness value, yet be stuck behind a wall in a local optimum—quite far from the objective in terms of the path needed to be taken. Indeed, our experiments with such a fitness-based evolutionary algorithm (Sect. 7) produced the expected failure, demonstrated in Fig. 2.

Reaching the global optimum implies the acceptance of *reduced* fitness over the course of searching for an optimal controller. This is at odds with the work-ings of an evolutionary algorithm, which in practice is driven to optimize the objective function rather than find the best way to reach the objective. Another way to frame this problem is that while an optimal solution may be representable by the controller, it may not be learnable given this simple objective function [3]. For EAs, learnability translates to *evolvability* [4].

One solution to this conflation problem was offered by [2] in the form of novelty search, which *ignores* the objective and searches for novelty (using a novelty metric that requires careful consideration, e.g., robot endpoint novelty). However, novelty for the sake of novelty alone lacks incentive for solutions that reach and stay at the objective. In [1] we offered an alternative view, namely, that the problem may lie with our ignorance of the *correct* objective function. We proposed that rather than exclusively seek out novelty or rely on a fixed objective-based metric to evaluate both the "journey" and the "destination" it

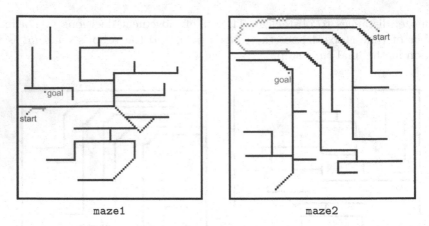

Fig. 2. Path (green) of a robot evolved by a standard evolutionary algorithm with fitness measured as distance-to-goal, evidencing how conflating the objective with the objective function leads to a non-optimal solution. (Color figure online)

might be more effective to separate the optimization of the solution from the optimization of the objective function. For example, just because the robot's objective is to reach the goal does not necessarily imply that the objective function is distance-to-goal.

Thus, we reasoned, we are now left with the task of finding this better objective function, either manually or—perhaps more intriguingly—through some automated means; after all, if we are searching for a good objective function, why not employ a search algorithm? We concluded in [1] with a final speculation, namely, that coevolution [5,6] might offer the key to simultaneously explore solution optimality and objective-function optimality.

In this paper we flesh out our previous speculations and offer preliminary evidence that solutions can be coevolved along with the objective functions that evaluate them. Our newly proposed algorithm is dubbed **SAFE**, for **S**olution **A**nd **F**itness **E**volution.

This paper describes an exploratory study of SAFE. We show that where a standard, fitness-based evolutionary algorithm fails, SAFE succeeds, on a par with (or slightly better) than novelty search.

We examine four algorithms in this work—(1) a standard genetic algorithm, (2) novelty search, (3) SAFE, and (4) random search—for their ability to solve maze problems. After recounting related work in Sect. 2, we present two key ideas, novelty search (Sect. 3) and coevolution (Sect. 4). We continue with a delineation of the robot simulator (Sect. 5). We then present SAFE (Sect. 6) followed by results involving maze-solving robots (Sect. 7). We end with a discussion and concluding remarks (Sect. 8).

2   Related Work

We believe that the idea presented herein—treating the objective function as an evolving entity—has not been previously examined as such. There are a number of related, but distinct, research directions summarized below.

Fitness approximation explores the ways one can estimate the fitness function by constructing an approximate model. This can be quite useful when an explicit fitness function does not exist, or the evaluation of the fitness is computationally expensive. An excellent review on the topic by [7] divides fitness-approximation approaches into three categories: (1) *Problem approximation* tries to replace the original statement of the problem by one which is approximately the same but easier to solve (e.g., evaluate a turbine blade using computational fluid dynamics—CFD—simulations rather than in a wind tunnel); (2) *functional approximation* constructs an alternate expression for the objective function, e.g., instead of a CFD simulation of a blade define an explicit mathematical model (see also work on surrogate fitness functions [8,9]); (3) *evolutionary approximation*, including fitness inheritance—wherein offspring fitness is estimated from parent fitness, and fitness imitation—where individuals are clustered into groups and only a group representative is evaluated whereupon it serves as a basis for estimating the fitness of the others in the cluster. Interestingly, coevolution has been used in fitness approximation [10].

The objective function conflation problem is also distinct from the topic of *dynamic fitness*, wherein the problem itself (along with the global optimum and/or the fitness landscape) changes during run-time [11]. The prediction of weather patterns is in line with this challenge. Further, *multi-objective optimization*, where there is more than one objective to optimize at a time, is also distinct [12] given that it is typically up to the user to define a multi-objective fitness function. Multi-objective Pareto front methods are perhaps closest to the concept we propose in that they seek not to make fixed assumptions about what makes a 'good' solution/objective [12]. However, a Pareto front does not separate the objective from the objective function, and ultimately leaves the challenge of picking the 'optimal' solution up to the user, traditionally from a set of candidate 'non-dominated' solutions at the front. Ultimately, the challenges of dynamic fitness and multi-objective optimization are each vulnerable to the conflation of objective and objective function—and might benefit from a SAFE-inspired approach.

More generally, the topic of *open-ended evolution* comes to mind. This type of evolution occurs in nature, admitting no externally imposed fitness criterion, but rather an implicit, emergent, dynamical one (that could arguably be summed up as survivability) [13–15]. Indeed, the original novelty-search paper began with: "This paper establishes a link between the challenge of solving highly ambitious problems in machine learning and the goal of reproducing the dynamics of open-ended evolution in artificial life." [2] While open-ended evolution is a popular topic in the field of artificial life, we prefer to keep the focus herein on computational problem solving.

3   Novelty Search

Novelty search was introduced by [2] and applied to the examination of maze problems where robot controllers were sought as candidate solutions. Recall that a controller defines the behavior of a robot, yielding a path through the maze concluding in an endpoint position. The key idea was that instead of rewarding endpoint closeness to objective, individual solutions would be considered valuable if their endpoint diverged from prior solution endpoints (i.e., the system was driven to identify solutions that led the robot to new parts of the maze, regardless of objective closeness).

The novelty metric employed should be carefully defined based on the problem at hand. In [2], the novelty metric was defined as the Euclidean distance between the endpoints of two individuals. Note that this is essentially a *phenotypic*, or outcome-based, novelty measure. Differently, a *genotypic* novelty metric would examine differences between the architecture of the solutions themselves (e.g., the robot controllers). We will revisit this in Sect. 6, where we present a genotypic novelty metric for SAFE objective functions.

In later work, novelty was combined with other objectives (e.g., performance). For example, in [16], "an existing creature evolution platform is extended with multi-objective search that balances drives for both novelty and performance". However, this does not involve an evolving objective function, but rather, "The suggested approach is to provide evolution with both a novelty objective ... and a local competition objective". To wit, the objectives are provided by the user, as opposed to SAFE, wherein they are evolved. Appendix F of [17] includes a comprehensive up-to-date list of references to research involving novelty search.

Effectively, novelty search is identical to a standard evolutionary algorithm, with the fitness function replaced by the novelty metric. We seek novel behaviors rather than the objective, hoping that the former will ultimately lead to the latter. The endpoint of a candidate solution is compared to its cohorts in the current population and to an archive of past individuals whose behavioral endpoints were highly novel when they emerged. The novelty score is the average of the distances to the k nearest neighbors (k was set to 15 both by [2] and by us; see Table 1 in Sect. 7 for a summary of parameters).

For future problem flexibility, we implemented the novelty archive differently than [2]. There, an individual robot's endpoint entered the archive if its novelty score was above a certain threshold (which was adjusted dynamically). We chose to implement a fixed-size archive, with new points entering the archive until it fills up. Once full, a newly found point that exceeds the archive's current minimum will replace that minimum; otherwise it will not enter the archive. Thus, the archive saves the most novel points (at the time of their emergence). Our choice of a different archive implementation stems from our initial forays into other problems (e.g., function optimization)—to be reported in the future—where no threshold value was readily apparent.

To summarize, a robot (controller's) novelty score is computed by having the robot wander the maze for the allotted number of steps (300—see Table 1). Then, its endpoint is compared to all endpoints of its cohorts in the current generation *and* to all endpoints in the archive. The final score is then the average of the $k = 15$ nearest neighbors.

4 Coevolution

Coevolution refers to the simultaneous evolution of two or more species with coupled fitness [6]. Strongly related to the concept of symbiosis, coevolution can be mutualistic, parasitic, or commensalistic [18]: (1) In mutualism, different species exist in a relationship in which each individual (fitness) benefits from the activity of the other; (2) in parasitism, an organism lives on or in another organism, causing it harm; and (3) in commensalism, members of one species gain benefits while those of the other species neither benefit nor are harmed. Interestingly, though the idea of coevolution originates (at least) with Darwin—who spoke of "coadaptations of organic beings to each other" [19]—it is arguably somewhat less pervasive in the field of evolutionary computation than one might expect.

A cooperative (mutualistic) coevolutionary algorithm involves a number of independently evolving species, which come together to obtain problem solutions. The fitness of an individual depends on its ability to collaborate with individuals from other species [5,6,20,21].

In a competitive (parasitic) coevolutionary algorithm the fitness of an individual is based on direct competition with individuals of other species, which in turn evolve separately in their own populations. Increased fitness of one of the species implies a reduction in the fitness of the other species [22].

We will discuss the particulars of coevolution—specifically, commensalistic coevolution—in Sect. 6, when we present the SAFE algorithm. To our knowledge, this is the first introduction of a commensalistic coevolutionary algorithm.

5 The Robot and the Maze

We implemented a robotic simulator, similar to the one used by [2], involving a robot moving in a two-dimensional environment that contains walls, with a controller dictating its actions, based on inputs from several sensors. The major difference from [2] is that we did not use NEAT—NeuroEvolution of Augmenting Topologies—for the controller, as we wanted full access to all code aspects, and we wanted to make our controller as interpretable as possible.

The maze environment is a two-dimensional grid of size 100×100 cells, where each cell can either be empty or (part of a) wall (refer to Fig. 1). Two cells are designated as the start and goal. The robot can move horizontally or vertically, but not diagonally. The geometry is taxicab (Manhattan), i.e., the distance between two points is the sum of the absolute differences of their Cartesian coordinates.

The robot has 8 sensors (Fig. 3):

– $dist_{up}$, $dist_{down}$, $dist_{left}$, $dist_{right}$: distances to the obstacle (or boundary) in the upward, downward, leftward, and rightward direction, respectively.

– $goal_{up}$, $goal_{down}$, $goal_{left}$, $goal_{right}$: indicators of whether the goal is towards the upward, downward, leftward, or rightward direction, respectively.

The *dist* values are computed by examining the robot's current position with respect to the nearest obstacle or boundary in the appropriate direction. The *goal* values are computed with respect to the robot's current position (the robot can sense the goal "beacon" through walls). If the goal is in the northeast quadrant of the robot, $goal_{up}$ and $goal_{right}$ are set to the maze height and width, respectively, while the other two *goal* sensors are set to 0. Similarly, the sensors are set when the goal is northwest, southeast, or southwest of the robot (Fig. 3).

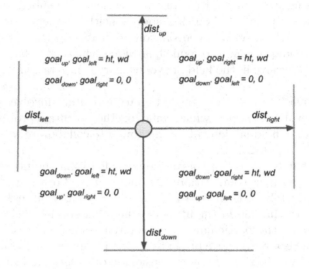

Fig. 3. The robot's 4 *dist* sensors indicate distance to nearest obstacle (wall or boundary). The 4 *goal* sensors indicate direction of goal: 2 are always set to *ht* and *wd* (maze height and width), and 2 are set to 0.

The robot moves for an allotted number of steps (set to 300, see Table 1). Movement is controlled by two variables, horizontal – h and vertical – v. The robot moves one cell in the horizontal position and one cell in the vertical position at every step, as follows:

1. Compute robot's 8 sensor values.
2. Set h and v to:

$$h = p_1\,dist_{right} + p_2\,goal_{right} + p_3\,dist_{left} + p_4\,goal_{left}$$
$$+ p_5\,dist_{down} + p_6\,goal_{down} + p_7\,dist_{up} + p_8\,goal_{up}$$

$$v = p_9\,dist_{right} + p_{10}\,goal_{right} + p_{11}\,dist_{left} + p_{12}\,goal_{left}$$
$$+ p_{13}\,dist_{down} + p_{14}\,goal_{down} + p_{15}\,dist_{up} + p_{16}\,goal_{up}$$

3. If $h \geq 0$ ($h < 0$) then move right (left), provided adjacent cell is neither boundary nor wall (otherwise don't move).
4. If $v \geq 0$ ($v < 0$) then move down (up), provided adjacent cell is neither boundary nor wall (otherwise don't move).

The solution search space is engendered by the vector $[p_1, \ldots, p_{16}] \in \mathbb{R} \cap [-1, 1]$, where each set of $[p_1, \ldots, p_{16}]$ values constitutes a robot's control vector, which determines its behavior. In this domain, a successful vector will drive the robot to the goal. The algorithms we examine below all aim to find a successful control vector.

6   SAFE: Solution and Fitness Evolution

SAFE is a coevolutionary algorithm that maintains two coevolving populations: a population of candidate solutions and a population of candidate objective functions (see Fig. 4). The evolution of each population is identical to a standard, single-population evolutionary algorithm—except where fitness computation is concerned. Below we describe the various components of the system: population composition, initialization, selection, elitism, crossover, mutation, and fitness computation.

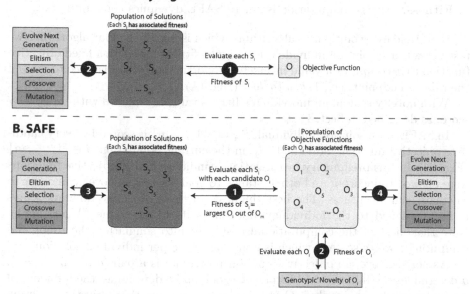

Fig. 4. A single generation of a standard evolutionary algorithm vs. a single generation of SAFE. The numbered circles identify sequential steps in the respective algorithms.

Populations. An individual in the solutions population is a list of 16 real values, each in the range $[-1, 1]$. These values are the robot parameters, p_i, described in Sect. 5.

An individual in the objective-functions population is a list of 2 real values $[a, b]$, each in the range $[0, 1]$, whose usage is described below. Population sizes and other parameters are given in Table 1 (Sect. 7).

Initialization. For every evolutionary run: both populations are initialized to random (fixed-length) lists, wherein each component value is in the appropriate range.

Selection. Tournament selection with tournament size 5, i.e., choose 5 individuals at random from the population and return the individual with the best fitness as the selected one.

Elitism. The 2 individuals with the highest fitness in a generation are copied ("cloned") into the next generation unchanged.

Crossover. Standard single-point crossover, i.e., select a random crossover point and swap two parent genomes beyond this point to create two offspring. The crossover rate is the probability with which crossover between two selected parents occurs. (Note that for an objective-function individual, which comprises two values, if crossover occurs, it will always be, ipso facto, at the same position.)

Mutation. Mutation is done with probability 0.4 (per individual in the population) by selecting a random gene (of the 16 or 2, respectively) and replacing it with a new random value in the appropriate range.

Fitness. Fitness computation is where SAFE dynamics come into play.

In a standard evolutionary algorithm (which is one of the four algorithms we test below) each solution individual is assigned fitness based on a fixed objective function that computes a value inversely proportional to the distance to goal of the robot's endpoint, i.e., $1/distToGoal$ (Fig. 4A).

With novelty search, an individual's fitness value is replaced with its novelty score, as described in Sect. 3.

In SAFE, each solution individual, S_i, $i \in \{1, \ldots, n\}$ is scored by every candidate objective-function individual, O_j, in the current population, $j \in \{1, \ldots, m\}$ (Fig. 4B). In this preliminary investigation, candidate SAFE objective functions were allowed to incorporate both 'distance to goal' as well as novelty in order to calculate solution fitness. The best (highest) of these objective function scores is then assigned to the individual solution as its fitness value (Algorithm 1). Note that most of the computational cost goes into simulating the robot and computing novelty scores—which is done only once per individual solution.

As noted above, an objective-function individual is a pair $[a, b]$; specifically, a determines the influence of 'distance to goal' and b determines the influence of 'phenotypic solution novelty'. $O_j(S_i)$ is the fitness score that objective function O_j assigns to solution S_i, given as:

$$O_j(S_i) = a \times \frac{1}{distToGoal_i} + b \times noveltyScore_i, \tag{1}$$

Algorithm 1. Compute fitness values of solutions population

1: $n \leftarrow$ size of solutions population
2: $m \leftarrow$ size of objective-functions population
3: **for** $i \leftarrow 1$ to n **do**
4:     simulate robot with S_i, and derive $endPosition_i$, $distToGoal_i$
5: **for** $i \leftarrow 1$ to n **do**
6:     compute $noveltyScore_i$, based on $endPosition$ and $archive$
7: **for** $i \leftarrow 1$ to n **do**
8:     **for** $j \leftarrow 1$ to m **do**
9:         $f_j \leftarrow O_j(S_i)$ # O_j uses $distToGoal$ & $noveltyScore$ (Equation 1)
10:     $solutionFitness_i \leftarrow \max f_j$

where $distToGoal_i$ is the distance to goal of robotic controller (solution) i's endpoint, and $noveltyScore_i$ is the novelty score of solution i. Rather than use one or the other, we let evolution discover an effective mix of the two objective function components (as opposed, e.g., to [23], wherein a weighted combination of fitness and novelty was used with pre-determined weights).

As for the objective-functions population, determining the quality of an evolving objective function places us in uncharted waters. Such an individual is not a solution to a problem, but rather the "guide"—or "path"—to a solution. As such, it is not clear what comprises a good measure of success.

We experimented with various forms of mutualistic coevolution, where the fitness of an objective function depends on its ability to ascribe fitness to solutions in the solutions population, such that better solutions (i.e., closer to the objective) receive higher fitness values, and worse solutions receive lower fitness values. In particular, we considered evaluating objective functions based on the correlation (i.e., Pearson's or Spearman's) between the solution fitness scores it generated and distance to the actual objective. While seemingly intuitive, this approach yielded unsatisfactory results, finding that this implementation ended up reinforcing the same local minima problem encountered when using a traditional evolutionary algorithm. Despite this failure, we plan to continue investigating the possibility of a mutualistic coevolutionary approach, where the fitness of objective functions is dependent in some way on the population of candidate solutions.

In the working version of SAFE presented here, we turned to a commensalistic coevolution strategy, where the objective functions' fitness does not depend on the population of solutions. Instead, it relies on *genotypic* novelty, based on the objective-function individual's two-valued genome, $[a, b]$. The distance between two objective functions—$[a_1, b_1]$, $[a_2, b_2]$—is simply the Euclidean distance of their genomes, given as: $\sqrt{(a_1 - a_2)^2 + (b_1 - b_2)^2}$. Note the contrast between genotypic novelty here and phenotypic (i.e., outcome-based) novelty used to evaluate solutions (both by novelty search and by SAFE).

Each generation, every candidate objective function is compared to its cohorts in the current population of objective functions and to an archive of past individuals whose behaviors were highly novel when they emerged. The novelty score is the average of the distances to the k ($= 15$) nearest neighbors, and is used in computing objective-function fitness (Algorithm 2).

Algorithm 2. Compute fitness values of objective-functions population

1: $m \leftarrow$ size of objective-functions population
2: **for** $i \leftarrow 1$ to m **do**
3:     compute $noveltyScoreObj_i$
4:     $objectiveFitness_i \leftarrow noveltyScoreObj_i$

To clarify, we recap SAFE's double use of the idea of novelty. First, solutions evolve not via pure goal-directedness or through pure phenotypic novelty, but rather through a combination of the two. The exact nature of this combination is determined by the evolving objective functions, which themselves use novelty (albeit genotypic) to explore the objective-function search space.

The code for SAFE and all the experiments carried out in this paper is available at https://github.com/EpistasisLab/.

7   Results

To test our new framework, we performed eight sets of experiments, using four algorithms, each set to search for robotic controllers that solve the two mazes of Fig. 1:

1. Standard evolutionary algorithm (fitness is $1/distToGoal$).
2. Novelty search.
3. SAFE.
4. Random search.

Table 1 summarizes the run parameters.

Random search serves as a yardstick to ensure evolution is tackling a non-trivial task. It is done by drawing 100,000 solutions at random and outputting the best one. 100,000 equals maximal number of generations × population size, which are used by the previous algorithms. This ensures a fair comparison in terms of resources.

The results are shown in Table 2. A solution is deemed successful if the robot gets to within a distance of 2 from the goal. We note that SAFE is on a par with novelty search on maze1, and able to discover slightly more solutions for the harder maze2. SAFE also appeared to find a solution that reached the goal within slightly fewer generations, however this finding cannot be deemed significant. Figure 5 shows sample solutions found by SAFE (contrast this with the standard evolutionary algorithm, which always gets stuck in a local minimum, as exemplified in Fig. 2).

Table 1. Algorithm parameters. Unless stated, a relevant parameter not shown is that listed under 'Standard EA' (e.g., the generation count for novelty search is that of the standard EA).

Description	Value
Standard EA	
Number of evolutionary runs	500
Maximal number of generations	500
Size of solutions population	200
Type of selection	Tournament
Tournament size	5
Type of crossover	Single-point
Crossover rate	0.8
Probability of mutation (solutions)	0.4
Number of top individuals copied (elitism)	2
Maximal no. steps taken by robot	300
Stop if distance to goal $\leq$	2
Novelty search	
Average over k nearest neighbors, $k =$	15
Archive size	1000
SAFE	
Size of objective-functions population	200
Probability of mutation (objective functions)	0.4
Random search	
Number of random individuals drawn	100,000

Table 2. Results of 500 evolutionary runs per each algorithm. Success: number of successful runs, where the algorithm discovered a solution that gets the robot to a distance of 2 or less from the goal. Generations: average number of generations (standard deviation) to success.

Algorithm	Maze	Success	Generations
Standard EA	maze1	0	—
	maze2	0	—
Novelty	maze1	322	145 (132)
	maze2	10	249 (163)
SAFE	maze1	328	141 (130)
	maze2	17	227 (152)
Random	maze1	9	—
	maze2	0	—

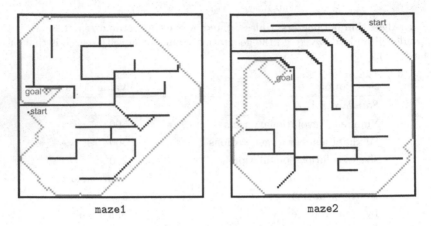

maze1   maze2

Fig. 5. Solutions to the maze problems, evolved by SAFE.

As explained in Sect. 6, the evolving objective function is a real-valued tuple $[a, b]$, with a being the distance-to-goal coefficient, and b being the novelty coefficient. Examining all successful objective functions evolved by SAFE for maze1 (those that led to a maze solution), we observed that the average of a was 0.85 ($sd = 0.16$) and the average of b was 0.99 ($sd = 0.01$); for maze2, average a was 0.83 ($sd = 0.18$) and average b was 0.98 ($sd = 0.02$). Evolution is thus seen to strike a fairly even balance between being goal-oriented and seeking novelty, favoring the latter to a small extent. This motivated a final experiment, wherein we ran the standard evolutionary algorithm, but with the fitness function of SAFE (Eq. 1) and fixed $a = b = 0.5$. The number of successful runs was 307 for maze1 and 17 for maze2, on par with SAFE and novelty search. Of course, SAFE's task is more difficult, given that it is not handed the objective function but must evolve it.

Recall from Sect. 5 that the robot's movement is controlled by two variables, horizontal $- h$ and vertical $- v$, each defined as a weighted sum of the eight sensor values: $dist_{right}$, $goal_{right}$, $dist_{left}$, $goal_{left}$, $dist_{down}$, $goal_{down}$, $dist_{up}$, $goal_{up}$. An evolved solution is a vector of 16 real values, $[p_1, \ldots, p_{16}]$, where $[p_1, \ldots, p_8]$ are the weights used by h, and $[p_9, \ldots, p_{16}]$ are the weights used by v. Given that the model is fairly white box in nature (as opposed, e.g., to a neural network's black-box essence), we might attempt to examine the evolved solutions—whose summary is shown in Table 3. For maze1 we note that h's *left* (absolute) sensor values are quite low, perhaps because the start position is close to the left wall, so there is no point in going left. Also, the *right* values are high for both h and v, so apparently right is a good direction. For maze2 we observe that moving left seems to be favored, and—to a lesser degree—down.

Table 3. Evolved robotic control parameters—$[p_1, \ldots, p_{16}]$—of successful solutions. Each value is the average (standard deviation) of all successful runs. Parameters are shown alongside the sensors they weight.

p	Sensor	maze1	maze2
Horizontal			
p_1	$dist_{right}$	0.68 (0.26)	−0.27 (0.33)
p_2	$goal_{right}$	−0.56 (0.29)	0.04 (0.38)
p_3	$dist_{left}$	0.11 (0.39)	−0.82 (0.18)
p_4	$goal_{left}$	−0.04 (0.29)	0.52 (0.4)
p_5	$dist_{down}$	−0.65 (0.25)	0.58 (0.42)
p_6	$goal_{down}$	−0.73 (0.2)	0.38 (0.45)
p_7	$dist_{up}$	0.45 (0.38)	0.11 (0.2)
p_8	$goal_{up}$	0.33 (0.28)	−0.37 (0.37)
Vertical			
p_9	$dist_{right}$	0.65 (0.23)	0.1 (0.1)
p_{10}	$goal_{right}$	0.62 (0.31)	−0.09 (0.45)
p_{11}	$dist_{left}$	−0.4 (0.46)	0.78 (0.15)
p_{12}	$goal_{left}$	−0.55 (0.31)	0.54 (0.29)
p_{13}	$dist_{down}$	−0.24 (0.46)	0.11 (0.17)
p_{14}	$goal_{down}$	0.03 (0.38)	−0.15 (0.5)
p_{15}	$dist_{up}$	−0.64 (0.34)	−0.21 (0.39)
p_{16}	$goal_{up}$	0.41 (0.3)	−0.15 (0.52)

8   Discussion and Concluding Remarks

Aiming to confront the optimization conflation problem—where the objective is conflated with the objective function—we separated these two entities into two populations, and presented SAFE, a coevolutionary algorithm to evolve the two simultaneously. We showed that both the "journey" and the "destination" can be discovered together, i.e., the solution and the objective function needed to find it.

Our aim herein has been to provide preliminary proof for a novel idea. As such, we did not perform extensive experiments on parameter combinations, although recent research suggests that parameters should not matter too much [24].

Stanley [17] recently wrote that, "While a good algorithm is sometimes one that performs well, sometimes a good algorithm is instead *one that leads to other algorithms and new frontiers*." While ours is but a first foray into new territory we hope it leads to new frontiers. Indeed, there are several exploratory avenues that come to mind:

– First, we would like to add other and more complex problem domains, hoping to reinforce the promising results presented herein and identify areas where the SAFE concept is not only competitive but clearly advantageous. It is of particular interest to demonstrate whether SAFE can adapt itself to problems where little prior knowledge exists regarding either the objective or the best path to said objective.
– The coevolutionary dynamics engendered by SAFE are likely to be quite interesting and worthy of study in and of themselves. The impact of elements including elitism approach, run parameters, and convergence criteria are all worthy of future consideration.
– Our evolving objective function comprises two components, distance-to-objective and novelty. We could consider other components that might be added to the mix, e.g., the robot's distances to walls, its trajectory (behavior) through the environment [25,26], and so forth.
– We used a simple measure to drive objective-function evolution—genotypic novelty. Other, possibly better measures might be designed. As noted in Sect. 6 we did preliminary work on mutualistic fitness scores for objective functions, work which—though it did not pan out—we plan to continue.
– Explore the concept of whether having an objective function that changes over the course of evolutionary search may itself be an important aspect of evolvability in certain deceptive domains.

Considering the "coadaptations of organic beings to each other" [19], is it not natural to view the objective function as a constantly shifting, evolving entity? We speculate that leveraging a combination of open-endedness and coevolutionary search may offer a path to solving a variety of future problems where traditional approaches fail, characterized by deceptive landscapes, high dimensionality, and a lack of prior knowledge.

Acknowledgements. This work was supported by National Institutes of Health (USA) grants AI116794, LM010098, and LM012601.

References

1. Sipper, M., Urbanowicz, R.J., Moore, J.H.: To know the objective is not (necessarily) to know the objective function. BioData Min. **11**(1), 21 (2018)
2. Lehman, J., Stanley, K.O.: Exploiting open-endedness to solve problems through the search for novelty. In: Proceedings of the Eleventh International Conference on Artificial Life (ALIFE). MIT Press, Cambridge (2008)
3. Domingos, P.: A few useful things to know about machine learning. Commun. ACM **55**(10), 78–87 (2012)
4. Wagner, G.P., Altenberg, L.: Perspective: complex adaptations and the evolution of evolvability. Evolution **50**(3), 967–976 (1996)
5. Zaritsky, A., Sipper, M.: Coevolving solutions to the shortest common superstring problem. Biosystems **76**(1), 209–216 (2004)
6. Pena-Reyes, C.A., Sipper, M.: Fuzzy CoCo: a cooperative-coevolutionary approach to fuzzy modeling. IEEE Trans. Fuzzy Syst. **9**(5), 727–737 (2001)

7. Jin, Y.: A comprehensive survey of fitness approximation in evolutionary computation. Soft Comput. **9**(1), 3–12 (2005)
8. Buche, D., Schraudolph, N.N., Koumoutsakos, P.: Accelerating evolutionary algorithms with Gaussian process fitness function models. IEEE Trans. Syst. Man Cybern. Part C (Appl. Rev.) **35**(2), 183–194 (2005)
9. Brownlee, A.E.I., Regnier-Coudert, O., McCall, J.A.W., Massie, S.: Using a Markov network as a surrogate fitness function in a genetic algorithm. In: Proceedings of the IEEE Congress on Evolutionary Computation. pp. 1–8, July 2010
10. Schmidt, M.D., Lipson, H.: Coevolution of fitness predictors. IEEE Trans. Evol. Comput. **12**(6), 736–749 (2008)
11. Grefenstette, J.J.: Evolvability in dynamic fitness landscapes: a genetic algorithm approach. In: Proceedings of the 1999 Congress on Evolutionary Computation, CEC 1999, vol. 3, pp. 2031–2038. IEEE (1999)
12. Deb, K., Agrawal, S., Pratap, A., Meyarivan, T.: A fast elitist non-dominated sorting genetic algorithm for multi-objective optimization: NSGA-II. In: Schoenauer, M., et al. (eds.) PPSN 2000. LNCS, vol. 1917, pp. 849–858. Springer, Heidelberg (2000). https://doi.org/10.1007/3-540-45356-3_83
13. Sipper, M.: If the milieu is reasonable: lessons from nature on creating life. J. Transfigural Math. **3**(1), 7–22 (1997)
14. Sipper, M.: Machine Nature: The Coming Age of Bio-Inspired Computing. McGraw-Hill, New York (2002)
15. Banzhaf, W., et al.: Defining and simulating open-ended novelty: requirements, guidelines, and challenges. Theory Biosci. **135**(3), 131–161 (2016)
16. Lehman, J., Stanley, K.O.: Evolving a diversity of virtual creatures through novelty search and local competition. In: Proceedings of the 13th Annual Conference on Genetic and Evolutionary Computation, pp. 211–218. ACM (2011)
17. Stanley, K.O.: Art in the sciences of the artificial. Leonardo **51**(2), 165–172 (2018)
18. Wikipedia: Symbiosis (2018). https://en.wikipedia.org/wiki/Symbiosis
19. Darwin, C.R.: On the Origin of Species by Means of Natural Selection, or the Preservation of Favoured Races in the Struggle for Life. John Murray, London (1859)
20. Potter, M.A., De Jong, K.A.: Cooperative coevolution: an architecture for evolving coadapted subcomponents. Evol. Comput. **8**(1), 1–29 (2000)
21. Dick, G., Yao, X.: Model representation and cooperative coevolution for finite-state machine evolution. In: 2014 IEEE Congress on Evolutionary Computation (CEC), pp. 2700–2707. IEEE, Piscataway (2014)
22. Hillis, W.: Co-evolving parasites improve simulated evolution as an optimization procedure. Physica D: Nonlinear Phenomena **42**(1), 228–234 (1990)
23. Cuccu, G., Gomez, F.: When novelty is not enough. In: Di Chio, C., et al. (eds.) EvoApplications 2011. LNCS, vol. 6624, pp. 234–243. Springer, Heidelberg (2011). https://doi.org/10.1007/978-3-642-20525-5_24
24. Sipper, M., Fu, W., Ahuja, K., Moore, J.H.: Investigating the parameter space of evolutionary algorithms. BioData Min. **11**(2), 1–14 (2018)
25. Gomez, F.J.: Sustaining diversity using behavioral information distance. In: Proceedings of the 11th Annual Conference on Genetic and Evolutionary Computation, GECCO 2009, pp. 113–120. ACM, New York (2009)
26. Doncieux, S., Mouret, J.B.: Behavioral diversity with multiple behavioral distances. In: Proceedings of the 2013 IEEE Congress on Evolutionary Computation (CEC), pp. 1427–1434. IEEE (2013)

A Model of External Memory for Navigation in Partially Observable Visual Reinforcement Learning Tasks

Robert J. Smith[✉] and Malcolm I. Heywood

Dalhousie University, Halifax, NS, Canada
{rsmith,mheywood}@cs.dal.ca

Abstract. Visual reinforcement learning implies that, decision making policies are identified under delayed rewards from an environment. Moreover, state information takes the form of high-dimensional data, such as video. In addition, although the video might characterize a 3D world in high resolution, partial observability will place significant limits on what the agent can actually perceive of the world. This means that the agent also has to: (1) provide efficient encodings of state, (2) store the encodings of state efficiently in some form of memory, (3) recall such memories after arbitrary delays for decision making. In this work, we demonstrate how an external memory model facilitates decision making in the complex world of multi-agent 'deathmatches' in the ViZDoom first person shooter environment. The ViZDoom environment provides a complex environment of multiple rooms and resources in which agents are spawned from multiple different locations. A unique approach is adopted to defining external memory for genetic programming agents in which: (1) the state of memory is shared across all programs. (2) Writing is formulated as a probabilistic process, resulting in different regions of memory having short- versus long-term memory. (3) Read operations are indexed, enabling programs to identify regions of external memory with specific temporal properties. We demonstrate that agents purposefully navigate the world when external memory is provided, whereas those without external memory are limited to merely 'flight or fight' behaviour.

Keywords: External memory · Visual reinforcement learning ·
First person shooter · Partially observable · Tangled program graphs

1 Introduction

Visual reinforcement learning represents a recent development in which reinforcement learning agents learn directly from high dimensional visual information alone. This means that rather than requiring a human to design high-quality task specific features, the learning agent can develop its own application-specific features, and in the process still match or better the performance of a human on

© Springer Nature Switzerland AG 2019
L. Sekanina et al. (Eds.): EuroGP 2019, LNCS 11451, pp. 162–177, 2019.
https://doi.org/10.1007/978-3-030-16670-0_11

the same task. To date, advances of this form have been dominated by agents employing some form of deep learning (e.g. [1,2]). In addition, two works from genetic programming (GP) have demonstrated that GP specifically designed to discover emergent low level representations [3,4] or GP augmented with matrix processing operations [5] can also address visual reinforcement learning tasks, at a fraction of the complexity of the deep learning solutions.

In all the above cases, solutions represent reactive agents applied to tasks that, for the most part, are described by complete information. Specifically, the Arcade Learning Environment – on which the above advances to visual reinforcement learning have been based – emulates titles from the Atari 2600 console [6]. As such, there is little or no need to retain temporal properties. Conversely, more modern gaming titles generally imply that: (1) the visual environment is much more expressive (3D rather than 2D), resulting in much more variety in the opponent/object types, and (2) the 'fields of view' through which the player perceives the game are detailed, but limited, resulting in high levels of partial observability. This has rekindled an interest in memory mechanisms so that the visual reinforcement learning agent can potentially develop its own policies for navigating the partially observable world. Specifically, several works demonstrate that merely augmenting previous deep learning architectures with recurrent connections is insufficient. Instead, external memory is necessary, with explicit 'read' and 'write' operations supported [7–9]. The resulting combination of both short-term recurrent and long-term external memory has been particularly effective at scaling deep learning to the challenge of path finding in partially observable gaming tasks [10].

In this work, we are interested in identifying a suitable memory framework so that GP can also discover pathfinding strategies in partially observable, high-dimensional, visual reinforcement learning tasks. Section 2 reviews previous research with GP memory models, with a bias to those that have informed the design of our approach. Section 3 summarizes the Tangled Program Graph (TPG) approach to visual reinforcement learning [3,4]. Section 4 details the specifics of the proposed short and long-term memory models. Experiments under the ViZDoom environment indicate that without long-term memory, only reactive 'flight-or-flight' behaviours are evolved (Sect. 5). Conversely, the combined short and long-term memory model develops strategies for navigating the partially observable world. Indeed, the pathfinding strategy is systematic, with external memory employed to return the agent to a favoured region of the world, irrespective of where the agent is initialized. Section 6 concludes the work.

2   Background

For the purposes of this review, we recognize two forms of memory as employed with GP: scalar and indexed. Scalar memory is synonymous with register references in linear GP. Thus, linear GP might manipulate code defined in terms of a 'register-to-register' operation: $R[x] := R[x]\langle op \rangle R[y]$ in which $R[\cdot]$ is a register, $x, y \in \{0, \ldots, R_{max} - 1\}$ represent indexes to registers and $\langle op \rangle$ is a two argument operator. Several authors pioneered the use of scalar memory in linear GP [11,12]. In addition, we can also distinguish between stateful and stateless operation in linear GP. *Stateful* operation implies that after execution of the program, the values of the registers are not reset. Instead, execution at state $s(t+1)$ commences with the register values as left from the previous execution of the program. Stateful operation is potentially useful in tasks requiring some form of sequence learning, e.g. intrusion detection [13] or financial time series [14].

Indexed memory implies that addresses are specified with an explicit read and write operation (i.e. provision of 'load' and 'store' operations in linear GP). Teller investigated various formal properties that might result from supporting indexed memory [15]. Teller also demonstrated the evolution of indexed memory as a basic data structure [16], a topic extensively investigated by Langdon [17]. Other early developments included the evolution of 'mental models', such as recalling the content of discrete worlds defined as 4×4 toroidal grids [18,19]. To do so, a two phase approach to evolving memory content was assumed [18,19]. In phase 1, the agent can only write to memory, i.e. the agent can navigate the entire world noting points of interest without reading from memory. In phase 2, the agent is rewarded for systematically revisiting all the points of interest using only the information from memory. More recent applications of indexed memory have demonstrated its utility for obstacle avoidance in Khepera robots [20], and the application to financial time series classification tasks [14].

In this work, we will assume support for both scalar memory and indexed memory. However, the TPG framework composes solutions as graphs of teams of programs. Thus, scalar memory implies that state is 'local' to each program. In order to make indexed memory 'global' all programs should be capable of issuing read and write instructions. To do so, we build on an earlier observation in which the state of indexed memory is never reset, thus common to all individuals [21]. Our motivation being to encourage synchronization on how memory is employed between the TPG agents as evolution progresses.

In short, the first person perspective of the environment in which agents have to operate implies that: (1) both the task environment is explored and suitable memory content is developed simultaneously. (2) navigation polices are developed for operation under high dimensional continuous state spaces under partial observability. (3) memory content is developed despite the stochastic nature of some interactions (i.e. the location of encounters with opponents or the location from which agents are re-spawn).

3   Tangled Program Graphs

Tangled program graphs (TPG) represent a flexible framework for the emergent co-discovery of modular relationships between programs. Earlier research demonstrated that TPG is particularly effective at discovering simple agent solutions to visual reinforcement learning benchmarks from the Arcade Learning Environment [3,4], as well as agents able to simultaneously play multiple titles [4,22].

In the following we review the basic composition of TPG in terms of three components: bid-based GP, coevolution of GP teams, and evolving graphs.

3.1 Bid-Based GP

Bid-based GP [23] defines an individual in terms of a program, bgp, producing a single output, b, and a discrete action, a. An individual's action is selected from the set of atomic actions of the task domain, $a \in \mathcal{A}$ (e.g. the enumeration of joystick directions to 8 specific directions). A Bid-based GP program returns a single output post execution, representing the 'confidence' in suggesting its action, or the 'bid', b. In effect, Bid-based GP separates defining context from defining an action. Bid-based GP individuals therefore never represent a solution on their own (because they only ever have a single action), but have to cooperate within the content of a team of programs (Sect. 3.2).

In this work, a linear representation is assumed for its capacity to implicitly support stateful scalar memory (Sect. 2). Programs are defined in terms of a simple register machine [11,12,24] in which instructions may reference up to three register fields (target, source1 and source2) dependent on the opcode $\langle op_i \rangle$ acting on register content, Table 1. Mode bits switch between two sources of address (register-input, register-memory) and register-register instructions. Opcodes supported in this work are defined in Table 1. We note that no attempt is made to introduce application-specific operations.

Table 1. Relation between opcodes and operands. Register-Register instructions imply an address range in the form of registers alone, or $x, y \in \{0, \dots, R_{max} - 1\}$. Register-Input and Register-Memory references introduce different ranges for the y operand: Register-Input $y \in \{0, \dots, N - 1\}$, Register-Memory $y \in \{0, \dots, M - 1\}$. N is the number of pixels in the input, and M is the number of indexible locations in external memory (Sect. 4). ln and exp are (protected) natural logarithm and exponential operators (e.g. absolute value of the operand is assumed) and NaN is trapped in the case of division.

Opcode	Instruction
$\langle op_0 \rangle \in \{<\}$	IF $R[x] \langle op_0 \rangle R[y]$ THEN $R[x] = -R[x]$
$\langle op_1 \rangle \in \{cosine, ln, exp\}$	$R[x] = \langle op_1 \rangle (R[y])$
$\langle op_2 \rangle \in \{+, -, \div, \times\}$	$R[x] = R[x] \langle op_2 \rangle R[y]$

3.2 Coevolution of GP Teams

Programs are cooperatively coevolved using a symbiotic two population model representing teams and programs [23]. The Program population provides a pool of Bid-based GP individuals, whereas each member of the Team population defines a team in terms of a set of pointers to members of the program population.

Initialization begins by first constructing P teams, each with ω unique Bid-based GP individuals.[1]

Fitness is only expressed for individuals in the Team population. Evaluation of an individual from the team population, tm_i implies that for the current state of the environment, $s(t)$, all Bid-based GP individuals with membership in this team ($bgp_j \in tm_i$) have their programs executed. Whichever Bid-based GP individual has the largest program output 'wins' the right to suggest its action.

Under reinforcement learning tasks the action represents the agent's decision, and results in a change of task environment state, $s(t + 1)$. The process of evaluating team tm_i continues until a game specific end state is encountered (e.g., Ms.Pac-Man is 'eaten' by a ghost) or a computational budget is encountered ($t = \tau_{max}$). Thus, for any state, $s(t)$, there can only be one winning program per team. Teams are therefore ultimately rewarded for discovering 'good' decompositions. The process of team evaluation repeats until all teams in the Team population have completed their fitness assessment, i.e. game score. The bottom Gap teams are then deleted and any members of the Program population that are not indexed by any team can also be deleted. This results in a Program population whose size varies (variation operators can introduce new individuals, Sect. 3.3).

3.3   Evolving Graphs of Teams of Programs

As noted in Sect. 3.2, the worst Gap teams are deleted at each generation, where this is subject to the constraint that these represent 'root teams' ($\equiv$ agents). At generation 0, this will always be the case, however, as the generations increase, teams will begin to incorporate other teams, thus #agents $\leq$ #teams (or equivalently #root teams $\leq P$). This process of adaptation is achieved through the action of the selection and variation operators (Algorithm 1). In short, only teams representing root teams are candidates for variation (Step 1). After the worst performing agents are deleted, then Gap new root teams are introduced. The process begins by cloning another root team (selected with uniform probability), Step 3a, where this implies that the parent and offspring will compete at the next generation. Steps 3b and 3c probabilistically add or subtract from the (offspring) root team complement of programs. Step 3d applies a test for modifying programs within the (offspring) root team. Note that all variation tests are iterative [23], resulting in multiple programs being modified and multiple modifications potentially being applied to the same offspring. This is necessary in order to promote a variable length representation as there is no crossover operator employed.

Modifications to the action of a program can take one of two forms: (1) changing the current action, a, to another atomic action (Step 3(d)ivB), or (2) changing the current action, a, to a pointer to another root team (Step 3(d)ivC). Note also, that various constraints are enforced during the application of variation operators, including: (1) team complement cannot decrease below two, (2)

---

[1] Given the larger state space than encountered in the original TPG work, we begin with larger teams, i.e. a state space of $\approx$78,000 versus $\approx$1,300 in [3,4].

Algorithm 1. Selection and variation. 'IDroots' returns the subset of teams corresponding to root teams. 'DelAgents' deletes the worst performing root teams (leaving any teams that the root team indexed). 'CloneRoot' chooses one member of AP with uniform probability and clones it. 'Del/AddProg' modify the complement of programs within a team. 'SelectProg' identifies one Bid-based GP individual from a team's complement for modification. 'ModInstr' modifies a program through iterative application of multiple mutation operators [23].

1. $AP = \text{IDroots}(P)$
2. $AP' = \text{DelAgents}(AP, Gap)$ ▷ Remove the worst performing Gap agents
3. for $(i := 1 \text{ to } Gap)$
   (a) agentR = CloneRoot(AP)
   (b) DelProg(agentR, p_d)   ▷ Apply until test false
   (c) AddProg(agentR, p_a)
   (d) while $(rnd(0, 1] < p_m)$ ▷ Modify programs within new agent
       i. prog = SelectProg(AgentR)
       ii. prog' = CloneProg(prog)
       iii. prog' = ModInstr(prog', $p_{del}, p_{add}, p_{mut}, p_{swp}$)
       iv. IF $(rnd(0, 1] < p_{nm})$ ▷ Modify action of a program
           A. IF $(rnd(0, 1] < p_{atomic})$
           B. THEN ChangeAction(prog', $\mathcal{A}$)
           C. ELSE ChangeAction(prog', AP)
   (e) $P = P \cup \text{AgentR}$ ▷ Insert new root team into population

there must be at least two different actions across the complement of programs within each team, (3) each team must have at least one program with an atomic action.

Evaluation of an agent always starts with the root team. As per Sect. 3.2, all programs within the team are evaluated and the program with the largest output under state $s(t)$ wins the right to suggest its action, a. If this is an atomic action ($a \in \mathcal{A}$), the agent's decision at time step t is determined. Conversely, if the action is a pointer to another team ($a \in AP$), then program evaluation is repeated at the new team under state $s(t)$. Such a process could result in an infinite loop, e.g. $tm_R \rightarrow tm_3 \rightarrow tm_{15} \rightarrow tm_5 \rightarrow tm_3$. We avoid this by marking the teams visited during any given state evaluation, which in this case would request the runner up program bid at tm_5. If this corresponded to an atomic action, then the agent's decision is again resolved; otherwise, evaluation of the indicated team would continue. In short, by marking teams evaluated and requiring each team to have at least one program with an atomic action, we avoid the halting problem. Further details of the TPG algorithm are available in [4].

4   Memory Model

The ability of TPG to scale to visual reinforcement learning tasks comes from the capacity to discover unique decompositions of the task with minimal assumptions [3,4]. The underlying representation for a 'program' in TPG is linear GP, thus *scalar* memory is supported in linear GP without additional special termi- nals/instructions [24]. Additionally, each program is *stateful*, implying that each instance of a program retains register state, $\mathcal{R}$, from the last instance at which it was executed [13,14]. However, a TPG agent is not a single program, but a collection of programs (Sect. 3). This implies that any state associated with scalar memory is 'myopic' in the sense that it is unaware of state retained by other programs in the same TPG individual.

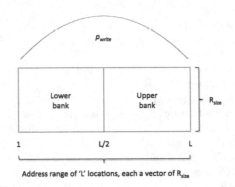

Fig. 1. View of external memory, $\mathcal{M}$, as perceived from the perspective of a write operation, Write($\mathcal{R}$). A write operation (initiated by program p) probabilistically dis- tributes (p_{write}) the contents of its register set $\mathcal{R}$ over the upper and lower 'banks' of the $R_{max} \times L$ locations. Locations in the region of $L/2$ are more likely to be written to, whereas locations towards 1 and L are less likely to be written to (Algorithm 2).

Designing an 'external' source of memory for TPG implies that any program should be able to read from and write to such a memory, hereafter external memory $\mathcal{M}$. Each TPG agent will share the *same* external memory, i.e. the underlying motivation is that convergence occurs through sharing $\mathcal{M}$ content. Thus, the state of $\mathcal{M}$ is not reset between training episodes or the evaluation of different TPG agents. It follows that $\mathcal{M}$ content is only explicitly initialized to 'null' values in the case of the single 'cold start' event, thereafter $\mathcal{M}$ content is a function of all the write operations performed during the course of evolution.

The data written to external memory takes the form of the content of the $\mathcal{R}$ registers associated with the program performing the write operation (Sect. 3), where all programs have the same number of registers, R_{max}. For the purposes of a write operation, external memory $\mathcal{M}$ will be defined in terms of an $R_{max} \times L$ matrix, Fig. 1. Writes will operate probabilistically, distributing the content of the registers, $\mathcal{R}$, across the L columns of external memory $\mathcal{M}$, as per Algorithm 2. This means that columns in the 'middle' of external memory are updated most frequently (short-term memory) whereas columns towards the beginning or end are updated less frequently (long-term memory). Section 5.1 discuss particular parameter choices for L and β.

Algorithm 2. Write function for External memory $\mathcal{M}$. Function called by a write instruction of the form: Write($\mathcal{R}$) where $\mathcal{R}$ is the vector of register content $\{R[0], R[R_{max} - 1]\}$ of the program when the write instruction is encountered. Step 1 identifies the mid point of memory $\mathcal{M}$, effectively dividing memory into upper and lower memory banks (Figure 1). Step 2 sets up the indexing for each bank such that the likelihood of performing a write decreases as a function of the distribution defined in Step 2a. Step 2(a)i defines the inner loop in terms of the number of registers that can source data for a write. Step 2(a)iA tests for a write to the upper memory bank and Step 2(a)iB repeats the process for the lower bank.

WriteExternal($\mathcal{R}, \mathcal{M}$)

1. mid $= \frac{L}{2}$
2. for (offset $:= 0 <$ mid)
   (a) $p_{write} = 0.25 - (\beta \times \text{offset})^2$
       i. for ($i := 0 < R_{max}$)
          A. IF (rnd$[0, 1) \leq p_{write}$)
             THEN ($\mathcal{M}[\text{mid} + \text{offset}][i] = \mathcal{R}[i]$)
          B. IF (rnd$[0, 1) \leq p_{write}$)
             THEN ($\mathcal{M}[\text{mid} - \text{offset}][i] = \mathcal{R}[i]$)

Read operations will be supported through a different mechanism. For a read, external memory $\mathcal{M}$ is treated as a single vector of addresses in the same way that inputs from the frame buffer are consecutively indexed. Figure 2 summarizes this relationship. Thus, mode bits select between one of three forms of each instruction, depending on whether the last source operand is a register reference, an external state reference, $s(t)$, or a reference to external memory, $\mathcal{M}$ (Sect. 3.1).

In summary, write operations distribute the content of the $\mathcal{R}$ registers associated with the program performing the write across the L columns of external memory $\mathcal{M}$. There is no attempt to perform content based addressing or index based addressing. The goal is to support short and long-term retention of register state. Conversely, a read operation assumes index based addressing. This means that a read is free to establish whether the material sourced should represent a long or short-term property. Naturally, it is also assumed that the content of the $\mathcal{R}$ registers is a representative summary of internal state of the TPG agent.

Fig. 2. Instructions may index registers alone (a register-register operation) or read a value from one of two external sources: state or external memory. State, $s(t)$, represents the content of the frame buffer at time step t. External memory, $\mathcal{M}$, represents the content of memory accessible to every program in any TPG graph. Both state and external memory are indexed sequentially for a read operation.

5   Empirical Evaluation

As noted in the introduction, our goal is to facilitate navigation and path finding under very high dimensional visual reinforcement learning environments with partial observability. To date, the most applicable results we are aware of are based on a modified version of the Quake environment in which (deep learning) agents are spawned in a specific location and required to 'capture the flag' of the opponent team [10]. Smaller studies have also been performed using 'T-maze' environments [7–9].

In this work, the ViZDoom environment will be assumed [25], in which case the learning agent is randomly spawned in one of 21 start (spawn) locations. A first person shooter field of view is enforced at all times, and the world consists of rooms and corridors populated with different objects that could potentially benefit the agent. Examples include Medical kits and two types of armour as well as a host of weaponry (with different attributes) that are located in specific places. Given that the location of objects is fixed, but the start state of the agent is variable, memory might enable the agent to reliably locate specific objects as part of its game strategy. However, interactions with opponent agents are stochastic, so hindering the ability to build informative memory models/content. State, $s(t)$ at any point in time is described in terms of a 320×240 pixel frame buffer, thus the input dimensionality for N in Fig. 2 is 76,800. The initial RGB format is concatenated into a single 24-bit decimal number [26]. The set of atomic actions consists of nine discrete actions: Forward, Backward, Turn Left/Right, Strafe Left/Right, Look Up/Down, and Shoot. Evolution is driven by alternating between sets of task scenarios [26] and sets of deathmatch episodes.

5.1   Parameterization

External memory is parameterized in terms of the number of registers a program indexes (R_{max}, Sect. 4) and the number of columns, L (Fig. 1). Thus, from the perspective of a read operation, Fig. 2, there are $M = R_{max} \times L$ indexible locations. In this work all our experiments assume $L = 100$ therefore the range of indexes supported during a read operation is $M = 800$. The probability distribution defined in Step 2a of Algorithm 2 is parameterized with $\beta = 0.01$. This means that for 'offset' values ≈ 0 the likelihood of a write operation is 25% (or 4 out of $R_{max} = 8$ registers will be written to the shortest term memory location as indexed about $\approx \frac{L}{2}$). However, as 'offset' $\rightarrow L$ (and respectively 1), then the likelihood of performing a write tends to 1%. Overall a write operation will result in 17% of $\mathcal{M}$ changing value under this parameterization. No claims are made regarding the relative optimality of such a parameter choice. Table 2 summarizes the parameters assumed for TPG (Sect. 3.3) and Table 1 details the instruction set.

The initial population resulting from the parameterization of Table 2 results in some 2.6 million instructions (across all possible 360 teams) of which $\approx 98,000$ read from $\mathcal{M}$ and $\approx 53,000$ write. Evolution is performed by alternating between six sets of ten task scenarios [26] and six sets of ten deathmatch episodes. After these sets are concluded, evolution moves to a purely generational mode, where 500 generations of deathmatch episodes are completed.

Table 2. TPG Parameterization. For the most part this follows an earlier work deploying TPG in ViZDoom without External Memory and restricted forms of environment that did not require significant amounts of navigation [26].

Team Population		Program Population	
Parameter	Value	Parameter	Value
Team Population Size (P)	360	Max. Instructions	1024
Gap	50% of Root Teams	Prob. Delete Instr. (P_{del})	0.5
Initial Prog. per Team (ω)	12	Prob. Add Instr. (P_{add})	0.5
P_d	0.7	Prob. Mutate Instr. (P_{mut})	1.0
P_a	0.7	Prob. Swap Instr. (P_{swp})	1.0
P_m	0.2	P_{nm}, P_{atomic}	0.2, 0.5

5.2 Results

Post training evaluation will compare the best four agents from ViZDoom 'deathmatches' in which agents of a given type compete for survival in the ViZDoom environment. Three agent types will be considered: default Bots available in ViZDoom, TPG without external memory (TPG), TPG with external memory (M-TPG).[2] Each of the three agent types assume the same fitness function (ViZDoom game score). In the following we first perform a qualitative comparison for TPG with and without memory relative to the default ViZDoom agents. This is then followed by a quantitative analysis for the role of external memory in facilitating the behaviours of the M-TPG agents.

Static Analysis. Twenty deathmatches are performed between the top four agents of each run with a tournament duration of 15 min of play (agents re-spawn after a death from any of twenty-one randomly chosen spawn sites). Table 3 summarizes the kills for each pairing of agent. Assuming a confidence of 95% ($\alpha = 0.05$) and introducing a Bonferroni correction of 3 to reflect the three hypothesis, leaves a target post hoc threshold of 0.0167 for significance. It is apparent that both TPG formulations provide significant improvements over the Bots provided in ViZDoom and TPG with external memory (M-TPG) is significantly better than TPG without.

We can also characterize various (static) properties of the *best* M-TPG agent as follows: 29 teams organized 289 programs, or a total of 186,439 instructions of which only a fraction are executed per TPG agent decision (an implicit property of TPG [4]) and an average of 9.95 programs per team (an average of 645.1 instructions per program). In terms of read references, there were 34,955 read instructions of which ≈85.5% read from state, $s(t)$, and ≈14.4% read from external memory. Write operations are far less frequent, with 553 write operations to external memory. Moreover, write operations only appeared in 24 programs

---

[2] Both TPG and M-TPG support per program stateful scalar memory, or a limited form of memory in which programs are unaware of each other's state.

Table 3. Static qualitative comparison of deathmatch kills for ViZDoom Bot, and TPG with and without external memory. The t-Test returns the p-value for the significance of the difference between the two means under dissimilar variance and non-paired tests.

Competition pairing		t-Test significance
TPG: 12.625 ± 6.01	Bot: 9.35 ± 4.29	$p = 1.1 \times 10^{-4}$
M-TPG: 14.8 ± 5.3	Bot: 8.6 ± 3.83	$p = 1.9 \times 10^{-14}$
M-TPG: 11.56 ± 5.78	TPG: 9.16 ± 6.1	$p = 0.012$

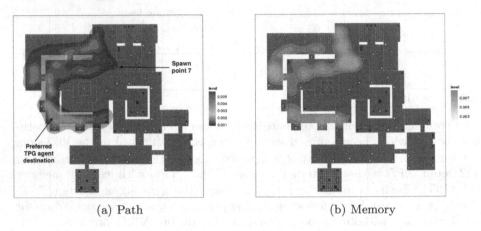

(a) Path   (b) Memory

Fig. 3. Heat map for navigated path frequency and memory access frequency from spawn point 7. (Figure best viewed in colour.)

($\approx 8.3\%$ of the total complement of programs), and all these programs appeared in 7 of the 29 teams comprising the M-TPG agent. In short, writing to external memory appears to be a specialist task undertaken by specialist programs and teams. Note, intron removal was not performed, so further simplifications are likely.

Behavioural Analysis of TPG with Memory. Post training, we can compare the paths that TPG agents took during the course of a deathmatch relative to each spawn point. In the following we will assume spawn points 7 and 9 for illustrative purposes. Specifically, we can build heat maps to characterize the typical paths taken by M-TPG agents and when they read/write to memory. Figures 3 and 4 summarize this for the case of our best M-TPG agent. In short, the M-TPG agent has learnt to navigate from the different spawn points to a specific common corridor. This corridor provides access to 'Plasma gun', 'Super Shotgun', and 'Chaingun'. In addition, the corridor itself facilitates the definition of specific fields of view, therefore making it more difficult for opponents to set up ambushes, but easier for the M-TPG agent to ambush opponents.

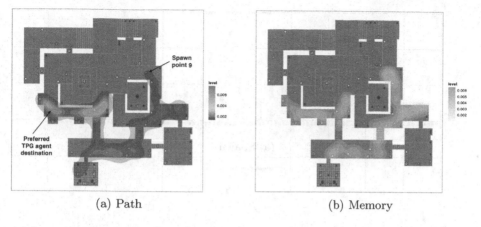

(a) Path   (b) Memory

Fig. 4. Heat map for navigated path frequency and memory access frequency from spawn point 9. (Figure best viewed in colour.)

It is also apparent that the selection of paths is quite systematic. In the case of spawn point 7, the M-TPG agent either uses the upper corridor to circumnavigate the central main room as much as possible (Fig. 3), or skirts around the side of the central room (reducing the likelihood of being shot in the back). It is also apparent that depending on where the M-TPG agent is, it will access external memory with different frequencies (subplot (b)). Indeed, one of the heat map 'hot spots' is at the site of the spawn point, likely indicating that the agent is attempting to orient itself.

Spawn point 9 enables the M-TPG agent to make much more use of corridor constructs, thus avoiding the central main room (where it is easier to be shot from behind). The M-TPG agent, however, always navigates to the same preferred location as with spawn point 7. Moreover, when navigating corridors, the agent tends to use a specific side, again minimizing the number blind spots. Also as observed for Spawn point 7, a higher number of references to external memory occur around the location of the initial spawn point.

Figure 5 provides a characterization for which locations in memory are referenced during a read on spawn points 7 and 6. Several general observations can be made: (1) Spawn point 9 appears to utilize a subset of the 'memories' utilized by Spawn point 7. (2) References are very sparse, with very specific locations being repeatedly read from. (3) Both short and long-term memory[3] are referenced, however, Spawn point 7 appears to make use of more memory scales between the two extremes.

---

[3] Short-term memory is located at indexes near '50', long-term at indexes near '1' and '100'.

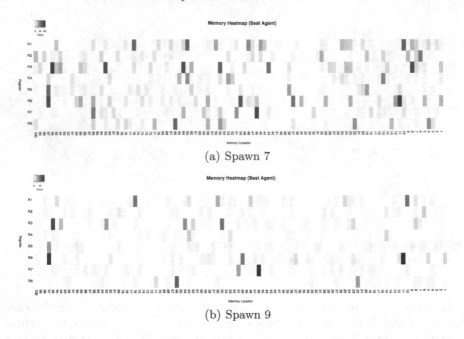

(a) Spawn 7

(b) Spawn 9

Fig. 5. Comparing heat maps for memory accesses for different spawn locations.

Behavioural Analysis of TPG Without Memory. Figure 6 summarizes TPG behaviour relative to spawn point 7, again using a heat map to characterize the most frequently visited locations. Most activity is around the spawn point itself, with a wide distribution of 'islands of behaviour' across the remainder of the map. Further visual analysis during play was able to qualify this as follows: (1) Agents are very reactive, they either see something to engage with and then either go to it, or move away from it (aka 'fight or flight'). (2) Most activity is around the spawn point as it is a relatively open spawn point and therefore has many opponents to engage with. (3) In cases when the agent does not die in the region of the spawn point, it will wander in the direction of 'objects' which may or may not be other agents. The unfortunate side effect of the latter is that the TPG agent then gets stuck, spending a considerable time 'hugging' an object such as a barrel. These appear as the 'islands' in the heat map, and generally result in the TPG agent then being successfully attacked by an opponent. In short, without external memory a reactive behaviour results in which the only reason that different locations are reached is accidental encounters with other agents/objects.

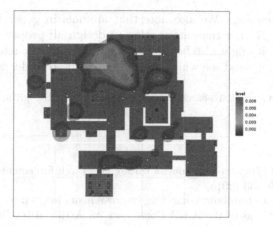

Fig. 6. Case of heat map for TPG behaviour without external memory relative to spawn point 7.

6   Conclusions

Deep learning approaches to external memory use the low dimensional representation discovered by deep learning as the source of content for (external) indexed memory [7–10]. In this work, we do not explicitly engineer a low dimensional representation through a deep learning architecture, although the original visual state space is defined in terms of a very high dimensional space ($\approx$78,000).[4] Instead we provide the opportunity to write register state probabilistically over short and long time scales. Read operations assume that memory is indexed, thus the long versus short-term distribution of data in external memory can be discovered. The external memory is common to all agents and stateful, thus providing a common context for content and addressing patterns.

The TPG framework is able to discover a cohesive model of memory operation involving hundreds of programs. Moreover, although all programs might at some point read from external memory, only a very specialist subset of programs write to memory. This division of duties was not designed in a priori, but an emergent phenomena. Indeed, the resulting TPG agents demonstrate purposeful behaviours in which they explicitly navigate the world in order to reach a preferred region of the ViZDoom environment, irrespective of spawn point. Conversely, TPG agents evolved without external memory are limited to reactive behaviours alone, i.e. they either pursue or evade other agents that they encounter.

We also note that the resulting external memory is very sparsely populated, i.e. very specific locations account for the majority of read operations. Biologically, research has also found evidence for sparse encodings in long-term memory, e.g. [27]. This is particularly interesting behavioural property, and a potential

---

[4] Conversely, deep learning solutions down sampled to a 84 × 84 state space.

topic for future research. We also note that animals in general support many forms of memory. Rather than attempting to design all properties into a single model of memory, it might also be useful to support multiple interfaces to different banks of memory and see which are ultimately used under what conditions.

Acknowledgments. This research was supported by NSERC grant CRDJ 499792.

References

1. Mnih, V., et al.: Human-level control through deep reinforcement learning. Nature **518**(7540), 529–533 (2015)
2. Hessel, M., et al.: Rainbow: combining improvements in deep reinforcement learning. In: Proceedings of the AAAI Conference on Artificial Intelligence, pp. 3215–3222 (2018)
3. Kelly, S., Heywood, M.I.: Emergent tangled graph representations for Atari game playing agents. In: McDermott, J., Castelli, M., Sekanina, L., Haasdijk, E., García-Sánchez, P. (eds.) EuroGP 2017. LNCS, vol. 10196, pp. 64–79. Springer, Cham (2017). https://doi.org/10.1007/978-3-319-55696-3_5
4. Kelly, S., Heywood, M.I.: Emergent solutions to high-dimensional multitask reinforcement learning. Evol. Comput. **26**(3), 347–380 (2018)
5. Wilson, D.G., Cussat-Blanc, S., Luga, H., Miller, J.F.: Evolving simple programs for playing Atari games. In: ACM Genetic and Evolutionary Computation Conference, pp. 229–236 (2018)
6. Bellemare, M.G., Naddaf, Y., Veness, J., Bowling, M.: The arcade learning environment: an evaluation platform for general agents. J. Artif. Intell. Res. **47**, 253–279 (2013)
7. Graves, A., Wayne, G., Danihelka, I.: Neural Turing machines. CoRR abs/1410.5401 (2014)
8. Greve, R.B., Jacobsen, E.J., Risi, S.: Evolving neural Turing machines for reward-based learning. In: ACM Genetic and Evolutionary Computation Conference, pp. 117–124 (2016)
9. Merrild, J., Rasmussen, M.A., Risi, S.: HyperNTM: evolving scalable neural Turing machines through HyperNEAT. In: Sim, K., Kaufmann, P. (eds.) EvoApplications 2018. LNCS, vol. 10784, pp. 750–766. Springer, Cham (2018). https://doi.org/10.1007/978-3-319-77538-8_50
10. Jaderberg, M., et al.: Human-level performance in first-person multiplayer games with population-based deep reinforcement learning. CoRR abs/1807.01281 (2018)
11. Nordin, P.: A compiling genetic programming system that directly manipulates the machine code. In: Kinnear, K.E. (ed.) Advances in Genetic Programming, pp. 311–332. MIT Press, Amsterdam (1994)
12. Huelsbergen, L.: Toward simulated evolution of machine language iteration. In: Proceedings of the Annual Conference on Genetic Programming, pp. 315–320 (1996)
13. Haddadi, F., Kayacik, H.G., Zincir-Heywood, A.N., Heywood, M.I.: Malicious automatically generated domain name detection using stateful-SBB. In: Esparcia-Alcázar, A.I. (ed.) EvoApplications 2013. LNCS, vol. 7835, pp. 529–539. Springer, Heidelberg (2013). https://doi.org/10.1007/978-3-642-37192-9_53

14. Agapitos, A., Brabazon, A., O'Neill, M.: Genetic programming with memory for financial trading. In: Squillero, G., Burelli, P. (eds.) EvoApplications 2016. LNCS, vol. 9597, pp. 19–34. Springer, Cham (2016). https://doi.org/10.1007/978-3-319-31204-0_2

15. Teller, A.: Turing completeness in the language of genetic programming with indexed memory. In: IEEE Congress on Evolutionary Computation, pp. 136–141 (1994)

16. Teller, A.: The evolution of mental models. In: Kinnear, K.E. (ed.) Advances in Genetic Programming, pp. 199–220. MIT Press, Amsterdam (1994)

17. Langdon, W.B.: Genetic Programming and Data Structures. Kluwer Academic, Dordrecht (1998)

18. Andre, D.: Evolution of mapmaking ability: strategies for the evolution of learning, planning, and memory using genetic programming. In: IEEE World Congress on Computational Intelligence, pp. 250–255 (1994)

19. Brave, S.: The evolution of memory and mental models using genetic programming. In: Proceedings of the Annual Conference on Genetic Programming (1996)

20. Nordin, P., Banzhaf, W., Brameier, M.: Evolution of world model for a minature robot using genetic programming. Robot. Auton. Syst. **25**, 105–116 (1998)

21. Spector, L., Luke, S.: Cultural transmission of information in genetic programming. In: Annual Conference on Genetic Programming, pp. 209–214 (1996)

22. Kelly, S., Heywood, M.I.: Multi-task learning in Atari video games with emergent tangled program graphs. In: ACM Genetic and Evolutionary Computation Conference, pp. 195–202 (2017)

23. Lichodzijewski, P., Heywood, M.I.: Symbiosis, complexification and simplicity under GP. In: Proceedings of the ACM Genetic and Evolutionary Computation Conference, pp. 853–860 (2010)

24. Brameier, M., Banzhaf, W.: Linear Genetic Programming. Springer, New York (2007). https://doi.org/10.1007/978-0-387-31030-5

25. Kempka, M., Wydmuch, M., Runc, G., Toczek, J., Jaśkowski, W.: ViZDoom: a doom-based AI research platform for visual reinforcement learning. In: IEEE Conference on Computational Intelligence and Games, pp. 1–8 (2016)

26. Smith, R.J., Heywood, M.I.: Scaling tangled program graphs to visual reinforcement learning in ViZDoom. In: Castelli, M., Sekanina, L., Zhang, M., Cagnoni, S., García-Sánchez, P. (eds.) EuroGP 2018. LNCS, vol. 10781, pp. 135–150. Springer, Cham (2018). https://doi.org/10.1007/978-3-319-77553-1_9

27. Quiroga, R.Q., Kreiman, G., Koch, C., Fried, I.: Sparse but not 'grandmonther-cell' coding in the medial temporal lobe. Trends Cogn. Sci. **12**(3), 87–91 (2008)

Fault Detection and Classification for Induction Motors Using Genetic Programming

Yu Zhang[1], Ting Hu[1(✉)], Xiaodong Liang[2], Mohammad Zawad Ali[2],
and Md. Nasmus Sakib Khan Shabbir[2]

[1] Department of Computer Science, Memorial University,
St. John's, NL A1B 3X5, Canada
{yu.zhang,ting.hu}@mun.ca
[2] Department of Electrical and Computer Engineering, Memorial University,
St. John's, NL A1B 3X7, Canada
{xliang,mzali,mnskshabbir}@mun.ca

Abstract. Induction motors are the workhorse in various industry sectors, and their accurate fault detection is essential to ensure reliable operation of critical industrial processes. Since various types of mechanical and electrical faults could occur, induction motor fault diagnosis can be interpreted as a multi-label classification problem. The current and vibration input data collected by monitoring a motor often require signal processing to extract features that can better characterize these waveforms. However, some extracted features may not be relevant to the classification, feature selection is thus necessary. Given such challenges, in recent years, machine learning methods, including decision trees and support vector machines, are increasingly applied to detect and classify induction motor faults. Genetic programming (GP), as a powerful automatic learning algorithm with its abilities of embedded feature selection and multi-label classification, has not been explored to solve this problem. In this paper, we propose a linear GP (LGP) algorithm to search predictive models for motor fault detection and classification. Our method is able to evolve multi-label classifiers with high accuracies using experimentally collected data in the lab by monitoring two induction motors. We also compare the results of the LGP algorithm to other commonly used machine learning algorithms, and are able to show its superior performance on both feature selection and classification.

Keywords: Genetic programming · Feature selection · Classification ·
Fault detection · Induction motor

1 Introduction

Three-phase induction motors are widely used in industrial systems given their compactness, ruggedness, reliability features, and lower maintenance. However,

© Springer Nature Switzerland AG 2019
L. Sekanina et al. (Eds.): EuroGP 2019, LNCS 11451, pp. 178–193, 2019.
https://doi.org/10.1007/978-3-030-16670-0_12

like any type of electromechanical devices, induction motors can experience various electrical and mechanical faults during operation, which can severely interrupt critical industrial processes and even lead to catastrophic failures. Early detection of faults in induction motors are essential to prevent unexpected interruption of industrial processes and ensure reliable operation.

Three-phase current and three-dimension vibration waveforms collected by monitoring an induction motor can be used to train predictive models that are able to classify different fault states of the motor. Since these waveforms do not provide the most precise and distinguishing information of the status of a motor, signal processing techniques are often used to extract a number of statistical features to characterize the waveforms. Therefore, the fault detection and classification of induction motors can be formulated as a multi-label classification problem with the necessity of feature selection.

Such a problem definition fits well in the realm of machine learning. Therefore, machine learning algorithms have been increasingly applied to solve this problem [1–4]. However, most of these existing studies divide this multi-label classification problem into multiple binary classifications, and often employ external filter feature selection methods that are independent of the classification algorithms.

Genetic programming (GP), as a branch of evolutionary computing, has emerged as a powerful tool to solve machine learning problems [5,6]. This is not only because GP can automatically evolve complex predictive models that map the input instances to the expected outcome [7], but also because of its stochastic and robust nature of the search for diversity and novelty [8–11]. GP has been applied to solve classification and regression problems in Physics, Economics, and Biology [12–15]. To the best of our knowledge, GP has not been explored extensively to solve the problem of induction motor fault detection.

The nature of GP makes it a very promising approach to solving multi-label classification problem for fault detection of induction motors [16–18]. First, it can discover novel non-linear models for high-dimensional data by constructing executable computer programs using arithmetic functions, logical functions, and branching statements. Second, the automatic feature selection of GP is embedded in the process of model evolution. This intrinsic selection of relevant features distinguishes GP from many approaches that manually select features using domain-expertise, or perform feature selection and construct classification models in separate stages. Third, the stochastic population-based search property of evolutionary algorithms allows generating diverse high-quality classification models, which enriches the analysis of model interpretation and feature importance.

In this research, we design a GP algorithm, particularly using a linear GP representation [19], to learn predictive models for induction motor fault detection, formulated as a multi-label classification problem. Training data are collected by manipulating and monitoring two real induction motors in lab experiments. In addition, our proposed LGP method is compared with other similar or commonly used machine learning algorithms, including tree-based GP [20], support

Table 1. The types of faults applied to the two motors in the lab. Multi-fault states are represented using a "+" sign to combine multiple single-faults, and n-BRB stands for the BRB faults on n rotor bar(s).

Motor 1 state	Motor 2 state
Healthy	Healthy
UNB	UV
BF	1-BRB
BF + UNB	2-BRB
BF + UNB + UV	3-BRB
BF + 1-BRB	UV + 3-BRB

vector machines [21], decision trees [22], random forests [23], and naive Bayes classifiers [24]. Our LGP algorithm is shown to be able to achieve superior classification performance.

2   Methods

In this section, we first describe the experimental data collection for induction motor faults, then explain the signal processing for extracting characteristic features of the collected waveform data, and finally present the design and configuration of our LGP algorithm.

2.1   Experiments and Data Collection for Induction Motor Faults

For this study, two induction motors are used in the lab experiments, and various single- or multi- electrical and mechanical faults are manually applied to the motors. Most faults introduced to Motor 1 are mechanical, whereas Motor 2 electrical. The three-phase stator currents and the three-dimension vibration signals are recorded and used for training predictive models that can classify various motor fault states. The different motor fault states, including both single- and multi-faults, are summarized in Table 1. Five fault states are considered for each motor. Most of the manually introduced motor faults are irreversible, and the data are collected in the lab by progressively damaging a motor. The labels in Table 1 for single faults and healthy state are explained as follows:

(1) Healthy: a healthy motor
(2) BF: bearing fault, realized by the sand blasting process on the bearing to cause roughness
(3) UV: unbalanced voltage, realized by inserting a resistor in one phase of the power supply
(4) UNB: unbalance condition of shaft rotation, realized by adding extra weight on part of the pulley
(5) BRB: broken rotor bar, realized by drilling a hole on a rotor bar.

Table 2. Statistical features of the current and vibration signals. We use 3000 time-step sample points to calculate each feature variable. In the table, i is the index of a time step, N is the total number of time steps ($N = 3000$), and x_i denotes the value of the current/vibration at the i-th time step.

Features	Definitions
Mean	$\bar{X} = \frac{1}{N} \sum_{i=1}^{N} x_i, i = 1, 2, \ldots, N$
Median	$X_{\text{median}} = \frac{1}{2}(x_{\lfloor (N+1)/2 \rfloor} + x_{\lfloor N/2 \rfloor + 1})$
Standard Deviation	$\sigma = \sqrt{\frac{1}{N} \sum_{i=1}^{N} (x_i - \bar{X})^2}$
Median Absolute Deviation	$X_{\text{Median\_AD}} = \text{median}(|x_i - X_{\text{median}}|)$
Mean Absolute Deviation	$X_{\text{Mean\_AD}} = \frac{1}{N} \sum_{i=1}^{N} |x_i - \bar{X}|$
L1 Norm	$\|L\|_1 = \sum_{i=1}^{N} |x_i|$
L2 Norm	$\|L\|_2 = \sqrt{\sum_{i=1}^{N} |x_i|^2}$
Maximum Norm	$\|L\|_\infty = \max\{|x_i|\}, i = 1, 2, \ldots, N$

The problem of fault detection and classification of induction motors is formulated as a multi-label classification problem. One *class label* corresponds to a particular *motor state*. The input samples are motor current and vibration signals. Our goal is to learn predictive models that can classify the correct fault states by monitoring an induction motor.

2.2 Signal Processing and Feature Extraction

The signal data collected by monitoring a motor at various states shown in Table 1 are the three-phase current and three-dimension vibration. To characterize the signatures of these waveform input data, a signal processing method, matching pursuit (MP), is utilized to extract motor state features [25, 26]. MP decomposes a signal into a linear expansion of waveforms that are selected from a redundant dictionary of functions to best match original signal [27, 28]. As a result, eight statistical features can be derived, including mean, median, standard deviation, median absolute deviation, mean absolute deviation, L1 norm, L2 norm, and the maximum norm. A total number of 840 samples were collected for each motor, and each motor state class has 120 samples. Table 2 gives the definitions of the extracted features.

2.3 Multi-label Classification by Linear Genetic Programming

Linear genetic programming (LGP) algorithm defines evolutionary programs as linear sequences of executable instructions [19, 20]. In traditional tree-based GP, an individual expressed as a syntax tree can only have one root, therefore, it can only generate a single output value, stored in the root node. In contrast, LGP allows designating multiple registers as outputs. This expands its ability to solve multi-label classification problem conveniently, as well as to discover the complex non-linear relationship among input variables.

For classification tasks, a typical LGP setup is to assign a single output register, meaning that the evolved solutions are binary classification models. Nevertheless, LGP algorithms can be conveniently modified to enable simultaneously producing multiple output labels. This allows us to evolve multi-label classification models. As discussed in Sect. 2.1, each induction motor can have one of the six possible states (one healthy state, five faulty states). Instead of performing six binary classifications, we design an LGP algorithm for a six-label classification.

Program Representation. Each genetic program is a candidate classifier, whose instructions can be either assignment or branch statements. In an assignment statement, the expression yields a value stored in calculation registers by applying arithmetic operations or trigonometric functions. A branch statement changes the execution of a program. We use *if-greater-then* to skip one subsequent instruction when the condition in the *if* statement is evaluated to *false*. Input registers receive the feature values of a training sample, and they can only serve as an operand on the right-hand side of an assignment statement. Calculation registers are provided to enhance computational capacity and can be used in both left- and right-hand sides of an assignment statement. Constant registers store real numbers randomly generated within a given range. In our method, each motor state is numerically coded as $i \in \{0, 1, 2, 3, 4, 5\}$. A program incorporates six output registers, and they constitute an *output register vector* $\mathbf{R} \in \mathcal{R}^6$. To guarantee that the return values of output registers are classification probabilities, we apply the *sigmoid* function to map register values to probabilities. The final prediction result of a genetic program is the index of the register that has the highest probability value, which corresponds to one of the six possible states, formulated as follows,

$$\mathbf{R} = (r[0], r[1], r[2], r[3], r[4], r[5]), \tag{1}$$

$$p(state = i|X) = sigmoid(r[i]), \tag{2}$$

$$Output = i \text{ with } max(p(state = i|X)), \tag{3}$$

where $r[i]$ is the ith output register, $X = (x_1, x_2, ..., x_8)$ is a training sample input containing eight feature values, $p(state = i|X)$ means that the probability that X belongs to state i. Figure 1 shows a schematic example of the LGP program.

To guarantee that each LGP program contains six output registers for fitness evaluation, we follow four steps to generate LGP programs in the initial population. First, we randomly choose a length l for a program to be initialized. Next, we generate six instructions whose return registers are the designated six output registers, in order to make sure that each output register has at least one effective instruction. Then, we randomly produce $l - 6$ instructions as in conventional settings. The final step is to randomly order the generated l instructions. Figure 2 shows the process of such a program generation process.

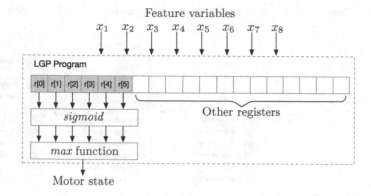

Fig. 1. A schematic example of the LGP program for the motor fault detection and classification. The value in the i-th output register corresponds to the probability that a sample X belongs to the i-th fault state. While evaluating an LGP program on a data sample, the program generates six probability values, and the label with the maximum probability will be the final output.

Genetic Variation. Since conventional genetic variations can violate our requirement that a program must produce six effective outputs, we design specific genetic variation operations to ensure all programs are valid. For a better explanation, we define a complete *output set* as a set containing six assignment instructions, each of which has a distinct output register as the return. A valid program must have a complete output set.

Crossover. When two parents are picked to swap instruction segments, it is possible that a child can have an incomplete output set. To prevent that, we maintain a *output-register table* for each program. The table uses six entries to record the number of each element of the output set for the program. When performing crossover, we first exchange segments between parents and then update their children's output-register tables. If a child's output-register table has a zero-valued entry, we abandon the offspring and randomly choose one of the parents to replace this child.

Mutation. We adopt micro mutation as a second variation, which alters an element of a randomly selected instruction such as the operation or an operand/return register in the selected instruction. Since a mutated program is possible, again, to have an incomplete output set, we first perform micro mutation to a parent, and If the mutant's output-register table has any zero-valued entries, we discard this mutant.

Fitness Function. The fitness of a program measures the accuracy of the classifier it represents. In this study, we employ the metrics of *precision* and *recall* by constructing the confusion matrix of true/false positives and negatives,

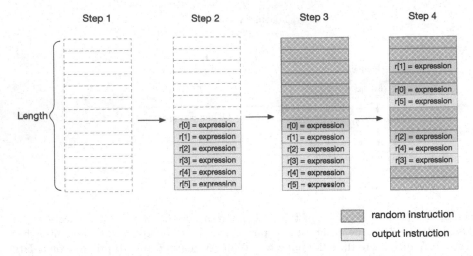

Fig. 2. The diagram of generating LGP programs for initialization. In step 1, we randomly decide the length of a program. In step 2, we generate six instructions using each of the six designated output registers as their returns. In step 3, we generate the remaining instructions randomly. In step 4, we randomly order the sequence of instructions.

and define the fitness function by computing their harmonic mean, known as the *F-measure* [29,30], defined as

$$Precision = \frac{T_P}{T_P + F_P}, \tag{4}$$

$$Recall = \frac{T_P}{T_P + F_N}, \tag{5}$$

$$F = 2 \cdot \frac{Precision \cdot Recall}{Precision + Recall}, \tag{6}$$

where T_P and F_P denote the numbers of true and false positive cases, respectively, and F_N represents the number of false negative cases. Here, we define that given a sample of label l_i, a true positive case is a prediction of the sample as l_i; otherwise the prediction is a false negative case. When a prediction is l_i, and the sample is of label l_j ($i \neq j$), such a case is a false positive of class l_i. We construct a confusion matrix for each class and compute their F-measure values, and the fitness function is the *average* of F-measure across all six classes. The goal of evolution is to maximize this averaged F-measure, which evaluates different and conflicting characteristics in the classification process.

2.4   Experiment Setup

The objectives of our experiments are:

- to show that the LGP classification algorithm can accurately predict motor states on real-world data;

Table 3. The parameter settings for the LGP classification algorithm

Parameter	Setting
Population size	1000
Function set	$+, -, \times, \div$ (protected), sin, cos, IF $>$
Terminal set	Motor features, random constants
Program length	[6, 200]
Fitness function	F-measure
Number of calculation registers	20
Number of feature registers	8
Number of constant registers	5
Constant bound	$[-2, 2]$
Crossover rate	0.9
Mutation rate	0.1
Crossover operator	Single-point crossover
Mutation operator	Micro mutation
Parent selection	Tournament with size 5
Survival selection	Truncation
Number of generations	1000
Number of runs	200

- to show that the LGP algorithm can automatically select the most relevant features for classification;
- to compare the proposed method with various machine learning techniques that have been applied to solve the same problem.

In the study, we implement our LGP algorithm in C programming language and compile the C-code by GCC 7.1. Table 3 shows the main parameter settings for the experiments. For each motor, the algorithm is run 200 times, and each GP program is evaluated after intron removal, i.e., removing structurally non-effective instructions. We adopt a five-fold cross-validation scheme to detect over-fitting. That is, we divide the training data into five equally-sized subsets, and in each cross-validation iteration, we use one subset for validation and train the LGP algorithm on other subsets. Therefore, each run of the algorithm produces five best classifiers on the five validation subsets. In total, we obtain 1000 best classifiers for Motor 1 and 2, respectively. For comparisons, we further implement the tree-based GP in C programming language, and we implement four other machine learning approaches using the *scikit-learn* library in Python [31], including support vector machines, decision trees, random forests, and naive Bayes classifiers. We ensure the same cross-validation partitions are used for all the counterpart algorithms.

3 Results

3.1 Best Evolved Classification Models

We first investigate the 1000 best classifiers found by the LGP algorithm for both motors. Although we collect both current and vibration signals of the motors in the lab experiments, the results using these two different signals as inputs are of equal quality. Thus, for simplicity, in this section, we discuss the results using current signals as the input. Figure 3 illustrates the distribution of the fitness of the best classifiers on the stator current training and test data for Motor 1 and 2, respectively. For Motor 1, we see that the majority of the 1000 best classifiers on the training set have fitness values in the range of [0.4, 0.6]. For motor 2, the majority of the best fitness values on the training set fall between [0.3, 0.6]. Figure 3 also shows the distribution of the fitness of the best classifiers on test data sets. We see that the majority of the fitness in Motor 1 also ranges from [0.35, 0.6], and in Motor 2, the interval shifts to [0.4, 0.75].

In addition to fitness, i.e., F-measure, we further look at other statistics of the best evolved classifiers. Figure 4 shows five metrics using stator current signals as input data. The best classifiers of Motor 1 can achieve a maximum accuracy as high as 0.973. For Motor 2, the best accuracy is 0.981. This suggests that our LGP algorithm performs equally well at identifying both mechanical (Motor 1) and electrical faults (Motor 2).

3.2 Feature Importance Assessment

GP can construct highly non-linear models of multiple features for the fault detection problem. Given the stochastic features of evolutionary algorithms, GP can automatically select a subset of features that are relevant to the classification. This embedded feature selection can boost the performance of the classifiers produced by GP.

In the experiments, we also look into the fitness and the number of effective features among the best classifiers for the two motors. We define an *effective feature* as any feature that remains in a GP program after intron removal. Since the performance of classifiers using stator current and vibration data are comparable, we only discuss the results of using the stator current as input data.

Figure 5 shows the fitness in a relation to the number of effective features in the best classifiers. We see that they have a monotonic relationship, and the best fitness is achieved when four or five effective features are used in the classifiers for both motors. This suggests that the proposed LGP method automatically selects relevant features and decreases the dimension of the feature space.

We define the *importance* of a feature as its appearance frequency in the evolved 1000 best classifiers [32]. That is, for instance, if a feature is an effective feature that contributes to the outcome of an LGP program, in 600 out of the 1000 best evolved classifiers, its importance is computed as 0.6.

Figure 6 shows the importance scores of the eight features using either current and vibration signals as input in motor 1 and 2, respectively. We see that using

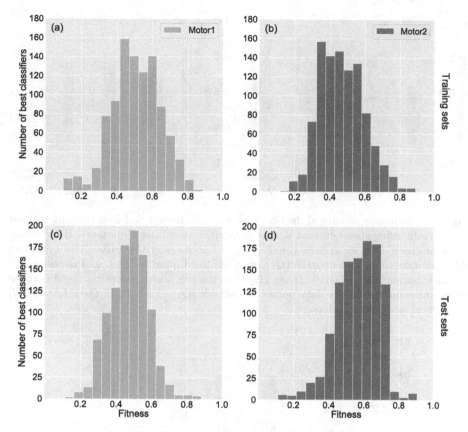

Fig. 3. Fitness distribution of the 1000 evolved best classifiers for Motor 1 (a, c) and 2 (b, d) on training (a, b) and test (c, d) data sets, respectively. The results shown here are generated using current signals as input.

the features extracted from the current signals, the top four features for both motor 1 and 2 are the same, i.e., mean, median, standard deviation, and L1 norm. Meanwhile, when the vibration signals are used to extract the features, the top four features for both motors are mean, median, standard deviation, and median absolute deviation.

3.3 Comparison with Other Machine Learning Algorithms

For comparison purposes, we implement five different machine learning approaches using the same data and cross-validation strategy. The compared methods include tree-based GP (TGP), support vector machines (SVM), decision trees (DT), random forests (RF), and naive Bayes classifiers (NB).

We perform a two-round training-comparison process. In the first round, we train all six methods using the full set of eight features using either the stator current or the vibration input data. In the second round, we only provide

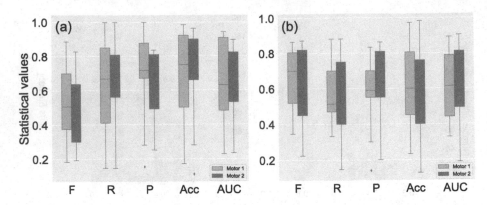

Fig. 4. Performance metrics of the best classifiers of Motor 1 and 2 on training (a) and test (b) sets, using stator current signals as input. F stands for F-measure, R for recall, P for precision, Acc for accuracy, and AUC for area under the curve. The box plots show the minimum, first quartile (Q_1), median, third quartile (Q_3), and maximum of the distributions. Interquartile range (IQR) is the box width, denoting the difference between the quartiles. Any data points fall outside $[Q_1 - 1.5 \times IQR, Q_3 + 1.5 \times IQR]$ are plotted as outliers in dots.

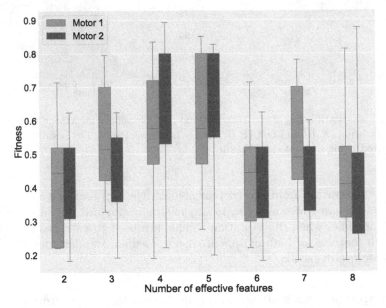

Fig. 5. The correlation of the fitness and the number of effective features in the 1000 best evolved classifiers. The results are produced using the stator current inputs for both motors.

the most important four features ranked by our LGP algorithm to train classifiers using the other machine learning methods. Then we compare the results of the first-round (without feature selection) and the second-round (with feature selection).

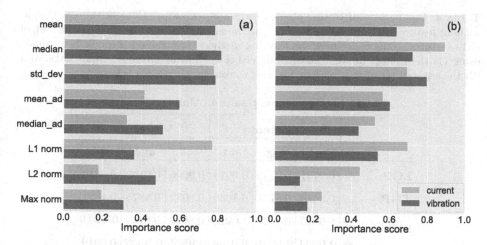

Fig. 6. The importance scores of the features in Motor 1 (a) and Motor 2 (b). Each bar denotes the importance score of a feature, defined as the occurrence frequency of a feature in the 1000 best classifiers. In the figure, *mean_ad* is the abbreviation for mean absolute deviation, and *median_ad* is the abbreviation for median absolute deviation.

Table 4 shows the statistics of the testing classification accuracy using either the stator current or the vibration input data for six algorithms. We test the statistical significance of the comparison results in Table 4 using a pairwise Wilcoxon rank-sum test with a confidence level of 95% [33, 34]. In Table 4, the results are printed boldface if they are statistically significant. An underline means that LGP performs significantly better than other methods without feature selection. The table shows that for both motors, LGP is significantly better compared to other methods without feature selection. Moreover, the features selected by our LGP algorithm significantly improve the accuracies of all other four common machine learning algorithms. This can be observed for both motors using either stator current or vibration input data, which suggests the effectiveness of both classifier construction and feature selection in our LGP algorithm.

4 Discussion

In this study, we design a linear genetic programming (LGP) algorithm for the fault detection and classification of induction motors, formulated as a multi-label classification problem. We collect experimental data by manipulating and monitoring two real induction motors in the lab. Five different mechanical and electrical faults are manually applied to each motor in order to collect their stator current and vibration signals that can be used to identify each of the fault states through training classification models. Using signal processing techniques, the original waveform signal data of stator current and vibration are subsequently characterized by eight extracted statistic features. Our LGP algorithm is able to take these input features, to evolve multi-label classifiers, and to predict the healthy/fault state of an induction motor with a high accuracy.

Table 4. Comparison of testing classification accuracies for six algorithms with and without feature selection (FS). Two GP methods (LGP and TGP) use the full set of eight features, while other machine learning methods are trained through two rounds using (1) the full set of eight features on either current or vibration input data, and (2) the selected set of the top four features ranked by our LGP algorithm.

(a) Comparison results on Motor 1

		Current			Vibration		
		best.	*avg.*	*std.*	*best.*	*avg.*	*std.*
LGP		<u>0.9731</u>	<u>0.9239</u>	0.0913	<u>0.9906</u>	<u>0.9195</u>	0.0844
TGP		0.9419	0.8727	0.1036	0.9032	0.8375	0.1391
SVM	w/o FS	0.9019	0.8837	**0.0238**	0.8625	0.8330	**0.0219**
	w FS	**0.9921**	**0.9219**	0.0409	**0.9855**	**0.9037**	0.0919
DT	w/o FS	0.8290	0.7663	**0.0138**	0.8220	0.8016	**0.0332**
	w FS	**0.9078**	**0.8557**	0.0394	**0.9531**	**0.9216**	0.0350
RF	w/o FS	0.8333	0.7885	**0.0762**	0.8461	0.7520	**0.0522**
	w FS	**0.9385**	**0.8058**	0.1098	**0.9432**	**0.8517**	0.0931
NB	w/o FS	0.8205	0.7993	0.0829	0.8377	0.7922	**0.0371**
	w FS	**0.9887**	**0.8993**	**0.0667**	**0.9677**	**0.9084**	0.0436

(b) Comparison results on Motor 2

		Current			Vibration		
		best.	*avg.*	*std.*	*best.*	*avg.*	*std.*
LGP		<u>0.9831</u>	<u>0.9029</u>	0.1017	<u>0.9528</u>	<u>0.8982</u>	0.0994
TGP		0.9619	0.8815	0.0928	0.9432	0.8761	0.1306
SVM	w/o FS	0.8015	0.7862	**0.0320**	0.7380	0.7037	**0.0419**
	w FS	**0.9822**	**0.9349**	0.0535	**0.9525**	**0.9245**	0.0677
DT	w/o FS	0.7014	0.6436	**0.0419**	0.8331	0.7421	0.0892
	w FS	**0.9063**	**0.8855**	0.1097	**0.8982**	**0.8770**	**0.0830**
RF	w/o FS	0.7318	0.6482	0.1090	0.8932	0.8048	0.0491
	w FS	**0.9251**	**0.8058**	**0.0839**	**0.9411**	**0.9007**	**0.0431**
NB	w/o FS	0.8387	0.7929	**0.0351**	0.8208	0.7814	**0.0514**
	w FS	**0.9717**	**0.8922**	0.0667	**0.9503**	**0.8496**	0.0710

With the intrinsic feature selection ability, our LGP algorithm can evaluate the importance of each feature based on its appearance frequency in the final evolved classifiers. This allows us to rank all eight features using either the stator current or the vibration input data, and assess their relevance to the classification.

Furthermore, we compare our LGP method with other machine learning approaches. Our results show that LGP performs the best using the full set of features, and is able to improve the performance of all other algorithms by selecting the more relevant features for their model training. We hope this study can showcase the great potentials of GP for machine learning problems arising from a wide range of application areas.

Acknowledgments. This research was supported by Newfoundland and Labrador Research and Development Corporation (RDC) Ignite Grant 5404.1942.101 and the Natural Science and Engineering Research Council (NSERC) of Canada Discovery Grant RGPIN-2016-04699 to TH.

References

1. Wu, S., Chow, T.W.: Induction machine fault detection using som-based RBF neural networks. IEEE Trans. Ind. Electron. **51**(1), 183–194 (2004)
2. Razavi-Far, R., Farajzadeh-Zanjani, M., Saif, M.: An integrated class-imbalanced learning scheme for diagnosing bearing defects in induction motors. IEEE Trans. Ind. Inform. **13**(6), 2758–2769 (2017)
3. Martin-Diaz, I., Morinigo-Sotelo, D., Duque-Perez, O., Romero-Troncoso, R.J.: An experimental comparative evaluation of machine learning techniques for motor fault diagnosis under various operating conditions. IEEE Trans. Ind. Appl. **54**(3), 2215–2224 (2018)
4. Godoy, W.F., da Silva, I.N., Goedtel, A., Palácios, R.H.C., Lopes, T.D.: Application of intelligent tools to detect and classify broken rotor bars in three-phase induction motors fed by an inverter. IET Electr. Power Appl. **10**(5), 430–439 (2016)
5. Poli, R., Langdon, W.B., McPhee, N.F.: A Field Guide to Genetic Programming (2008). http://lulu.com
6. Pappa, G.L., Ochoa, G., Hyde, M.R., Freitas, A.A., Woodward, J., Swan, J.: Contrasting meta-learning and hyper-heuristic research: the role of evolutionary algorithms. Genet. Program. Evol. Mach. **15**(1), 3–35 (2014)
7. Brameier, M., Banzhaf, W.: A comparison of linear genetic programming and neural networks in medical data mining. IEEE Trans. Evol. Comput. **5**(1), 17–26 (2001)
8. Agapitos, A., O'Neill, M., Brabazon, A.: Adaptive distance metrics for nearest neighbour classification based on genetic programming. In: Krawiec, K., Moraglio, A., Hu, T., Etaner-Uyar, A.Ş., Hu, B. (eds.) EuroGP 2013. LNCS, vol. 7831, pp. 1–12. Springer, Heidelberg (2013). https://doi.org/10.1007/978-3-642-37207-0_1
9. Guven, A.: Linear genetic programming for time-series modelling of daily flow rate. J. Earth Syst. Sci. **118**(2), 137–146 (2009)
10. Nguyen, S., Mei, Y., Zhang, M.: Genetic programming for production scheduling: a survey with a unified framework. Complex Intell. Syst. **3**(1), 41–66 (2017)

11. Parkins, A.D., Nandi, A.K.: Genetic programming techniques for hand written digit recognition. Sig. Process. **84**(12), 2345–2365 (2004)
12. Link, J., et al.: Application of genetic programming to high energy physics event selection. Nucl. Instr. Meth. Phys. Res. Sect. A: Accelerators Spectrometers Detectors Assoc. Equip. **551**(2–3), 504–527 (2005)
13. Chen, S.H., Yeh, C.H.: Evolving traders and the business school with genetic programming: a new architecture of the agent-based artificial stock market. J. Econ. Dyn. Control **25**(3–4), 363–393 (2001)
14. Liu, K.H., Xu, C.G.: A genetic programming-based approach to the classification of multiclass microarray datasets. Bioinformatics **25**(3), 331–337 (2009). https://doi.org/10.1093/bioinformatics/btn644
15. Hu, T., et al.: An evolutioanry learning and network approach to identifying key metabolites for osteoarthritis. PLoS Comput. Biol. **14**(3), e1005986 (2018)
16. Langdon, W.B., Poli, R.: Foundations of Genetic Programming. Springer, Heidelberg (2013)
17. Guo, H., Jack, L.B., Nandi, A.K.: Feature generation using genetic programming with application to fault classification. IEEE Trans. Syst. Man Cybern. Part B (Cybern.) **35**(1), 89–99 (2005)
18. Witczak, M., Obuchowicz, A., Korbicz, J.: Genetic programming based approaches to identification and fault diagnosis of non-linear dynamic systems. Int. J. Control **75**(13), 1012–1031 (2002). https://doi.org/10.1080/00207170210156224
19. Brameier, M.F., Banzhaf, W.: Linear Genetic Programming, 1st edn. Springer, Heidelberg (2010). https://doi.org/10.1007/978-0-387-31030-5
20. Koza, J.R.: Genetic Programming: On the Programming of Computers by Means of Natural Selection. MIT Press, Cambridge, MA (1992)
21. Cortes, C., Vapnik, V.: Support-vector networks. Mach. Learn. **20**(3), 273–297 (1995)
22. Breiman, L.: Classification and Regression Trees. Routledge, Abingdon (2017)
23. Liaw, A., Wiener, M., et al.: Classification and regression by randomForest. R News **2**(3), 18–22 (2002)
24. Rish, I.: An empirical study of the naive Bayes classifier. In: IJCAI 2001 Workshop on Empirical Methods in Artificial Intelligence, vol. 3, pp. 41–46. IBM, New York (2001)
25. Ali, M.Z., Shabbir, M.N.S.K., Liang, X., Zhang, Y., Hu, T.: Experimental investigation of machine learning based fault diagnosis for induction motors. In: Proceedings of 2018 IEEE Industry Applications Society (IAS) Annual Meeting, pp. 1–14. IEEE (2018)
26. Ali, M.Z., Shabbir, M.N.S.K., Liang, X., Zhang, Y., Hu, T.: Machine learning based fault diagnosis for single- and multi-faults in induction motors using measured stator currents and vibration signals. IEEE Trans. Ind. Appl. (2019, in press)
27. Li, J., Li, M., Yao, X., Wang, H.: An adaptive randomized orthogonal matching pursuit algorithm with sliding window for rolling bearing fault diagnosis. IEEE Access **6**, 41107–41117 (2018)
28. Mallat, S., Zhang, Z.: Matching pursuit with time-frequency dictionaries. Technical report, Courant Institute of Mathematical Sciences, New York, United States (1993)
29. Buckland, M., Gey, F.: The relationship between recall and precision. J. Am. Soc. Inf. Sci. **45**(1), 12–19 (1994)
30. Powers, D.M.: Evaluation: from precision, recall and F-measure to ROC, informedness, markedness and correlation. J. Mach. Learn. Technol. **2**, 37–63 (2011)

31. Pedregosa, F., et al.: Scikit-learn: machine learning in Python. J. Mach. Learn. Res. **12**, 2825–2830 (2011)
32. Hu, T., Oksanen, K., Zhang, W., Randell, E., Furey, A., Zhai, G.: Analyzing feature importance for metabolomics using genetic programming. In: Castelli, M., Sekanina, L., Zhang, M., Cagnoni, S., García-Sánchez, P. (eds.) EuroGP 2018. LNCS, vol. 10781, pp. 68–83. Springer, Cham (2018). https://doi.org/10.1007/978-3-319-77553-1_5
33. Wilcoxon, F.: Individual comparisons by ranking methods. Biometr. Bull. **1**(6), 80–83 (1945)
34. Wilcoxon, F., Katti, S., Wilcox, R.A.: Critical values and probability levels for the Wilcoxon rank sum test and the Wilcoxon signed rank test. Sel. Tables Math. Stat. **1**, 171–259 (1970)

Short Presentations

Fast DENSER: Efficient Deep NeuroEvolution

Filipe Assunção[✉][iD], Nuno Lourenço[iD], Penousal Machado[iD],
and Bernardete Ribeiro[iD]

CISUC, Department of Informatics Engineering, University of Coimbra,
Coimbra, Portugal
{fga,naml,machado,bribeiro}@dei.uc.pt

Abstract. The search for Artificial Neural Networks (ANNs) that are effective in solving a particular task is a long and time consuming trial-and-error process where we have to make decisions about the topology of the network, learning algorithm, and numerical parameters. To ease this process, we can resort to methods that seek to automatically optimise either the topology or simultaneously the topology and learning parameters of ANNs. The main issue of such approaches is that they require large amounts of computational resources, and take a long time to generate a solution that is considered acceptable for the problem at hand. The current paper extends Deep Evolutionary Network Structured Representation (DENSER): a general-purpose NeuroEvolution (NE) approach that combines the principles of Genetic Algorithms with Grammatical Evolution; to adapt DENSER to optimise networks of different structures, or to solve various problems the user only needs to change the grammar that is specified in a text human-readable format. The new method, Fast DENSER (F-DENSER), speeds up DENSER, and adds another representation-level that allows the connectivity of the layers to be evolved. The results demonstrate that F-DENSER has a speedup of 20 times when compared to the time DENSER takes to find the best solutions. Concerning the effectiveness of the approach, the results are highly competitive with the state-of-the-art, with the best performing network reporting an average test accuracy of 91.46% on CIFAR-10. This is particularly remarkable since the reduction in the running time does not compromise the performance of the generated solutions.

Keywords: Automatic Machine Learning ·
Convolutional Neural Networks · NeuroEvolution

1 Introduction

The increasing complexity of Machine Learning (ML) models has motivated the emergence of automatic (or semi-automatic) systems that seek to help humans in the design, optimisation and parameterisation of the models. In other words,

© Springer Nature Switzerland AG 2019
L. Sekanina et al. (Eds.): EuroGP 2019, LNCS 11451, pp. 197–212, 2019.
https://doi.org/10.1007/978-3-030-16670-0_13

such systems have the objective of working as black-box optimisers, empowering non-expert users with the ability to deploy ML models.

The goal of the current work is to automate the design and parameterisation of Artificial Neural Networks (ANNs). The tuning of this particular class of ML models involves numerous aspects: (i) pre-processing of the dataset, and definition of the data augmentation method; (ii) selection of the structure of the network (i.e., number and type of layers; number of neurons; activation functions; layers connectivity); and (iii) optimisation of the weights and bias values, and/or selection of the learning algorithm (and respective parameters). This task is even more challenging when addressing Deep Artificial Neural Networks (DANNs).

To automate the generation of ANNs we may resort to several techniques, such as, grid or random search, statistical methods (e.g., Bayesian optimisation), and/or we may use Evolutionary Computation (EC) to create ANNs. This is usually referred to as NeuroEvolution (NE). The bottleneck of NE is the time required by evolution, because EC techniques rely on a population of individuals that is evolved throughout a number generations. The evaluation of each individual – a potentially deep ANN – is consuming, which slows evolution.

The current work builds upon Deep Evolutionary Network Structured Representation (DENSER) [1,2]: a general purpose NE approach; The new version, known as Fast DENSER (F-DENSER) makes the following contributions:

- Introduces a graph-like representation, that enables the evolution of skip-connections and residual networks (Sect. 4.1);
- Speeds up the search of DANNs without compromising performance (Sect. 5.3, Number of Epochs);
- Investigates the simultaneous optimisation of the topology and learning (Sect. 5.3, Train Time).

The remainder of the paper is organised as follows. Section 2 briefly surveys works on Automated Machine Learning (AutoML), focusing on NE for Deep Learning (DL); Sects. 3 and 4 describe DENSER and F-DENSER, respectively; Sect. 5 details the conducted experiments and analyses the experimental results; Sect. 6 presents the conclusions and addresses future work.

2   Related Work

AutoML [3] concerns the set of approaches that seek to automate the design of ML models, from the pre-processing of the dataset to the selection and parameterisation of the model. Several approaches exist, ranging from grid [4] or random [5] search, to statistical [6] and EC [7] methods.

The current work focuses on NE, i.e., the application of EC to the design and optimisation of ANNs. The majority of the literature structures NE approaches according to the target of evolution: learning, topology, or both learning and topology simultaneously [8]. Our objective is to generate solutions for complex problems which often require deep networks. Therefore, our scope is not concerned particularly to what is being optimised, but rather to the scale of the networks. Our focus is towards methods capable of generating deep networks.

Koutník et al. [9] optimise the learning of deep networks, where the weights are indirectly encoded by Discrete Cosine Transform (DCT) coefficients that are evolved using CoSyNE [10]. According to Turner and Miller [11] "the choice of topology has a dramatic impact on the effectiveness of NE when only evolving weights; an issue not faced when manipulating both weights and topology"; therefore the authors defend that the best results are obtained by the simultaneous evolution of the learning and topology.

The current work addresses (without lost of generality) the optimisation of Convolutional Neural Networks (CNNs) for image classification tasks. Coevolution DeepNEAT (CoDeepNEAT) [7] extends NeuroEvolution of Augmenting Topologies (NEAT) [12] to the evolution of deep networks. The main difference between the two approaches is that each evolutionary unit encodes an entire layer instead of a single neuron. There are separate populations of modules and blueprints, which are evolved simultaneously (a blueprint connects several modules). Suganuma et al. [13] apply Cartesian Genetic Programming (CGP) to the evolution of CNNs. The authors a-priori define blocks that evolution can sequence and stack. The specification of such blocks biases search towards topologies that are known to work well, and makes the emergence of novel and unexpected architectures difficult. Real et al. [14] follow an approach that is close to the one of Suganuma et al.: the structure of the network is fixed and composed of normal and reduction cells; all normal cells have the same architecture, and all the reduction cells have the same architecture as well. The architecture of each of the cell types defines the target of evolution (i.e., to a certain point we are evolving two layer types).

The main limitation to the massive use of such systems is the time and computational resources needed for using them. In EC it is a common practice to evaluate large populations: the larger the population the more time is needed for evolution to complete. For example, CoDeepNEAT reports having used 100 GPUs to perform training, and Real et al. have used 450 GPUs and even with that massive amount of GPUs they required 7 days to perform each run. Suganuma et al. only use 2 GPUs, and each experiment takes about 5 days; however the bias of the search space reduces the number of individuals that need to be evaluated. Thus, the computational requirements are often a barrier to the widespread application of EC to DL, making the proposal of highly time effective and low computational approaches necessary. Our objective is to optimise DENSER so it is capable of finding appropriate solutions with limited resources, and within a reasonable amount of time.

3 DENSER

Deep Evolutionary Network Structured Representation (DENSER) is a NE approach that aims at enabling the evolution of the topology and learning of DANNs, i.e., DENSER optimises: (i) the number and placing of layers; (ii) the type of the layers (e.g., convolutional, pooling, fully-connected); (iii) the hyper-parameters of each of the layers (e.g., for convolutional layers the parameters to be optimised can be the number of filters, the shape of the filters, stride, padding type, activation function); (iv) the choice of the learning algorithm; and (v) the hyper-parameters of the learning algorithm (e.g., for Backpropagation we can optimise the learning rate and momentum). Theoretically, it also enables the setup of other hyper-parameters, such as the pre-processing of the dataset (e.g., resize, mean image), and data augmentation (e.g., cropping, rotation).

To enable a flexible and easy to adapt representation DENSER combines the principles of Genetic Algorithms (GAs) with those of Dynamic Structured Grammatical Evolution (DSGE) [15], with a genotype that has two genotypic levels. The outer-level encodes the macro-structure of the network, i.e., it encodes a linear sequence of evolutionary units (layers, learning and any other param-eterisations). The structure of the outer-level is specified by the user, and is responsible for defining the search space, and consequently what is to be opti-mised. An example of the structure of the outer-level for evolving CNNs is: [(features, 1, 10), (classification, 1, 4), (softmax, 1, 1), (learning, 1, 1)]; each tuple keeps information regarding the type of evolutionary unit stored in the outer-level position, and the minimum and maximum number of evolutionary units of that type, respectively. Therefore, the example outer-level structure can encode CNNs with a minimum of 3, and up to 15 layers: up to 10 convolution, pooling and/or batch-normalisation layers, followed by up to 4 fully-connected layers, and a final softmax layer (the reason why features stands for convo-lution and/or pooling layers, and why classification stands for fully-connected layers will be made clear next). In addition, there is a learning evolutionary unit, that handles the parameterisation of the learning algorithm. For each position in the outer-level there is a corresponding inner-level encoded by grammatical derivations; the used grammar must have one and only one production rule to expand the type of the evolutionary unit. Therefore, whilst the outer-level encodes the macro-structure, the inner-level is responsible for the parameters, their ranges, and type. Figure 3 shows a grammar for representing CNNs. It is now clear why the features outer-level type stands for convolutional, pooling, or batch-normalisation layers (lines 1–3), and why the classification type represents fully-connected layers (line 13). The numerical parameters (e.g., num-filters in line 4) are defined by a block that keeps information about its name, type (int or float), the number of values to generate, and the minimum and maximum bounds, respectively. The categorical parameters are defined by means of gram-matical derivation choices (e.g., activation function, line 16). The learning rule can be found in line 6 of Fig. 4.

Evolution in DENSER is promoted by genetic operators specifically designed to manipulate ANNs, that act on the two genotypic levels. The mutations add, replicate or remove evolutionary units, respecting the defined outer-structure, and within the lower and upper bounds for the evolutionary units (outer-level), and perturb the float and integer values, or change grammatical expansion pos-sibilities (inner-level). Two crossover operators are applied: one changes layers within the same module, and another swaps entire modules. A module is a set of evolutionary units that belong to the same index of the outer-level. The initial population is randomly generated, respecting the constraints on the minimum and maximum number of evolutionary layers defined in the outer-level structure.

4 Fast DENSER

Fast DENSER (F-DENSER) main goal is to speedup evolution without compromising the end results. It also seeks to make DENSER more flexible by enabling the layers of the evolved networks to be connected to any of the previous layers; this way it is easy to promote the emergence of skip connections or even residual networks. Next, we will discuss the changes made to DENSER in order to make it more flexible, and fast. The code for F-DENSER has been release as open-source, and can be found at https://github.com/fillassuncao/f-denser.

4.1 Representation

The individuals in DENSER are encoded as ordered linear sequences of feed-forward layers. Notwithstanding, in the original paper, Assunção et al. [1] define a merge-input parameter on the convolutional layers; the objective of this parameter is to merge the output of the convolutional layer with its input. The merge-input parameter takes part in many of the convolutional layers of the best performing networks; the problem is that this operator is layer-specific, and is not generalisable to pick any (or multiple) of the previous layers.

The core of the representation of F-DENSER is similar to DENSER: each evolutionary unit in the outer-level has a corresponding inner-level. The difference relies on the encoding of the connections between layers. Instead of assuming that between two consecutive evolutionary units encoding layers the connection is feed-forward, there is a list of connections for each layer specifying its inputs. This way, the encoding is close to a directed acyclic graph that is mapped into a DANN. Resembling what is done in graph-based approaches, like CGP [11], we introduce a new parameter: the maximum number of levels back, which sets a bound for the number of previous layers that can be used as input. To avoid disjoint graphs that generate invalid solutions, the layers have to be always connected to the previous layer.

Figure 1 depicts an example of the encoding of an individual, using the outer-level structure [(features, 1, 10), (classification, 1, 4), (softmax, 1, 1), (learning, 1, 1)], and the grammar of Fig. 3.

4.2 Initialisation

Previous experiments have demonstrated that when the initial population is generated entirely at random, during the initial generations the number of layers is trimmed. This is expected, as generating a high number of hyper-parameters for very deep networks would unlikely provide high performing networks. Therefore, in F-DENSER the initial population is still generated at random, but we impose a lower upper bound on the maximum number of layers of the outer-level structure, so that the initial individuals have a low number of layers, and evolution proceeds in a constructive way, by deepening the DANNs.

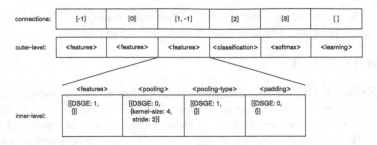

(a) Genotype. The connections define the list of inputs of each layer; -1 is the input of the network, and the remaining are the indexes of the outer-level (starting in 0). The connections are restricted to the evolutionary units that encode layers. We only represent the inner-level of one of the layers because of space constraints.

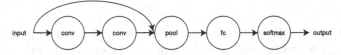

(b) Phenotype. We assume that the layers that do not have their inner-level specified correspond to convolution, convolution, and fully-connected, respectively.

Fig. 1. Example of an individual encoding a CNN (with the grammar of Fig. 3, and the learning production rule from the grammar of Fig. 4).

4.3   Genetic Operators

To promote evolution we only apply mutations to the candidate solutions. The mutations are specifically tailored for DANNs:

Add evolutionary unit – selects a position of the outer-level structure and adds a new valid evolutionary unit. The new evolutionary unit either replicates an existing evolutionary unit, or generates a new one at random (similarly to the initialisation procedure). Replicating an evolutionary unit means that if later in evolution one of the parameters of the replicas is changed, all copies are affected. The creation of new genotypic material cannot violate the constraint on the maximum length of the outer-level structure;

Remove evolutionary unit – chooses an evolutionary unit and removes it, assuring the new number of evolutionary units does not turn bellow the minimum bound. The connections to the layer that is discarded are removed, and not replaced by any other valid ones. The only exception is the connection to the previous layer: the layer previous to the one that is removed is added to the connections of the layer next to the one that is removed;

Add connection – picks at random a layer and adds another layer to its inputs. The inputs are restricted by the number of levels back;

Remove connection – similar to the add connection, but instead of adding a layer, removes one. The connection to the previous layer cannot be removed.

The remaining mutations act on the inner-level, and have no differences to the described in DENSER (same mutations of DSGE): they manipulate the real-values, and the integers associated to the grammatical expansion possibilities.

4.4 Evolution

F-DENSER uses a $(1 + \lambda)$ Evolutionary Strategy (ES). The rationale behind choosing an ES is related to reducing the number of individuals that need to be evaluated in each generation, and thus increase the speedup of the algorithm. The typical evolutionary run in DENSER uses a population of size 100. With $\lambda = 4$, in F-DENSER we need to assess the quality of 5 individuals in each generation, which translates into performing 1/20th of the evaluations per generation. The question lies in understanding the impact that having less individuals (e.g., diversity loss) has in the quality of the best found DANNs.

The best individual is re-evaluated in each generation (does not happen in DENSER). The train of ANNs is stochastic, because of the initial random setting of the weights, and thus different trains can obtain very different performances. The purpose of the re-evaluation is to minimise the effect of this randomness, and thus we seek to generate networks that are robust to different initialisations.

4.5 Evaluation

The usual methodology for evaluating the quality of the DANNs is based on assessing their performance on a given task, i.e., train the generated model either with fixed learning parameters, or with the evolved ones. The fitness of the candidate solution is the validation accuracy after train, since we will be dealing with a balanced dataset. However, we can use any other metric without the loss of generality of the framework.

The training of a network until convergence can be very time consuming. For that reason, many practitioners evaluate the networks with a low number of epochs. The problem with such small trains is that they difficult the optimisation of the learning parameters. The training for a small number of epochs also makes unfeasible the generation of networks that require no further training after evolution. In this paper we test and compare two methods for evaluating the generated DANNs: (i) we use the typical policy of a low number of training epochs; and (ii) we define a maximum GPU training time.

5 Experimentation

In this work we focus our experiments in the evolution of CNNs for the CIFAR-10. The experimental setup, and statistical tests are detailed in Sects. 5.1, and 5.2, respectively. Experiments are designed to compare the performance of F-DENSER and DENSER, the evaluation stop criteria, and the ability of F-DENSER to evolve the topology, learning, and layer's connectivity (Sect. 5.3).

Table 1. Experimental parameters.

Category	Parameter	F-DENSER	DENSER
	Number of runs	10	
	Number of generations	150	100
	Population size	5	100
	Crossover rate	–	70%
	Mutation rate	–	30%
Evolutionary Engine	Add layer rate	25%	–
	Remove layer rate	25%	–
	Add connection rate	15%	–
	Remove connection rate	15%	–
	DSGE-level rate	15%	–
	Tournament size	–	3
	Elite size	–	1
	Train set	42500 instances	
Dataset	Validation set	7500 instances	
	Test set	10000 instances	
	Number of epochs	10	
	Train time	10 min.	–
Train	Loss	Categorical Cross-entropy	
	Batch size	125	
	Learning rate	0.01	
	Momentum	0.9	
	Padding	4	
Data Augmentation	Random crop	4	
	Horizontal flipping	50%	

5.1　Experimental Setup

The parameters used for performing the experiments described next are summarised in Table 1. The table is organised into sections: (i) evolutionary engine –
parameters required by the evolutionary algorithm; (ii) dataset – partitioning of
the dataset; (iii) train – training stop conditions, and learning algorithm parameters used for the experiments conducted with a fixed learning policy; and (iv)
data augmentation – strategy for augmenting the dataset during training (never
evolved in the reported experiments).

Despite a higher number of generations in F-DENSER than in DENSER, the
number of evaluated individuals is a small fraction; whilst F-DENSER evaluates
750 individuals in 150 generations, DENSER evaluates 10000 individuals in 100
generations. To speedup search, F-DENSER applies a higher mutation rate, and
multiple mutations can be applied to the same individual in one generation.
The mutation rates concerning the number of layers are by individual, and the
remaining ones are by layer. The DSGE-level mutation rate concerns the per-
gene probability of changing any of the expansion possibilities or terminals.

The dataset is partitioned (in a stratified way) into three disjoint sets. The
train set is used for tuning the weights of the individuals, and the validation
set is used to assign fitness, and to perform early stop based on the validation
loss (on the experiments performed using time as the stop criteria). The test
set is kept out of evolution. On the experiments where the learning parameters
are evolved, the batch size, learning rate and momentum are not fixed. For the
experiments where the train stop criteria is GPU time it is important to mention
that we are using 8 GeForce GTX 1080 Ti GPUs. A run only uses one GPU.

$$\begin{aligned}
\text{<features>} ::=\ &\text{<convolution>} \mid \text{<convolution>} &&(1)\\
&\mid \text{<pooling>} \mid \text{<pooling>} &&(2)\\
&\mid \text{<batch-norm>} &&(3)\\
\text{<convolution>} ::=\ &\text{layer:conv [num-filters,int,1,32,256] [filter-shape,int,1,2,5]} &&(4)\\
&\text{[stride,int,1,1,3] <padding> <activation> <bias>} &&(5)\\
\text{<batch-norm>} ::=\ &\text{layer:batch-norm} &&(6)\\
\text{<merge-input>} ::=\ &\text{merge-input:True} &&(7)\\
&\mid \text{merge-input:False} &&(8)\\
\text{<pooling>} ::=\ &\text{<pool-type> [kernel-size,int,1,2,5]} &&(9)\\
&\text{[stride,int,1,1,3] <padding>} &&(10)\\
\text{<pool-type>} ::=\ &\text{layer:pool-avg} \mid \text{layer:pool-max} &&(11)\\
\text{<padding>} ::=\ &\text{padding:same} \mid \text{padding:valid} &&(12)\\
\text{<classification>} ::=\ &\text{<fully-connected>} &&(13)\\
\text{<fully-connected>} ::=\ &\text{layer:fc <activation>} &&(14)\\
&\text{[num-units,int,1,128,2048 <bias>} &&(15)\\
\text{<activation>} ::=\ &\text{act:linear} \mid \text{act:relu} \mid \text{act:sigmoid} &&(16)\\
\text{<bias>} ::=\ &\text{bias:True} \mid \text{bias:False} &&(17)\\
\text{<softmax>} ::=\ &\text{layer:fc act:softmax num-units:10 bias:True} &&(18)
\end{aligned}$$

Fig. 2. Grammar used by F-DENSER for evolving the topology (10 epochs).

$$\begin{aligned}
\text{<features>} ::=\ &\text{<convolution>} \mid \text{<pooling>} &&(1)\\
\text{<convolution>} ::=\ &\text{layer:conv [num-filters,int,1,32,256]} &&(2)\\
&\text{[filter-shape,int,1,2,5] [stride,int,1,1,3]} &&(3)\\
&\text{<padding> <activation> <bias>} &&(4)\\
&\text{<batch-normalisation> <merge-input>} &&(5)\\
\text{<batch-normalisation>} ::=\ &\text{batch-normalisation:True} &&(6)\\
&\mid \text{batch-normalisation:False} &&(7)
\end{aligned}$$

Fig. 3. Grammar used by DENSER for the evolution of the topology. The production rules that are equal to those of Fig. 2 are omitted.

The tests are conducted on the CIFAR-10 [16] dataset: composed of 50000 train and 10000 test real-world images. Each instance is a 32×32 RGB image, that belongs to one of ten classes. The CIFAR-10 has been used in the benchmark of numerous NE approaches for evolving CNNs, and thus is a good baseline.

5.2 Statistical Analysis

To check if the samples follow a Normal Distribution we use the Kolmogorov-Smirnov and Shapiro-Wilk tests, with $\alpha = 0.05$. The tests reveal that the data does not follow any distribution and thus the non-parametric Mann-Whitney U test ($\alpha = 0.05$) is used to perform the pairwise comparisons between the setups.

5.3 Experimental Results

We perform four sets of experiments: (i) we compare F-DENSER with DENSER on the optimisation of the topology of CNNs (fixed learning policy with 10 epochs

Table 2. Results of the evolution of CNNs with F-DENSER and DENSER. The accuracy results on evolution and evolution (re-trained) concern evolution and thus are based on the validation set. The last row reports results on the test set; the value within brackets is the accuracy on the test set using augmented versions of the instances, i.e., each image is augmented 100 times, and the assigned label is the average of maximum of the average of the confidence values.

	F-DENSER	DENSER	p-value
Evolution	84.54%	85.51%	0.308
Evolution (re-trained)	83.21%	83.62%	0.624
Evolution (no limit)	87.79% (89.28%)	88.19% (89.65%)	0.624 (0.497)

per evaluation); (ii) we compare the evolution based on the number of epochs to a maximum GPU train time; (iii) we test evolving simultaneously the topology, and learning of the CNNs; and (iv) we optimise the topology and learning of CNNs, but with the possibility of establishing connections to previous layers.

Number of Epochs

To compare the performance of F-DENSER with DENSER we conduct the same experiments of the original paper, i.e., we promote the search for the topology of CNNs, trained with Backpropagation with a fixed learning policy for 10 epochs (Table 1). The grammars used for both methods are slightly different, and are shown in Figs. 2 and 3, for F-DENSER and DENSER, respectively. The main differences between the two rely in the features and convolution production rules. The convolution layer in DENSER has more properties, namely, batch-normalisation and merge-input (discussed above). DENSER had to tailor the convolution layer to encompass these extra parameters. We decided to follow a more flexible approach, and thus disregard these parameters. It is possible to mimic the merge-input with the novel feature of F-DENSER that enables layers to have more than one input (discussed below).

The results are summarised in Table 2: evolution refers to the average of the 10 best results found throughout evolution (the best result of each run); re-trained is the average performance of each of the best CNNs re-trained 5 times; and no limit is the average classification accuracy of the best networks trained without limit on the number of epochs, instead a early stop (without improvement for 12 epochs) based on the validation loss is used. This table structure is kept for the remainder of the experiments. The analysis of the results indicates that the performance of the evolutionary results of DENSER is slightly superior to F-DENSER. Notwithstanding, this changes if the networks are re-trained 5 times, demonstrating that the re-evaluation of the elite during evolution generates networks that are more robust. The training for an unlimited number of epochs (early stop of 12 epochs) leads to better performances in both methods.

Nevertheless, the time complexity needed for obtaining the best solutions is very different. The average run time for each run of F-DENSER is of 55 hours,

<features> ::= <convolution> | <convolution>   (1)

        | <pooling> | <pooling>   (2)

        | <dropout> | <batch-norm>   (3)

<classification> ::= <fully-connected> | <dropout>   (4)

<dropout> ::=layer:dropput [rate,float,1,0,0.7]   (5)

<learning> ::= <bp> <early-stop> [batch_size,int,1,50,500] epochs:400   (6)

      | <rmsprop> <early-stop> [batch_size,int,1,50,500] epochs:400   (7)

      | <adam> <early-stop> [batch_size,int,1,50,500] epochs:400   (8)

<bp> ::= learning:gradient-descent [lr,float,1,0.0001,0.1]   (9)

    [momentum,float,1,0.68,0.99] [decay,float,1,0.000001,0.001]   (10)

    <nesterov>   (11)

<nesterov> ::= nesterov:True | nesterov:False   (12)

<adam> ::= learning:adam [lr,float,1,0.0001,0.1] [beta1,float,1,0.5,1]   (13)

    [beta2,float,1,0.5,1] [decay,float,1,0.000001,0.001]   (14)

<rmsprop> ::= learning:rmsprop [lr,float,1,0.0001,0.1]   (15)

    [rho,float,1,0.5,1] [decay,float,1,0.000001,0.001]   (16)

<early-stop> ::= [early_stop,int,1,5,20]   (17)

Fig. 4. Grammar used by F-DENSER for the evolution of the topology and learning (10 min). Production rules equal to those of Fig. 2 are omitted.

while DENSER requires an average of 1083 hours to complete 100 generations, i.e., there is a speedup of 20x when we use F-DENSER[1]. Thus, F-DENSER can find solutions that are highly competitive with the ones discovered by DENSER, but 20 times faster. Focusing the structure of the evolved networks, it is noticeable that F-DENSER generates CNNs that have less layers (12.5) than the ones generated by DENSER (16.8). This difference is statistically significant (p-value $= 0.0375$). We have conducted preliminary experiments with DENSER (on the same conditions) but with a smaller population, and the results show that when performing only 750 evaluations (same as F-DENSER), the results are bellow the ones obtained by F-DENSER and DENSER.

Train Time

From this point onward we focus on F-DENSER. We compare the two stop criteria for the optimisation of the topology (T) (evaluation for 10 epochs/minutes), and the simultaneous evolution of the topology and learning (T+L). The evaluation during 10 min enables the use of dropout layers and facilitates the tune of the learning rate decay. The learning policy includes early stop, and thus the training time can be inferior to the maximum granted. The used grammar is shown in Fig. 4, and adds dropout and learning rules to the previous grammar. The experiment focusing the evolution of the topology (10 min) disregards the production rules associated to learning, and uses the parameters of Table 1.

---

[1] F-DENSER and DENSER are run in the same machines, which have the following specifications: 1080 Ti GPUs, 64 GB of RAM, and an Intel Core i7-6850K CPU.

Table 3. Results of the evolution of the topology (T) and learning (L) parameters of CNNs with F-DENSER. The first p-value concerns the statistical result of the comparison between Epochs (T) and Time (T), and the second p-value the results of the comparison between Time (T) and Time (T+L).

	Epochs (T)	Time (T)	p-value	Time (T+L)	p-value
Evolution	84.54%	88.52%	0.000	88.29%	0.187
Evolution (re-trained)	83.21%	87.47%	0.000	87.56%	0.347
Evolution (no limit)	87.79%	87.65%	0.472	87.76%	0.308
	(89.28%)	(88.89%)	(0.103)	(88.90%)	(0.472)

The first two columns of Table 3 compare the evaluation based on the number of epochs to a maximum training time. A perusal analysis of the results indicates that evaluating the CNNs up to a maximum of 10 min generates the best results, and therefore without further training the results that come out of evolution are better with F-DENSER (statistically significant). Notwithstanding, we note that the evaluation on 10 epochs corresponds to varying training times, i.e., the networks can take much less or much more than 10 min to complete the 10 epochs; thus, we re-evaluate the best networks generated with 10 epochs for 10 min. Each network is evaluated 5 times, and the validation accuracy increases from 83.21% to 85.33% (the test accuracy is not recomputed because it already has no time limit). The opposite, i.e., re-evaluate the networks generated with 10 min during 10 epochs decreases the average validation accuracy from 87.47% to 76.60%. This shows that when evaluating the CNNs considering the number of epochs the outcome of evolution are networks that train fast, compared to networks that are capable to learn during more epochs. The results reported by both methods are competitive, and consequently we conduct the remaining experiments with the time as stop criteria because it enables longer trains, which makes the evolution of learning policies more efficient. When we use time as the stop criteria an average of 61.2 epochs are performed.

The results of the simultaneous evolution of the topology and learning are presented on the last columns of Table 3. In terms of performance, there are no statistical differences between evolving the topology, or simultaneously the topology and learning. The results of the evolution of the topology are as good as the simultaneous evolution of topology and learning because there is a first stage of defining a learning rate that is known to work well. This indicates that it is advantageous to evolve both. Despite the larger search space, given the same time the results are of the same magnitude. Therefore, it is not needed to a-priori define learning parameters we are not sure to be the most appropriate.

Connections to Previous Layers

To investigate the effect of allowing the layers to have more than one input we experiment with the number of levels back set to 5, i.e., we focus the simultaneous

Table 4. Results of the evolution of topology and learning (T+L) and topology, learning and backward connections (T+L+B) of CNNs with F-DENSER.

	Time (T+L)	Time (T+L+B)	p-value
Evolution	88.29%	87.73%	0.271
Evolution (re-trained)	87.56%	86.88%	0.187
Evolution (no limit)	87.76% (88.90%)	87.17% (88.32%)	0.472 (0.430)

evolution of the topology, learning, and layer's connectivity, and the features layers can connect to the 5 previous layers. We use the grammar of Fig. 4, and evaluate each individual for 10 min. The results are summarise in Table 4.

There are no statistical differences in performance between receiving multiple input connections to connecting only to the previous layer. Nonetheless, 6 out of the 10 best networks have at least a layer that receives more than one input. This shows that although there is no increase in performance, the best solutions tend to use this feature. On the other hand, there is a statistical difference in the number of layers (p-value $= 0.0375$): whilst the average number of layers of the feed-forward CNNs is 14.9, when establishing multiple connections to previous layers the average is 12.2. Therefore, despite the lack of performance gains, the evolved networks are simpler.

Discussion

Figure 5 depicts a box-plot of the test accuracy of all the experiments conducted with F-DENSER. The results on the evolution of the topology with the individuals evaluated for 10 epochs have a slightly higher median than when evaluated for 10 min, but the dispersion of the results (reported by the interquartile range) is also higher. In addition, as the evaluation for 10 min facilitates the evolution of learning policies, and as the differences in the test accuracies are not statistical we carried out the remainder of the experiments considering the evaluation for a maximum of 10 min. This approach also has the advantage of enabling a fair comparison between individuals by restricting their assess to computational resources equally. The comparison between the evolution of topology, and topology and learning supports the theory of Turner and Miller [11] that it is advantageous to evolve both the topology and learning simultaneously. In fact, the highest performing CNN in terms of validation accuracy (among all experiments) was generated when evolving both topology and learning – this network has an average test accuracy of 91.26% (i.e., an error of 8.74%). The experiment where a layer can establish connections to multiple previous layers reported a performance that is lower than the remainder tests. However, the comparison is not fair: we are searching larger search spaces, but evolution is conducted during the same search time (i.e., 150 generations). To test this hypothesis we perform more 50 generations for the evolution of the topology and learning, and topology, learning, and layer's connectivity. The new results report an average test clas-

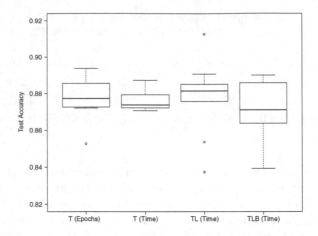

Fig. 5. Box-plot of the test accuracy of all the conducted experiments for the evolution of the topology (T), topology and learning (TL), and topology, learning, and backward connections (TLB). The stop condition is within brackets.

sification performance of 88.22% (89.33%), and 87.86% (88.92%), respectively. These results are slightly superior to those reported in Table 4, and show that longer evolutionary cycles may provide better results.

Assunção et al. [1] report that for CIFAR-10 the best CNN found with DENSER has an average error of 7.30% (with the same learning policy used during evolution), Stanley et. al. al. [7] also report an average error of 7.30%, and Suganuma et al. [13] an average error of 5.98% (with a learning rate policy different than the used during evolution). The average error reported by F-DENSER is comparable to those of Assunção et al. and Stanley et. al. (8.74% vs. 7.30%); most importantly F-DENSER generates a competitive result within a fraction of the time (and computational resources) of the other methods. The result of Suganuma et al. is on a different learning rate policy, that still had to be tried and fine-tuned after the end of evolution. Therefore, the conducted experiments show that the results of F-DENSER are comparable to the obtained by DENSER, with a significant time gain; it is also proven that F-DENSER can effectively evolve the topology and learning of CNNs. Further experiments need to be conducted in what regards the establishment of connections to previous layers to understand if more evolutionary time is needed, or even if the time needed for evaluating the individuals has to be greater so that the performance gains can be noticed. On the other hand, for the considered problem it may happen that there is no advantage in considering connections to previous layers.

The best result reported by F-DENSER is the outcome of the train with the same partitions used in evolution. The remainder of the state-of-the-art methods merge the train and validation sets, and thus the train is based on more instances. This makes it impossible to use early stop. The train of the best generated CNN with these conditions, and given the same number of epochs used during the trains without limit increases the test accuracy to 91.46% (i.e., error of 8.54%).

6 Conclusions and Future Work

The current paper proposes F-DENSER, a NE approach that extends DENSER. The introduced modifications expand the representation so that the connectivity of each layer is considered, and speedup evolution without compromising the overall performance. Experiments are designed to compare F-DENSER to DENSER, and to acknowledge the ability of F-DENSER to evolve only the topology, the topology and learning (simultaneously), or the topology, learning, and layers connectivity (simultaneously). The results show that considering the same conditions of DENSER (i.e., evolution of the topology, with the individuals evaluated for 10 epochs) F-DENSER generates results that have the same overall performance of DENSER, but in a fraction of the time; more precisely, there is a speedup of 20x. It is also demonstrated that F-DENSER is capable of simultaneously optimising the topology and learning of CNNs; this way, we do not need to conduct the search in two steps (manually find an adequate learning strategy and only then apply NE to the topology), but instead we can let evolution search for the topology, and learning strategy that is more adequate for that specific network and problem. These results evidence that assessing the quality of the individuals considering a time stopping criteria facilitates the optimisation of the learning strategy. The best result is obtained when simultaneously evolving the topology and learning: the generated CNN reports a test classification error of 8.54%, which is close to state-of-the-art results, but only requiring a fraction of the run time. In what regards the connection of a layer to previous layers we are aware that more experiments need to be conducted. Despite the encouraging results there is the need to understand if a time limit of 10 min is sufficient for evaluating such networks. Experiments are to be conducted for a longer number of generations, and on a wider set of benchmarks.

Acknowledgments. This work is partially funded by: Fundação para a Ciência e Tecnologia (FCT), Portugal, under the PhD grant SFRH/BD/114865/2016, and the project grant DSAIPA/DS/0022/2018 (GADgET), and is based upon work from COST Action CA15140: ImAppNIO, supported by COST (European Cooperation in Science and Technology): www.cost.eu.

References

1. Assunção, F., Lourenço, N., Machado, P., Ribeiro, B.: Evolving the topology of large scale deep neural networks. In: Castelli, M., Sekanina, L., Zhang, M., Cagnoni, S., García-Sánchez, P. (eds.) Genetic Programming, pp. 19–34. Springer International Publishing, Cham (2018). https://doi.org/10.1007/978-3-319-77553-1_2
2. Assunção, F., Lourenço, N., Machado, P., Ribeiro, B.: DENSER: deep evolutionary network structured representation. Genet. Program. Evolvable Mach. (2018). https://doi.org/10.1007/s10710-018-9339-y
3. Guyon, I., et al.: A brief review of the ChaLearn AutoML challenge: any-time any-dataset learning without human intervention. In: AutoML@ICML. JMLR Workshop and Conference Proceedings, vol. 64, pp. 21–30 (2016)

4. Duan, K.-B., Keerthi, S.S.: Which is the best multiclass SVM method? An empirical study. In: Oza, N.C., Polikar, R., Kittler, J., Roli, F. (eds.) MCS 2005. LNCS, vol. 3541, pp. 278–285. Springer, Heidelberg (2005). https://doi.org/10.1007/11494683_28
5. Bergstra, J., Bengio, Y.: Random search for hyper-parameter optimization. J. Mach. Learn. Res. **13**, 281–305 (2012)
6. Bergstra, J., Yamins, D., Cox, D.D.: Making a science of model search: Hyperparameter optimization in hundreds of dimensions for vision architectures. In: JMLR Workshop and Conference Proceedings, ICML (1), vol. 28, pp. 115–123 (2013)
7. Miikkulainen, R., et al.: Evolving deep neural networks. CoRR abs/1703.00548 (2017)
8. Floreano, D., Dürr, P., Mattiussi, C.: Neuroevolution: from architectures to learning. Evol. Intell. **1**(1), 47–62 (2008)
9. Koutník, J., Cuccu, G., Schmidhuber, J., Gomez, F.J.: Evolving large-scale neural networks for vision-based TORCS. In: FDG, pp. 206–212. Society for the Advancement of the Science of Digital Games (2013)
10. Gomez, F.J., Schmidhuber, J., Miikkulainen, R.: Accelerated neural evolution through cooperatively coevolved synapses. J. Mach. Learn. Res. **9**, 937–965 (2008)
11. Turner, A.J., Miller, J.F.: The importance of topology evolution in neuroevolution: a case study using cartesian genetic programming of artificial neural networks. In: Bramer, M., Petridis, M. (eds.) Research and Development in Intelligent Systems XXX, pp. 213–226. Springer, Cham (2013). https://doi.org/10.1007/978-3-319-02621-3_15
12. Stanley, K.O., Miikkulainen, R.: Evolving neural networks through augmenting topologies. Evol. Comput. **10**(2), 99–127 (2002)
13. Suganuma, M., Shirakawa, S., Nagao, T.: A genetic programming approach to designing convolutional neural network architectures. In: Proceedings of the Genetic and Evolutionary Computation Conference, pp. 497–504. ACM (2017)
14. Real, E., Aggarwal, A., Huang, Y., Le, Q.V.: Regularized evolution for image classifier architecture search. arXiv preprint arXiv:1802.01548 (2018)
15. Lourenço, N., Assunção, F., Pereira, F.B., Costa, E., Machado, P.: Structured grammatical evolution: a dynamic approach. In: Ryan, C., O'Neill, M., Collins, J. (eds.) Handbook of Grammatical Evolution, pp. 137–161. Springer, Cham (2018). https://doi.org/10.1007/978-3-319-78717-6_6
16. Krizhevsky, A., Hinton, G.: Learning multiple layers of features from tiny images (2009)

A Vectorial Approach to Genetic Programming

Irene Azzali[1]($\boxtimes$) [iD], Leonardo Vanneschi[2] [iD], Sara Silva[3,4] [iD], Illya Bakurov[2] [iD],
and Mario Giacobini[1] [iD]

[1] DAMU - Data Analysis and Modeling Unit, Department of Veterinary Sciences,
University of Torino, Turin, Italy
{irene.azzali,mario.giacobini}@unito.it
[2] NOVA Information Management School (NOVA IMS),
Universidade Nova de Lisboa, Campus de Campolide, 1070-312 Lisbon, Portugal
{lvannesc,ibakurov}@novaims.unl.pt
[3] LASIGE, Faculdade de Ciências, Universidade de Lisboa, Lisbon, Portugal
sara@fc.ul.pt
[4] BioISI – Biosystems & Integrative Sciences Institute, Faculdade de Ciências,
Universidade de Lisboa, Lisbon, Portugal

Abstract. Among the various typologies of problems to which Genetic Programming (GP) has been applied since its origins, symbolic regression is one of the most popular. A common situation consists in the prediction of a target time series based on scalar features and other time series variables collected from multiple subjects. To manage this problem with GP data needs a *panel* representation where each observation corresponds to a collection on a subject at a precise time instant. However, representing data in this form may imply a loss of information: for instance, the algorithm may not be able to recognize observations belonging to the same subject and their recording order. To maintain the source of knowledge supplied by ordered sequences as time series, we propose a new approach to GP that keeps instances of the same observation together in a vector, introducing vectorial variables as terminals. This new representation allows aggregate functions in the primitive GP set, included with the purpose of describing the behaviour of vectorial variables. In this work, we perform a comparative analysis of vectorial GP (VE-GP) against standard GP (ST-GP). Experiments are conducted on different benchmark problems to highlight the advantages of this new approach.

Keywords: Genetic programming · Vector-based representation ·
Panel data regression

1 Introduction

Among the several existing typologies of problems to which Genetic Programming (GP) [1] can be applied, symbolic regression is undoubtedly one of the most

© Springer Nature Switzerland AG 2019
L. Sekanina et al. (Eds.): EuroGP 2019, LNCS 11451, pp. 213–227, 2019.
https://doi.org/10.1007/978-3-030-16670-0_14

popular. The objective of symbolic regression is to find a function that describes the relationship between inputs and corresponding outputs, developing a model that can be used to make predictions on new inputs. Of great importance, belonging to the family of symbolic regression, is *panel data forecasting*.

A panel dataset is a collection of observations for multiple subjects at different equal-spaced time intervals [2]. Therefore, if the independent variables among the M observations measured are $X_1^i, \ldots, X_N^i$ where $i = 1, \ldots, M$ and $X_K^i, \ldots, X_N^i$ change in time ($X_{j_t}^i$ with $i = 1, \ldots, M$, $j = K, \ldots, N$ and $t = 1, \ldots, T$ denoting time series variables) and Y is a dependent variable (Y_t^i with $t = 1, \ldots, T$ denoting a target time series variable), we can express the dataset as

$$\{X_1^i, \ldots, X_{K-1}^i, X_{K_1}^i, \ldots, X_{K_T}^i, \ldots, X_{N_1}^i, \ldots, X_{N_T}^i, Y_1^i, \ldots, Y_T^i\}$$

where $i = 1, \ldots, M$ refers to the subject being observed. The interest of panel data regression lies in predicting dependent variables which are hard to measure. To clarify, let us consider the example panel reported in Table 1. In this example, individual characteristics are collected for different persons and years in order to predict the income. The standard GP approach can be easily applied to panel data regression; however, there can be a potential disadvantage. Data instances (fitness cases) are treated independently. Therefore the algorithm is not able to recognize that two (as in lines 1 and 2 in Table 1), or more, observations belong to the same individual. This situation may result in a loss of knowledge regarding the time series, that may instead have been useful to effectively model the target.

Table 1. Example standard panel dataset.

Person ID	Age	Sex	Year	Income
1	27	1	2015	1600
1	28	1	2016	1500
2	42	2	2015	1900
2	43	2	2016	2000
2	44	2	2017	2100
3	34	1	2015	3300

The idea behind this work is to design a novel GP system, that we call *Vectorial GP* (VE-GP), able to exploit the source of information provided by the additional dimension of time of panel datasets. To make the algorithm consider the whole time-series, we aggregate related data instances referring to the same entity in a vectorial representation, so that variables that change in time become *vectorial variables*. Therefore, the panel dataset represented in Table 1 is transformed into the representation reported in Table 2. In this configuration, a GP tree can be composed either by scalar terminals (to represent features such as Sex and Person ID) or by vectorial terminals (for instance to represent Ages and

Table 2. Example vectorial panel dataset.

Person ID	Ages	Sex	Years	*Incomes*
1	[27, 28]	1	[2015, 2016]	[1600, 1500]
2	[42, 43, 44]	2	[2015, 2016, 2017]	[1900, 2000, 2100]
3	34	1	2015	3300

Years). Moreover, the technique we propose adds new functions to the primitive set, defined with the purpose of describing the behaviour of temporal, or more generally vectorial, variables. From now on, we use the term *time series* to indicate an observation of a vectorial variable, also called *time series variable*. Therefore, a time series is a sequence of recorded values, belonging to one entity and represented as a vector.

The paper is organized as follows: in Sect. 2, we present previous and related work and we motivate the novelty of our approach. In Sect. 3, we describe VE-GP focusing on the new techniques and approaches we propose in the different blocks of a GP structure. Test problems and experimental setting used to explore the performance of VE-GP are presented in Sect. 4. Sections 5 and 6 describe the result obtained, comparing VE-GP with the standard GP and discuss the possible limitations of the new approach. Finally, in Sect. 7, we conclude with final considerations, and we give ideas for possible future works.

2   Previous and Related Work

Working with time series in GP has always been a challenging problem due to the inherent difficulty of handling this type of data. Common strategies include feature extractions to reduce the series into scalar features [3] or element by element treatment, where each entry of the series is an independent terminal [4]. However, some previous works explored the idea of keeping the native data type of time series, the vector. Holladay et al. [5] introduced a vector-based GP to predict the feature vector of fixed length signals. In this paper, vectors were possible inputs rather than scalars, and the primitive set included domain dependent functions that act on both of them. A more recent work of Bartashevich et al. [6] allows vectors as primitives and include vectorial functions such as the cross and dot product to build GP individuals. Other approaches have been proposed to preserve the ordered essence of time series such as Vera et al. [7]. In this latter work the authors investigated the serial processing of data where the time sequence is presented to the algorithm in series so that the elements of a sequence are processed in the same order as they are recorded.

Starting from the idea of [5] we move further to provide a more exhaustive vector-based GP. Our VE-GP is less problem-specific and aims at providing an algorithm able to deal with any naturally ordered variable of any length. Moreover, in VE-GP we have included new structures that advance the search space of the considered problems. VE-GP elevates the capabilities of previously proposed vector-based approaches and provides a more sophisticated technique.

3   Vectorial GP

VE-GP is built on top of the GPLAB toolbox [8]. GPLAB includes most of the traditional features usually found in many GP systems. We have chosen it because its highly modular structure makes it a particularly versatile, generalist and easily extendable tool, highly suited for testing new elements and techniques. Moreover, it is written in MATLAB, which provides a particularly appropriate environment to manage vectors. In the following paragraphs we describe the primitive set and other particular elements of VE-GP.

Functions of Arity One. Since VE-GP is specific for time series variables, we have integrated the set of classical arity one functions with aggregate functions. Standard aggregate functions collapse the whole time series variable into a single value of more significant meaning. They can be included in the primitive set when we deal with time series prediction based on past time series variables. We even face problems where the time series target flows simultaneously to the time series predictors, which means that the time instants, corresponding to the entries of a vector, are the same for both the target and the predictors. Therefore, specially meant for this latter problem, we have added cumulative aggregate functions. These operators applied on a time series return a vector whose entries are the aggregate values of only previous time values. These versions of the aggregate function, the standard and the cumulative ones, allow GP to foresee any kind of time series, from the ones that take place during the recording of data to the future ones. All the arity one functions can be also easily applied to scalars, considering them as a vector of 1×1 dimension. The primitives of arity one used by VE-GP are described in Table 3.

Functions of Arity Two. Concerning arity two we have included new functions inspired by classical vector operations. These functions can manage vectors of different lengths completing the shortest one with the null-element of the function involved. In the case of a scalar and a vector as inputs we have provided a specific evaluation in order not to consider scalars as 1×1 vectors. The primitives of arity two used by VE-GP are described in Table 4. The functions VSUMW, V_W, VprW and VdivW are called standard arity two functions.

Parametric Aggregate Functions. We have introduced parametric aggregate functions that apply the referring aggregate function only to the values belonging to the time window described by parameters. Regarding standard aggregate functions, the parameters p and q define respectively the initial and final position of the range to be considered. Therefore the standard aggregate function is applied to the input values of position $p, \ldots, q-1, q$. To have an admissible range

Table 3. Description of functions of arity one. The columns represent the primitive function (first column), MATLAB name (second column) and the outcome of the function (third column) applied on a given vector v (in this example, v=[1, 2.5, 4.3, 0.7]).

Primitive function (pf)	MATLAB name	pf(v)
Mean	V_mean	$V\_mean([1, 2.5, 4.3, 0.7]) = 2.1$
Max, Min	V_max, V_min	$V\_max([1, 2.5, 4.3, 0.7]) = 4.3$ $V\_min([1, 2.5, 4.3, 0.7]) = 0.7$
Sum	V_sum	$V\_sum([1, 2.5, 4.3, 0.7]) = 8.5$
Mode	V_mode	$V\_mode([1, 2.5, 4.3, 0.7]) = 0.7$
Length	V_length	$V\_length([1, 2.5, 4.3, 0.7]) = 4$
2-Norm	V_2norm	$V\_2norm([1, 2.5, 4.3, 0.7]) = 5.1$
Cumulative mean	C_mean	$C\_mean([1, 2.5, 4.3, 0.7]) = [1, 1.8, 2.6, 2.1]$
Cumulative sum	C_sum	$C\_sum([1, 2.5, 4.3, 0.7]) = [1, 3.5, 7.8, 8.5]$
Cumulative max, min	C_max, C_min	$C\_max([1, 2.5, 4.3, 0.7]) = [1, 2.5, 4.3, 4.3]$ $C\_min([1, 2.5, 4.3, 0.7]) = [1, 1, 1, 0.7]$
Exp, Log, Cos, Sin, Abs, Square, Cube, Sqrt	V_exp, V_log*, V_cos, V_sin, V_abs, V_2, V_3, V_sqrt* *(protected version as [1])	$V\_exp([1, 2.5, 4.3, 0.7]) = [2.7, 12.2, 73.7, 2.0]$ $V\_log([1, 2.5, 4.3, 0.7]) = [0, 0.9, 1.5, -0.4]$ $V\_cos([1, 2.5, 4.3, 0.7]) = [0.5, -0.8, -0.4, 0.8]$ $V\_sin([1, 2.5, 4.3, 0.7]) = [0.8, 0.6, -0.9, 0.6]$ $V\_abs([1, 2.5, 4.3, 0.7]) = [1, 2.5, 4.3, 0.7]$ $V\_2([1, 2.5, 4.3, 0.7]) = [1, 6.3, 18.5, 0.5]$ $V\_3([1, 2.5, 4.3, 0.7]) = [1, 15.6, 79.5, 0.3]$ $V\_sqrt([1, 2.5, 4.3, 0.7]) = [1, 1.6, 2.1, 0.8]$

$p < q$. Concerning cumulative aggregate functions, we remind that the output is a vector. The $i-$ith entry of the output depends on the values belonging to the window described by parameters p and q. In this case, p defines how far to look back from the i position determining the initial value of the range, while q defines how many values to consider. Thus, the $i-$ith entry of the output is the aggregate function applied on input values of position $i-(p-1), \ldots, i-(p-1)+q-1$. To have admissible range in this case $p > q$. The primitives of the parametric aggregate function used by VE-GP are described in Table 5.

It is noteworthy that many of the new functions are not replicable by the standard GP.

Table 4. Description of functions of arity two. The columns represent the primitive function (first column), MATLAB name (second column), the outcome of the function (third column) applied on a given scalar s and vector v1 and the outcome of the function (fourth column) applied on two vectors v1 and v2 (in this example, $s = 0.2$, $v1 = [1, 2.5, 4.3]$, $v2 = [0.1, 3.2, 4, 1.1]$).

Primitive function (pf)	MATLAB name	pf(s,v1)	pf(v1,v2)
Element-wise sum	VSUMW	VSUMW(0.2, [1, 2.5, 4.3]) = [1.2, 2.7, 4.5]	VSUMW([1, 2.5, 4.3], [0.1, 3.2, 4, 1.1]) = [1.1, 5.7, 8.3, 1.1]
Element-wise	V_W	V_W(0.2, [1, 2.5, 4.3]) = [−0.8, −2.3, −4.1]	V_W([1, 2.5, 4.3], [0.1, 3.2, 4, 1.1]) = [0.9, −0.7, 0.3 − 1.1]
Element-wise product	VprW	VprW(0.2, [1, 2.5, 4.3]) = [0.2, 0.5, 0.9]	VprW([1, 2.5, 4.3], [0.1, 3.2, 4, 1.1]) = [0.1, 8, 17.2, 1.1]
Scalar	VscalprW	VscalprW(0.2, [1, 2.5, 4.3]) = 1.6	VscalprW([1, 2.5, 4.3], [0.1, 3.2, 4, 1.1]) = 26.4
Element-wise division	VdivW* *(protected version as [1])	VdivW(0.2, [1, 2.5, 4.3]) = [0.2, 0.08, 0.05]	VdivW([1, 2.5, 4.3], [0.1, 3.2, 4, 1.1]) = [10, 0.8, 1.1, 0.9]
Scalar Division	VscaldivW* *(protected version as [1])	VscaldivW(0.2, [1, 2.5, 4.3]) = 0.3	VscaldivW([1, 2.5, 4.3], [0.1, 3.2, 4, 1.1]) = 12.8

Initialization. Given the new representation in VE-GP, several challenges arise. Firstly, a big number of scalar inputs can cause a poor initial representation of new functions and terminals, as such barely used during the evolutionary process. Secondly, it is possible to obtain final solutions whose output is a scalar and not a vector because many of the integrated functions collapse a vector into a scalar. We have designed a different initialization strategy which releases unique and innovative characteristics of VE-GP during the evolution. The strategy is resumed in the procedure described in Fig. 1.

The motivation behind the second rule is to ensure a representative amount of trees in the initial population whose output is not a scalar. Similarly, the third rule ensures trees where the new functions are meaningfully used. In our opinion, the initialization strategy that we propose does not introduce significant bias in the evolutionary process, contrarily, it aims to aid VE-GP to free its full potential during the evolution.

We have furthermore initialized the values of the parameters for the aggregate functions. Because a time series variable can have a different length among fitness cases, we have randomly set p and q between 1 and the maximum time series length for all the parametric aggregate functions.

Table 5. Description of parametric aggregate functions. The columns represent the primitive function (first column), MATLAB name (second column), the outcome of the function (third column) applied on a given vector. For standard functions p is the initial position while q is the final position of the range, instead for cumulative functions p is the number of backward steps to be made in order to determine the initial position of the range while q is the amplitude of the range (in this example, v=[1, 2.5, 4.3, 0.7, 1.6], $(p,q) = (2,3)$ for standard functions, $(p,q) = (3,2)$ for cumulative functions).

Primitive function (pf)	MATLAB name	$pf_{p,q}(\mathbf{v})$
Mean in [p,q]	$\texttt{V\_mean}_{p,q}$	$\texttt{V\_mean}_{2,3}([1, 2.5, 4.3, 0.7, 1.6]) = 3.4$
Max in [p,q], Min in [p,q]	$\texttt{V\_max}_{p,q}$, $\texttt{V\_min}_{p,q}$	$\texttt{V\_max}_{2,3}([1, 2.5, 4.3, 0.7, 1.6]) = 4.3$ $\texttt{V\_min}_{2,3}([1, 2.5, 4.3, 0.7, 1.6]) = 2.5$
Sum in [p,q]	$\texttt{V\_sum}_{p,q}$	$\texttt{V\_sum}_{2,3}([1, 2.5, 4.3, 0.7, 1.6]) = 6.8$
Cumulative mean in [p,q]	$\texttt{V\_Cmean}_{p,q}$	$\texttt{V\_Cmean}_{3,2}([1, 2.5, 4.3, 0.7, 1.6]) = [0, 1, 1.8, 3.4, 2.5]$
Cumulative max in [p,q], Cumulative min in [p,q]	$\texttt{V\_Cmax}_{p,q}$, $\texttt{V\_Cmin}_{p,q}$	$\texttt{V\_Cmax}_{3,2}([1, 2.5, 4.3, 0.7, 1.6]) = [0, 1, 2.5, 4.3, 4.3]$ $\texttt{V\_Cmin}_{3,2}([1, 2.5, 4.3, 0.7, 1.6]) = [0, 1, 1, 2.5, 0.7]$
Cumulative sum in [p,q]	$\texttt{V\_Csum}_{p,q}$	$\texttt{V\_Csum}_{3,2}([1, 2.5, 4.3, 0.7, 1.6]) = [0, 0, 3.5, 6.8, 5]$

Create an empty population P (the initial population) of size N.

1. Generate $n1$ trees in P using Ramped Half-and-Half initialization algorithm (RHH) [1];
2. Generate $n2$ trees in P using RHH such that each tree always generates an output which is a vector:
   (a) randomly generate a tree t by means of RHH;
   (b) randomly select a standard primitive function pf of arity two;
   (c) randomly select a vector-terminal v;
   (d) create the following tree, using post-fix notation, *(pf t v)*;
3. Generate $n3$ trees in P using RHH where aggregate primitive functions are forced to receive a vector-terminal as an input.

Fig. 1. Proposed initialization strategy.

Genetic Operators. VE-GP includes a new type of mutation. Besides the classical one there is the mutation of aggregate function parameters. This operator allows parameters to evolve so that the most informative window where to apply the relative aggregate function is found. Firstly, the algorithm searches the tree for parametric aggregate functions and randomly selects one. Secondly, a random parameter is chosen and mutated according to the procedure reported in Table 6. Every time a genetic operator requires a new tree, parameters are set according to the initialization default values.

Table 6. Parameters mutation.

Standard aggregate functions parameters p,q	Cumulative aggregate function parameters p,q
– Random selection of p or q; – If p randomly change it from 1 to q; – If q randomly change it from p to the maximum time series length	– Random selection of p or q; – If p randomly change it from 1 to the maximum time series length – If q randomly change it from 1 to p

4   Experiments

4.1   Benchmark Problems

We have tested the proposed VE-GP against a standard GP system (ST-GP) on four benchmark problems. To investigate the competitiveness of VE-GP we have chosen a first problem where the target does not involve the new primitive functions in order to see if VE-GP is penalized by having unnecessary functions and structures. Three more problems include some of the new functions in the target, and they are meant to show the performances of both algorithms considering that ST-GP can just try to approximate the new functions at its best.

Korns5. This benchmark problem is inspired by Korns problem number five for symbolic regression [9]. We have chosen to involve four variables of random numbers between -50 and 50 as the input, named $X1, X2, X3, X4$ respectively. Differently from the true Korns problem number five, the latter variable $X4$ for our experiment is a vector of length 10. The target expression is:

$$K5 = \text{VSUMW}(3.0, \text{VprW}(2.13, \text{V\_log}(X4))).$$

The dataset for VE-GP consists of 1000 instances, while for ST-GP it consists of 10000 instances because we have vertically untied the variable $X4$ and the target $K5$ to have the classical panel data representation.

Benchmark1. This benchmark problem is a new one that makes use of the aggregate functions implemented to produce the target. The vectorial dataset consists of 100 instances, where each instance is composed by four features: a random number between -10 and 10, another random number between -10 and 10, a vector of random numbers between 10 and 40 of length 10, and a vector of random numbers between -5 and 5 of length 10. Naming the variables in order as $X1, X2, X3$ and $X4$ the target reads as follows:

$$B1 = \text{V\_W}\Big(\text{VSUMW}(X4, \text{V\_mean}(X3)), \text{V\_W}(\text{VscalprW}(X3, X4), X2)\Big).$$

Conversely, for the scalar dataset, we have vertically untied the vectorial features so that it consists of 1000 instances.

Benchmark2. This benchmark problem involves the cumulative functions described in the previous section. The vectorial dataset is the same used for the previous benchmark B1, while the target is now

$$B2 = \text{VprW}\Big(\text{VdivW}(\text{VSUMW}(X3, X1), \text{V\_Cmean}(X4)), X2\Big).$$

Again for the scalar dataset we have vertically untied the vectorial variables.

Benchmark3. This benchmark problem includes parametric aggregate functions. Therefore the evolution of parameters is integrated in VE-GP. The variables involved are five: $X1$ is a vector of length 20 of random numbers between 10 and 30, $X2$ is a random number between 50 and 60, $X3$ is a random number between 5 and 10, $X4$ is a random number between -2 and 2, and $X5$ is a random number between 0 and 1. The target is:

$$B3 = \text{VSUMW}(\text{VprW}(\text{V\_Cmin}_{3,3}(X1), \text{VdivW}(X2, X3)), X4).$$

The dataset for VE-GP consists of 100 instances, while for ST-GP it consists of 1000 instances because, as for the other problems, we have vertically untied the vectorial variables.

4.2   Parameters and Statistical Test

The experimental parameters used in all the problems are provided in Tables 7 and 8. They were essentially the same for both ST-GP and VE-GP to facilitate the comparison between the techniques. We should remark that the choice of new terminal functions between cumulative or standard version depends on the chosen recording time of the time series involved. Fitness is calculated as the Root Mean Square Error (RMSE) between the output and the target. Since the output of trees built by VE-GP is supposed to be a vector, for this latter algorithm we have calculated the RMSE vertically disbanding both output and target; in this way the measures of fitness are ensured to be comparable between the two techniques. Moreover, when a VE-GP tree wrongly produces scalars as an output, each scalar is replicated until the length of the corresponding target to make it a vector. We have decided to penalize these trees by multiplying their fitness for a huge constant (100).

  We have tested both techniques on a total of 50 runs, each of which considers a different training and test data partition; from now on the term test set stands for unseen data. In particular, at the beginning of a VE-GP run, 70% of the instances are randomly selected as the training set, while the remaining ones form the test set. We have kept the same division for ST-GP with correspondence of the observations, so that time series are not split. We have performed a set of tests to analyse the statistical significance of the results. At first, the Kolmogorov-Smirnov test has shown that final fitness data are not normally distributed and hence we have opted for a rank-based statistic. Test decision for the null hypothesis of no difference in performance between ST-GP and VE-GP has been

Table 7. Standard GP parameters.

	K5	B1	B2	B3
Runs	50	50	50	50
Population	500	500	500	500
Generations	100	100	100	100
Training-Testing division	70%-30% of vectorial instances	70%-30% of vectorial instances	70%-30% of vectorial instances	70%-30% of vectorial instances
Genetic operators	Crossover, probability 0.9-Mutation, probability 0.1	Crossover, probability 0.9-Mutation, probability 0.1	Crossover, probability 0.9-Mutation, probability 0.1	Crossover, probability 0.9-Mutation, probability 0.1
Initialization	Ramped Half-and-Half [1], max depth 6	Ramped Half-and-Half [1], max depth 6	Ramped Half-and-Half [1], max depth 6	Ramped Half-and-Half [1], max depth 6
Functions set	plus, minus, times, protected div as [1]	plus, minus, times, protected div as [1]	plus, minus, times, protected div as [1]	plus, minus, times, protected div as [1]
Terminals set	Input variables, random numbers	Input variables, random numbers	Input variables, random numbers	Input variables, random numbers
Selection for reproduction	Lexicographic Parsimony Pressure [11], tournament size=10	Lexicographic Parsimony Pressure [11], tournament size=10	Lexicographic Parsimony Pressure [11], tournament size=10	Lexicographic Parsimony Pressure [11], tournament size=10
Elitism	Replication probability 0.1, best individual is kept	Replication probability 0.1, best individual is kept	Replication probability 0.1, best individual is kept	Replication probability 0.1, best individual is kept
Maximum depth	17	17	17	17

calculated with the Wilcoxon Rank Sum Test on the final test fitness. We have opted for it because it is a non parametric test and considers the median which is more robust than the mean to outliers. In order to quantify the assuming difference in performance between the two approaches, we have even used a Vargha Delaney A-test which is an index of effect size [10].

5   Results

In this section, we analyse the performance achieved by the two algorithms on the four problems. The evolution fitness plot (Fig. 2) shows the best fitness in each generation for the training and the test set, median of 50 runs. Besides evolution plots, there are boxplots based on the test fitness at the end of evolution. The statistical test comparing final test fitness between both techniques can be found in Table 9. In this table, p is the p-value of the Wilcoxon test with a 5% level of significance. The term A represents the value of the Vargha Delaney A-test. The test returns a number between 0 and 1, representing the probability that a randomly selected observation from the first sample is bigger than a randomly selected observation from the second sample. In our specific case, the first sample is formed by the best fitness found by ST-GP while the second sample is composed of the best fitness found by VE-GP. It is important to remember that fitness is measured via RMSE, therefore, the lower it is, the better performance it means. Vargha and Delaney in [10] provided a suggested threshold for interpreting the size of the difference: 0.5 means no difference at all, up to 0.56 indicates a small difference, up to 0.64 indicates medium and anything over 0.71 is large. The same intervals apply below 0.5.

Table 8. Vectorial GP parameters.

	K5	B1	B2	B3
Runs	50	50	50	50
Population	500	500	500	500
Generations	100	100	100	100
Training-Testing division	70%-30% of instances	70%-30% of instances	70%-30% of instances	70%-30% of instances
Genetic operators	Crossover, probability 0.9-Mutation, probability 0.1	Crossover, probability 0.9-Mutation, probability 0.1	Crossover, probability 0.9-Mutation, probability 0.1	Crossover, probability 0.5-Mutation, probability 0.1-Mutation of parameters, probability 0.4
Initialization	30% Ramped Half-and-Half [1], 70% Ramped Half-and-Half with forced initialization	30% Ramped Half-and-Half [1], 70% Ramped Half-and-Half with forced initialization	30% Ramped Half-and-Half [1], 70%Ramped Half-and-Half with forced initialization	30% Ramped Half-and-Half [1], 70%Ramped Half-and-Half with forced initialization
Functions set	VSUMW, V_W, VprW, VdivW, V_mean, V_max, V_min, V_sum	VSUMW, V_W, VprW, VdivW, V_mean, V_max, V_min, V_sum VscalprW	VSUMW, V_W, VprW, VdivW, V_Cmean, V_Cmax, V_Cmin, V_Csum	VSUMW, V_W, VprW, VdivW, V_Cmeanpq, V_Cmaxpq, V_Cminpq, V_Csumpq
Terminals set	Input variables, random numbers	Input variables, random numbers	Input variables, random numbers	Input variables, random numbers
Selection for reproduction	Lexicographic Parsimony Pressure [11], tournament size=10	Lexicographic Parsimony Pressure [11], tournament size=10	Lexicographic Parsimony Pressure [11], tournament size=10	Lexicographic Parsimony Pressure [11], tournament size=10
Elitism	Replication probability 0.1, best individual is kept	Replication probability 0.1, best individual is kept	Replication probability 0.1, best individual is kept	Replication probability 0.1, best individual is kept
Maximum depth	17	17	17	17

Table 9. Statistical results of final test fitness comparison.

K5	B1	B2	B3
$p = 0.14$	$p = 6.79 \cdot 10^{-18}$	$p = 1.08 \cdot 10^{-9}$	$p = 1.34 \cdot 10^{-12}$
$A = 0.41$	$A = 1$	$A = 0.85$	$A = 0.91$

Firstly, if we consider the K5 benchmark, there is no significant disparity in performance between the algorithms. This confirms our expectation since the target of the problem does not involve the new functions; the difference between techniques, thus, it is just in data representation. The VE-GP algorithm moreover is not affected by unnecessary improvement of the initialization step and by the extension of the primitive set. Table 9 and Fig. 3 show differently that VE-GP outperforms ST-GP for B1, B2, and B3. Moreover, Fig. 2 reveals an increasing error for the ST-GP test set on both B1 and B2 problems which means that overfitting is occurring. Therefore ST-GP is not able to understand the underlying relationship between the data. This phenomenon does not happen to VE-GP that increases in fitness during generation for both training and test data. A notable observation that emerges from the B2 evolution plot is the growing amplitude of percentiles. This consideration stresses the fact that every time ST-GP tries to extract the implicit relationship between data it fails, remaining stuck to high error levels. Concerning B3, the difficulty of finding the correct window of time emerges even from VE-GP, where percentiles show the presence

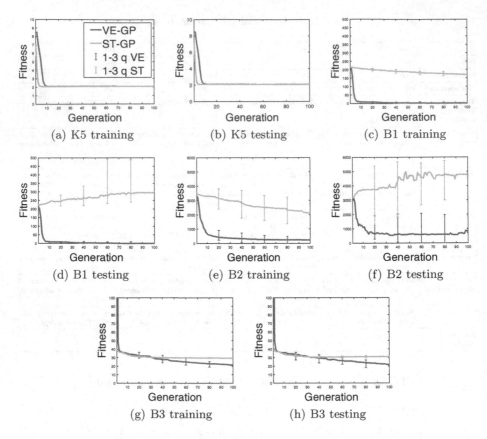

Fig. 2. ST-GP and VE-GP fitness evolution plots. The bars represent the first and the third quartile. K5 through (a) and (b), B1 through (c) and (d), B2 through (e) and (f) and B3 through (g) and (h).

of runs not able to overcome ST-GP in 50 generations. Nevertheless, at the end of the evolutionary process, every VE-GP model outperforms the ST-GP ones that stabilize at high error suggesting the idea of no future improvements.

6   Discussion

There is one issue that deserves attention. We have observed that VE-GP reveals its potential when the aggregate functions involved in the target generate variable values among fitness cases. To clarify the statement, we have considered the following benchmark problem

$$BBN = \mathtt{VSUMW}(-5.41, \mathtt{VprW}(4.9, \mathtt{VdivW}(\mathtt{VSUMW}(\mathtt{V\_max}(X4), X3), X1)))$$

where $X1$ and $X2$ are random numbers between 0 and 1, $X3$ is a vector of random numbers between -1 and 1 of length 10 and $X4$ is a vector of random numbers between 2 and 3 of length 10. In this case, the vectorial term

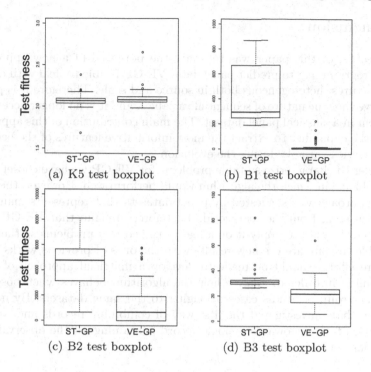

(a) K5 test boxplot   (b) B1 test boxplot

(c) B2 test boxplot   (d) B3 test boxplot

Fig. 3. ST-GP and VE-GP fitness boxplots for test set.

V_max($X4$) returns a value close to 3 for all the observations. Therefore ST-GP can approximate it with a constant despite the absence among primitives of the aggregate functions. We have investigated the performances of both GP method on this problem as before. The Wilcoxon rank test on the final fitness of the test set returns a p-value equal to 0.12 demonstrating that there is no difference in performances as expected.

The vector based approach of VE-GP suggests further developments and different applications from what we have treated in this paper. Although this work is specially meant for time-series vectors, the technique is in fact usable for every dataset that involves variables whose suitable representation is a vector. Moreover, we can even think about a matricial approach of genetic programming. Some variables, in fact, may change in space and time. Therefore, a matrix is a suitable representation for its records. Additionally, it is possible to group vectorial variables to catch intra-records dependencies. As for the VE-GP, we will introduce new functions inspired by classical matrix operations such as matrix row by column multiplication or determinant and we will enhance the different blocks of GP structure in order to exploit this new representation.

7   Conclusion

The objective of this paper was to study the potential of a new approach of genetic programming to predict panel data. VE-GP is able to deal with different data structures, heterogeneous both in source and scale. Therefore it is possible to preserve the true nature of sequential variables, and it is no longer necessary to untie them in a classical panel dataset. The main contribution of this approach is however the capability to extract the most informative features of the behaviour of a time series variable during the evolution.

In order to characterize suitable problems for VE-GP we have chosen benchmark problems in which the algorithm would perform well. However, the idea of VE-GP approach was suggested by panel datasets that represents many problems which stem from the real world. Therefore, to claim that VE-GP reveals advantages, we plan to apply it on already studied real problems so that it will be possible to compare the new results with the ones of previous works [12].

Future efforts should be made to develop a matricial approach of genetic programming in order to analyse how the algorithm behaves when posed with a different group of fitness cases belonging to the same datasets. By means of VE-GP we have investigated the first way of combining records and we would like to move further towards a single record containing all the observations as the training set.

Acknowledgments. This work was partially supported by FCT through funding of LASIGE Research Unit (UID/CEC/00408/2019), BioISI Research Unit (UID/MULTI/04046/2013), and projects INTERPHENO (PTDC/ASP-PLA/28-726/2017), PERSEIDS (PTDC/EMS-SIS/0642/2014), OPTOX (PTDC/CTA-AMB/30056/2017), BINDER (PTDC/CCI-INF/29168/2017), GADgET (DS-AIPA/DS/0022/2018) and PREDICT (PTDC/CCI-CIF/29877/2017).

References

1. Poli, R., Langdon, W., McPhee, N.: A Field Guide to Genetic Programming. Lulu Enterprises, UK Ltd. (2008). https://doi.org/10.1007/s10710-008-9073-y
2. Dermofal, D.: Time-series cross-sectional and panel data models. Spat. Anal. Soc. Sci. **32**, 141–157 (2015). https://doi.org/10.1017/CBO9781139051293.009
3. Guo, H., Jack, L.B., Nandi, A.K.: Automated feature extraction using genetic programming for bearing condition monitoring. In: Proceedings of the 14th IEEE Signal Processing Society Workshop Machine Learning for Signal Processing, pp. 519–528 (2004). https://doi.org/10.1109/MLSP.2004.1423015
4. De-Falco, I., Della-Cioppa, A., Tarantino, E.: A genetic programming system for time series prediction and its application to el niño forecast. Soft Comput.: Methodol. Appl.cations **32**, 151–162 (2005). https://doi.org/10.1007/3-540-32400-3_12
5. Holladay, K., Robbins, K.A.: Evolution of signal processing algorithm using vector based genetic programming. In: 15th International Conference on Digital Signal Processing, pp. 503–506 (2007). https://doi.org/10.1109/ICDSP.2007.4288629

6. Bartashevich, P., Bakurov, I., Mostaghim, S., Vanneschi, L.: Evolving PSO algorithm design in vector fields using geometric semantic GP. In: GECCO 2018: Proceedings of the Genetic and Evolutionary Computation Conference, pp. 262–263 (2018). https://doi.org/10.1145/3205651.3205760

7. Alfaro, E.C., Sharman, K., Esparcia-Alcázar, A.: Genetic programming and serial processing for time series classification. Evol. Comput. **22**, 265–285 (2013). https://doi.org/10.1162/EVCO_a_00110

8. Silva, S., Almeida, J.: GPLAB a genetic programming toolbox for MATLAB (2007). http://gplab.sourceforge.net/index.html

9. McDermott, J., O'Reilly, U.M., Luke, S., White, D.: A community-led effort towards improving experimentation in genetic programming. http://gpbenchmarks.org/

10. Vargha, A., Delaney, H.D.: A critique and improvement of the CL common language effect size statistics of McGraw and Wong. J. Educ. Behav. Stat. **25**(2), 101–132 (2000). https://doi.org/10.2307/1165329

11. Luke, S., Panait, L.: Lexicographic parsimony pressure. In: GECCO 2002: Proceedings of the Genetic and Evolutionary Computation Conference, pp. 829–836 (2002)

12. Bisanzio, D., et al.: Spatio-temporal patterns of distribution of West Nile virus vectors in eastern Piedmont region, Italy. Parasites Vectors 4, 230 (2011). https://doi.org/10.1186/1756-3305-4-230

Comparison of Genetic Programming Methods on Design of Cryptographic Boolean Functions

Jakub Husa[✉]

Faculty of Information Technology, IT4Innovations Centre of Excellence,
Brno University of Technology, Brno, Czech Republic
ihusa@fit.vutbr.cz

Abstract. The ever-increasing need for information security requires a constant refinement of contemporary ciphers. One of these are stream ciphers which secure data by utilizing a pseudo-randomly generated binary sequence. Generating a cryptographically secure sequence is not an easy task and requires a Boolean function possessing multiple cryptographic properties. One of the most successful ways of designing these functions is genetic programming. In this paper, we present a comparative study of three genetic programming methods, tree-based, Cartesian and linear, on the task of generating Boolean functions with an even number of inputs possessing good values of nonlinearity, balancedness, correlation immunity, and algebraic degree. Our results provide a comprehensive overview of how genetic programming methods compare when designing functions of different sizes, and we show that linear genetic programming, which has not been used for design of some of these functions before, is the best at dealing with increasing number of inputs, and creates desired functions with better reliability than the commonly used methods.

Keywords: Genetic programming · Cartesian Genetic programming · Linear Genetic programming · Cryptographic Boolean functions · Comparative study

1 Introduction

In 1882 Frank Miller, and later Gilbert Vernam, came up with the concept of a one-time pad and created a cipher which could under the right conditions be entirely unbreakable [1]. However for the cipher to work, it required the creation of pads of numbers which would be unique, truly random and could never be reused. These strict conditions made the cipher unfeasible for use in everyday life. However, its concept had survived and given birth to a family of stream ciphers.

These ciphers replace the one-time pads, with a single generator able to create a near infinite sequence so long that none of its parts would ever need

© Springer Nature Switzerland AG 2019
L. Sekanina et al. (Eds.): EuroGP 2019, LNCS 11451, pp. 228–244, 2019.
https://doi.org/10.1007/978-3-030-16670-0_15

to be reused. However, a deterministically generated sequence cannot be truly random. Instead, stream ciphers focus on making the sequence so complex that it can not be crypto-analyzed in feasible time. This requires a use of well designed specialized Boolean functions with cryptographic properties.

There are three main ways how to these functions can be created, random search, algebraic constructions, and heuristic methods (and their combinations) [2]. One of the most efficient heuristic methods are Evolutionary Algorithms. Inspired by the natural evolutionary process, these maintain a population of individuals each representing a potential solution. The individuals are then combined and mutated in a cycle guided by a fitness function, which determines the quality of each solution and steers the evolutionary process towards its goal.

In this paper, we focus on a specific subset of evolutionary algorithms called genetic programming (GP), which has been shown to provide great results in evolving Boolean functions of varying sizes and properties. The two most commonly used GP methods, tree-based, and Cartesian GP have already been the subject of multiple studies, while linear GP is usually overlooked. Another contribution of our paper is that we use two different population schemes for each GP method. To the best of our knowledge, this is a type of comparison none of the related works have performed before. Our aim is to provide a fair and comprehensive comparison of the individual GP methods and gain insight into what approach is most suited for each of the many various tasks.

The rest of this paper is organized as follows. Section 2 describes preliminaries of what Boolean functions are, how are they used in stream ciphers, and what are their cryptographic properties. In Sect. 3 we describe the various GP methods and go over related works that used GP to create Boolean functions in the past. In Sect. 4 we define our objectives, what fitness functions we use, how our experiments are set up, and how have the parameters used by each GP method been optimized. Section 5 shows the results of our experiments and highlights the most interesting findings. The work concludes with Sect. 6 which provides a summary and outlines the possible future works.

2   Preliminaries

Boolean function is a function $B^n \rightarrow B$ where $B \in \{0,1\}$ and $n \in \mathbb{N}$. In other words, it is a function which takes multiple binary inputs and provides a single binary output. The simplest way of representing a Boolean function is with a *truth table*, which assigns a specific output to every possible combination of inputs via a binary vector of length n^2, with n being the number of inputs [3].

Another way a binary function can be represented is the *Walsh spectrum*. It is defined as:

$$W_f(a) = \sum_{x \in \mathbb{F}_2^n} (-1)^{f(x) \oplus (x.a)} \qquad (1)$$

And shows the correlation (Hamming distance) between the examined function $f(x)$ and all linear functions $(a.x)$ [4]. This representation is useful for determining the function's nonlinearity (defined below). It can be calculated from the truth table representation using Fast Walsh–Hadamard transform in $O(n2^n)$, with n being the number of inputs [5].

The third way of representing Boolean functions is with the *algebraic normal form*, which uses a multivariate polynomial P defined as [3]:

$$P(x) = \oplus_{w \in \mathbb{F}_2^n} h(w).x^w \tag{2}$$

with $h(w)$ defined by the Möbius inversion principle [6]:

$$h(w) = \oplus_{x \preceq w} f(x), \text{ for any } w \in \mathbb{F}_2^n \tag{3}$$

In other words, algebraic normal form represents functions as a logical XOR of terms, which are themselves comprised of logical AND of the function's inputs. This representation allows to determine the function's algebraic degree (defined below), and for n inputs it can be calculated from the truth table representation in $O(n2^n)$.

2.1 Use of Boolean Functions

Boolean functions are used in stream ciphers when generating the pseudo-random sequence. The most commonly used type of generators use a Linear Feedback Shift Register (LFSR). It is comprised of a binary register of length m, initialized by a secret key and an initialization vector to some non-zero value. When generating a sequence, values in the register are shifted by one bit to the right. The right-most bit leaving the register is used as its output, and a new left-most bit is calculated by a (linear) generating polynomial from its current state. For a register of any common length, there is a well-known list of *primitive polynomials*, which will ensure that the generator will pass through every possible configuration except the state of all zeros before repeating itself, and thus generate a sequence with a period of $2^m - 1$. This means that for sufficiently long registers the sequence is effectively non-repeating [7].

LFSR generators are fast and easy to implement in both software and hardware and provide an output with good statistical properties [8,9]. However, its linear nature makes it susceptible to many cryptographic attacks. Given $2m$ bits of output, the initial setting of the register (and thus the secret key) can be reconstructed in $O(m^2)$ using the Berlekamp-Massey algorithm [10,11].

To ensure security the relationship between LFSR's internal state and its output needs to be obscured with a cryptographically sound Boolean function. Figure 1 shows the two main ways in which it can be applied, which determines what properties the function needs to possess.

The *combiner* model utilizes outputs of several short LFSRs of co-prime length and requires a Boolean function with a high degree of correlation immunity (defined below), equal to the number of LFSRs used [8]. The *filter* model utilizes several bits of a single long LFSR. This means that the Boolean function only requires a correlation immunity of 1, allowing for higher values of other cryptographic properties. However, the placement of input bits (tap positions) may open the generator to other types of cryptographic attacks, which are yet to be fully explored [12,13].

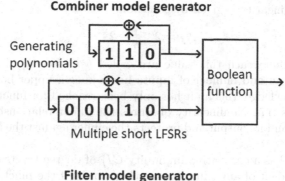

Combiner model generator

Generating polynomials

Boolean function

Multiple short LFSRs

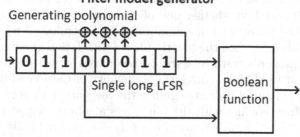

Filter model generator

Generating polynomial

Single long LFSR

Boolean function

Fig. 1. Use of Boolean functions in combiner and filter model of LFSR generators.

2.2 Property Definitions

In this paper, we focus on Boolean functions suitable for LFSR generators and possessing four cryptographic properties. Balancedness, nonlinearity, correlation immunity, and algebraic degree. Each property determines how difficult it will be to break the cipher using a specific type of attack. The individual properties are mutually conflicting, and when designing Boolean functions one always looks for a compromise. For a function to be directly applicable in a cipher it should also possess a high degree of Algebraic and Fast-Algebraic immunity, and have at least 13 inputs [14], though the ideal number of inputs is considered to be at least 20 [8]. But functions with fewer inputs and possessing only some of the cryptographic properties are still important, as they are used as building blocks for larger more secure functions constructed by analytical approaches [15].

A Boolean function is balanced if its truth table contains an equal number of ones and zeros, making its output statistically indistinguishable from a random sequence. An unbalanced function would cause a statistical relationship between plaintext and ciphertext of the secured messages, and make them vulnerable to attack by frequency analysis [8,16].

Linear Boolean function is a function that can be created by logical XOR of its inputs (including a constant 0 function). An affine function is either a linear function or its complement. Nonlinearity N_f of a function f is the minimal Hamming distance between the function's truth table and the truth table of any of the affine functions. For functions with an even number of inputs the maximum nonlinearity is given by equation:

$$N_f = 2^{n-1} - 2^{\frac{n}{2}-1} \tag{4}$$

Functions that reach this value are called *bent* functions. Nonlinearity of functions with an odd number of inputs has a known upper bound, but to the best of our knowledge, this limit has only been reached for functions with seven or fewer inputs [17]. Nonlinearity obscures the linear relationship between the state of LSFR and its output and protects the cipher against the Fast Correlation attack [16,18,19].

Function f has a correlation immunity CI_f of degree t if its output is statistically independent of any t inputs. In other words, if the function's truth table was split in half based on whether one of its inputs is 0 or 1, and both of the sub-tables would contain the same number of 1s, regardless of which input was used to create this split, then the function has a correlation immunity of the first degree. If the function's truth table could be quartered using a combination of any two inputs, it would have correlation immunity of the second degree, and so on. Correlation immunity protects against the Siegenthaler's correlation attack [20] and is in direct conflict with the function's algebraic degree Deg_f (defined below). If $t \in <2, n-1>$ then $Deg_f \leq n - CI_f - 1$, otherwise $Deg_f \leq n - CI_f$, where n is the number of function inputs. This limitation is known as Siegenthaler's Inequality [19]. If a function is both immune to correlation and balanced it is called *resilient*.

Algebraic degree Deg_f of function f is defined as the maximum number of elements in a single term when it's represented in the algebraic normal form (defined above). High algebraic degree protects the functions against the Berlekamp-Massey algorithm [10,11], Rønjom-Helleseth attack [21], and other algebraic attacks [22].

3   Genetic Programming Methods

Genetic programming is a subset of evolutionary algorithms focused on the evolution of executable structures. For Boolean functions, this usually means a logic equation used to calculate its truth table. This allows functions with many inputs to be stored in a compact form, and to be evolved with greater efficiency.

The most common form of GP is the Tree-based genetic programming (TGP). It represents its genotype using a syntactic tree consisting of two types of nodes. Internal (function) nodes that represent logical operators, and take their inputs from other function nodes or leaves. Leaf (terminal) nodes represent either the function's input variables or constants. TGP uses genotypes of variable length, but the maximum allowed depth of its tree may be restricted to prevent it from bloating to an unmanageable size [23].

Cartesian Genetic programming (CGP), represents the chromosome using a two-dimensional array of acyclically interconnected nodes. Each node contains a logical operator and several inputs that can be connected either to the function's input variables or a node in one of the previous columns. Because every node can be utilized in multiple data paths, performing genetic crossover is difficult, and offsprings are usually created utilizing only the mutation operator. In software implementations, the grid is often implemented as a single row, with no restrictions on how many levels-back can a node's input reach. The output is usually taken from a node specified by the last gene(s) of the genotype [24].

Linear Genetic programming (LGP) represents the chromosome as a linearly-executed sequence of instruction operating over a finite set of registers. Each instruction specifies a logical operator, several operands chosen from the set of registers and function inputs, and an output selected from the set of registers. The numbers of instructions and registers are mutually independent, and the values left in the registers after all instructions have been executed serve as the function output [25].

3.1   Related Works

The earliest application of evolutionary algorithms for the design of crypto-graphic Boolean functions focused on finding functions with high nonlinearity with a genetic algorithm [26]. TGP has first been used to design cryptographi-cally sound Boolean functions with 8 inputs [27], while CGP was used to search for bent Boolean functions with up 16 and later 18 inputs [28,29]. LGP was used to design bent functions with up to 24 inputs and has been shown to cope with increasing number of variables better than CGP [30].

There have been multiple studies comparing genetic algorithms, TGP, and CGP on the design of cryptographic Boolean functions applicable in stream ciphers, and functions with correlation immunity and minimal hamming weight useful in preventing side-channel attacks. All of these studies have shown that GP methods provide better results than the other evolutionary approaches [14,31–34].

In other works, TGP has been used to design algebraic constructions for combining existing bent Boolean functions into larger bent functions with up to 20 inputs [35], and a comparative study experimenting with designs of bent and resilient functions using CGP has shown that the latter may benefit from the use of various crossover operators [36].

4   Objectives

Our goal is to provide a comprehensive comparison of the three GP methods. For this reason, we choose several tasks of varying difficulty. To see how each method copes with growing search space we use Boolean functions of 6, 8, 10, and 12 inputs. To see how they handle increasingly restrictive criteria, we design four types of functions. Bent Boolean functions maximizing the nonlinearity property. Balanced functions with high nonlinearity. Resilient functions with correlation immunity of the first degree. And finally, resilient functions with high algebraic degree approaching the Siegenthaler's inequality (Siegenthaler functions).

For each type of a Boolean function, we define a fitness function that we try to maximize. Because the performance of evolutionary algorithms depends on how well is the fitness function able to guide its search, we use not only the raw values of cryptographic properties but define coefficients that show how close a solution is to meeting the specified criteria.

For balancedness we define coefficient:

$$BAL = 1 - \frac{|ONES - ZEROS|}{2^n} \tag{5}$$

Where ONES and ZEROS are the number of 1s and 0s in the function's truth table, and n is the number of function inputs. For Correlation immunity we define coefficient:

$$CRI = \frac{SPLIT}{n} \tag{6}$$

Where SPLIT is the number of inputs that can split the Boolean function's truth table and create two halves with an equivalent number of 1s. For algebraic degree we define coefficient:

$$DEG = max(1, \frac{Deg_f}{n-2}) \tag{7}$$

The value $n-2$ is one degree less than the limit determined by Siegenthaler's inequality to allow the evolved Boolean functions to also be balanced and highly nonlinear. Lastly, we define an acceptability coefficient:

$$ACC = \begin{cases} 1 & \text{if the function meets all criteria} \\ \frac{1}{2^{n-1}} & \text{otherwise} \end{cases} \tag{8}$$

This coefficient is used to reduce the fitness of Boolean functions that do not represent acceptable solutions to be within the range $<0,1>$, which is worse than the fitness of any acceptable solution.

Using the coefficients and the raw value of nonlinearity, we define the fitness function for each of the four types of functions being evolved:

$$F_{Bent} = N_f * ACC \tag{9}$$
$$F_{Balanced} = N_f * ACC * BAL \tag{10}$$
$$F_{Resilient} = N_f * ACC * BAL * CRI \tag{11}$$
$$F_{Siegenthaler} = N_f * ACC * BAL * CRI * DEG \tag{12}$$

Another possible way of combining the multiple criteria would be to utilize a multi-objective algorithm. However, related works suggest that this approach is not competitive when designing Boolean functions with cryptographic properties [33, 34], and so we leave this option out of the scope of this paper.

4.1   Experimental Setup

The various GP methods usually manage their population in different ways. To make our comparison fair, we perform all experiments using two different population schemes.

First scheme is the Steady-state Tournament (SST). Its initial population is generated randomly, and new individuals are created by randomly selecting thee individuals, and replacing the worst of them with a child of the better two. The child is created by a crossover followed by mutation. TGP uses standard subtree crossover that replaces a randomly selected function node (and its subtree) from one parent with a randomly selected function node (and its subtree) from the other parent. CGP and LGP use standard one-point crossover set to only split the chromosomes between whole nodes, respectively instructions.

Second scheme is the $(1 + \lambda)$ Evolution strategy (EST). Its initial population is randomly generated and new individuals are created in generations, by selecting the currently best individual and creating λ offsprings via mutation. If the parent and any of its offspring have the same fitness, the offsprings are preferred when choosing a new parent of the next generation, to promote fitness-neutral mutations. TGP mutates individuals by replacing randomly selected function node with a new, randomly generated subtree. CGP and LGP perform mutation by changing any of its individual genes to a randomized value with a small probability.

To increase the overall informative value of our comparison, we include a fourth evolutionary method. A genetic algorithm (GAL), whose chromosome represents the Boolean function's truth table directly, and not as an equation. It uses a one-point crossover and mutates individuals by flipping any number of individual bits in its chromosome to their opposite value, with a small probability.

Table 1. Fitness values desired for each type of function and number of inputs.

Inputs	Bent	Balanced	Resilient	Siegenthaler
6	28	24	24	24
8	120	112	112	112
10	496	480	480	480
12	2016	1984	1984	1984

For each evolved type of function, we choose a desired nonlinearity (and a corresponding fitness). For bent functions, we use the optimal value given by Eq. 4. For the rest of the functions we determine the desired nonlinearity experimentally by running each algorithm for 1 000 000 evaluations, and selecting the best value reached to be our goal, as shown in Table 1. The selected values are lower than some of the known lower bounds on maximum nonlinearity [37].

However, while we believe that these values can be reached using GP as well, the number of evaluations required would make our type comparison unfeasible.

All GP methods use {AND, OR, XOR, XNOR} as their set of operators, and had two logical constant {TRUE, FALSE} added to their Boolean function's inputs. The project is implemented in C++ using the Evolutionary Computation Framework[1], parallelized with Message Passing Interface[2]. During parallel execution, the main core maintains the population and performs the quick and easy tasks of selection, crossover, and mutation, while the computationally expensive task of obtaining each individual's truth table, cryptographic properties, and the resulting fitness, is passed to a number of worker cores.

4.2   Parameter Optimization

To determine each method's best performing setup, we use a one-at-a-time optimization method. GP methods were optimized on the task of evolving bent functions with 12 inputs. Because the size of a function's truth table is determined by the number of its inputs the chromosome length of GAL could not be optimized, and is included only for comparison. Because the ideal mutation rate is highly dependent on the chromosome length, it was optimized for each number of inputs separately. The examined ranges and the best values found are shown in Table 2.

For SST scheme, a medium sized population of around 15–25 individuals performed the best, with the exception of TGP method which performed even better

Table 2. Parameters optimized for each genetic programming method, the genetic algorithm with different number of inputs, and for both population schemes.

Optimized property	Examined range	TGP	CGP	LGP	GAL6	GAL8	GAL10	GAL12
$(1 + \lambda)$ Evolution Strategy								
Population	1 + 1–1000	1 + 5	1 + 5	1 + 5	1 + 5	1 + 5	1 + 5	1 + 5
Mutation rate	0.00025–1.0	1.0	0.035	0.025	0.04	0.008	0.0025	0.001
Chrom. length	31–4095	–	511	511	(64)	(256)	(1024)	(4096)
Tree depth	5–12	9	–	–	–	–	–	–
Free registers	5–100	–	–	15	–	–	–	–
Steady-State Tournament								
Population	3–1000	40	20	20	20	20	20	20
Mutation rate	0.00025–1.0	1.0	0.02	0.0125	0.04	0.008	0.0025	0.001
Chrom. length	31–4095	–	511	511	(64)	(256)	(1024)	(4096)
Tree depth	5–12	9	–	–	–	–	–	–
Free registers	5–100	–	–	15	–	–	–	–

---

[1] http://ecf.zemris.fer.hr/.

[2] https://www.open-mpi.org/.

with a larger population of around 35–50 individuals, possibly due to utilizing a different crossover operator than the other methods. This result is somewhat surprising, as TGP commonly performs the best with a population containing hundreds or thousands of individuals [38]. For EST scheme, the smallest populations always performed the best, but we have chosen a population with 5 offsprings to allow for a reasonable degree of parallelization.

For TGP method the genotype size was set as the maximum depth of its tree. For CGP and LGP we have worked with the number of nodes (respectively instructions) and restricted our search to genotype lengths equivalent to a fully grown TGP tree ($2^n - 1$). For CGP we have used a single-row implementation with no limit on each node's reach. For all GP methods, the optimal result was equivalent to a maximum of 511 logical operators, implying that this value depends on the task itself, rather than the method being used.

For CGP and LGP, the ideal mutation rate depended on the population scheme and was higher for EST, which uses mutation as its only genetic operator. For TGP and GAL, the mutation rate was the same for both population schemes. For LGP we have also optimized the number of available registers and found the optimal value to be around 15, including the final register whose value is used as the function output.

5   Results

The experiments include a multitude of tasks. We use four Evolutionary algorithms, GAL, TGP, CGP, and LGP, two population schemes SST and EST, evolve four types of Boolean functions, bent, balanced, resilient, and Siegenthaler, and consider use 6, 8, 10, and 12 inputs for each. Combination of all these setups results in 128 individual tasks. For each of these, we have performed 100 independent runs. Each run is limited to a maximum of 1 000 000 evaluation, and if it does not find a solution before this limit, the run is considered unsuccessful.

Our basis of comparison for the results is the number of fitness function evaluations required to find a Boolean function with the desired fitness. The complete set of results is shown in Tables 3, 4, 5, and 6 (one for each type of Boolean function).

All four tables order the experiments based on the number of inputs, the GP method, and population scheme. We consider the median outcome to be the most telling indicator of success and the first and third quartile as secondary indicators. These values are shown in columns "Q1", "Median", and "Q3", rounded to the nearest integer value. The last column labeled "Suc." shows how many of the 100 runs have been successful.

Our results show that the GAL, which was included to provide a comparison between GP and non-GP evolutionary algorithms, performs poorly and was only able to create bent functions of 6, and balanced functions of 6 and 8 inputs. It fails completely when required to create functions possessing correlation immunity, and for all other setups fails to provide even a single viable result.

Table 3. Experiment results of designing bent Boolean functions.

Method	Q1	Median	Q3	Suc.	Method	Q1	Median	Q3	Suc.
6 inputs									
GAL-EST	14372	31131	67776	100	GAL-SST	9915	30385	70250	100
TGP-EST	399	894	2157	100	TGP-SST	365	485	605	100
CGP-EST	161	331	642	100	CGP-SST	245	425	685	100
LGP-EST	298	634	1333	100	LGP-SST	305	535	1325	100
8 inputs									
GAL-EST	1000000	1000000	1000000	0	GAL-SST	1000000	1000000	1000000	0
TGP-EST	922	2179	3853	100	TGP-SST	565	765	1045	100
CGP-EST	290	531	1159	100	CGP-SST	440	615	1010	100
LGP-EST	624	1359	2496	100	LGP-SST	860	1755	4030	100
10 inputs									
GAL-EST	1000000	1000000	1000000	0	GAL-SST	1000000	1000000	1000000	0
TGP-EST	2016	4311	8710	100	TGP-SST	805	1085	1415	100
CGP-EST	488	934	1718	100	CGP-SST	675	1165	1945	100
LGP-EST	1000	1716	2970	100	LGP-SST	1525	2725	4975	100
12 inputs									
GAL-EST	1000000	1000000	1000000	0	GAL-SST	1000000	1000000	1000000	100
TGP-EST	3220	6879	12757	98	TGP-SST	1125	1385	1815	100
CGP-EST	1237	1994	3211	100	CGP-SST	960	1495	2345	100
LGP-EST	2165	3701	5899	100	LGP-SST	2185	3715	7045	100

Table 4. Experiment results of designing balanced Boolean functions.

Method	Q1	Median	Q3	Suc.	Method	Q1	Median	Q3	Suc.
6 inputs									
GAL-EST	26	41	76	100	GAL-SST	25	65	105	100
TGP-EST	91	204	457	100	TGP-SST	125	185	245	100
CGP-EST	35	61	131	100	CGP-SST	185	265	405	100
LGP-EST	80	124	329	100	LGP-SST	105	155	225	100
8 inputs									
GAL-EST	1451	3444	7216	100	GAL-SST	1805	3045	4735	100
TGP-EST	370	701	1143	100	TGP-SST	325	445	565	100
CGP-EST	115	209	404	100	CGP-SST	305	425	715	100
LGP-EST	195	364	720	100	LGP-SST	265	465	855	100
10 inputs									
GAL-EST	1000000	1000000	1000000	0	GAL-SST	1000000	1000000	1000000	0
TGP-EST	731	1301	2066	99	TGP-SST	525	645	805	100
CGP-EST	210	334	522	100	CGP-SST	305	425	715	100
LGP-EST	321	674	1028	100	LGP-SST	565	985	1815	100
12 inputs									
GAL-EST	1000000	1000000	1000000	0	GAL-SST	1000000	1000000	1000000	0
TGP-EST	1522	2289	4082	100	TGP-SST	765	965	1175	100
CGP-EST	416	619	1095	100	CGP-SST	505	715	965	100
LGP-EST	700	1191	1814	100	LGP-SST	975	1635	2515	100

Table 5. Experiment results of designing resilient Boolean functions.

Method	Q1	Median	Q3	Suc.	Method	Q1	Median	Q3	Suc.
6 inputs									
GAL-EST	1000000	1000000	1000000	0	GAL-SST	1000000	1000000	1000000	0
TGP-EST	195	359	677	100	TGP-SST	205	325	455	100
CGP-EST	116	217	341	100	CGP-SST	145	255	410	100
LGP-EST	217	474	928	100	LGP-SST	245	535	1335	100
8 inputs									
GAL-EST	1000000	1000000	1000000	0	GAL-SST	1000000	1000000	1000000	0
TGP-EST	625	966	1537	100	TGP-SST	525	625	805	100
CGP-EST	246	441	877	100	CGP-SST	325	565	935	100
LGP-EST	688	1139	1929	100	LGP-SST	480	1095	2590	100
10 inputs									
GAL-EST	1000000	1000000	1000000	0	GAL-SST	1000000	1000000	1000000	0
TGP-EST	1231	1706	2387	100	TGP-SST	805	1045	1245	100
CGP-EST	451	861	1377	100	CGP-SST	605	1005	1905	100
LGP-EST	822	1546	2647	100	LGP-SST	1185	2015	3870	100
12 inputs									
GAL-EST	1000000	1000000	1000000	0	GAL-SST	1000000	1000000	1000000	0
TGP-EST	2091	3106	4901	100	TGP-SST	1085	1365	1735	100
CGP-EST	886	1509	2472	100	CGP-SST	1045	1795	3070	100
LGP-EST	1311	2184	4352	100	LGP-SST	1460	2755	5415	100

Table 6. Experiment results of designing Siegenthaler Boolean functions.

Method	Q1	Median	Q3	Suc.	Method	Q1	Median	Q3	Suc.
6 inputs									
GAL-EST	1000000	1000000	1000000	0	GAL-SST	1000000	1000000	1000000	0
TGP-EST	785	1944	3550	100	TGP-SST	1145	1745	3335	100
CGP-EST	507	1104	1874	100	CGP-SST	745	1315	2670	100
LGP-EST	826	1686	3822	100	LGP-SST	1080	2025	6400	100
8 inputs									
GAL-EST	1000000	1000000	1000000	0	GAL-SST	1000000	1000000	1000000	0
TGP-EST	4336	8189	17444	100	TGP-SST	8615	18085	35935	100
CGP-EST	4897	9219	18703	100	CGP-SST	7995	15465	41335	100
LGP-EST	7359	14892	25307	100	LGP-SST	8000	19975	52290	100
10 inputs									
GAL-EST	1000000	1000000	1000000	0	GAL-SST	1000000	1000000	1000000	0
TGP-EST	19424	56054	196702	85	TGP-SST	50765	116905	327265	93
CGP-EST	26188	63944	125741	100	CGP-SST	36360	76195	236620	95
LGP-EST	33291	67382	130232	100	LGP-SST	37445	77385	253780	96
12 inputs									
GAL-EST	1000000	1000000	1000000	0	GAL-SST	1000000	1000000	1000000	0
TGP-EST	81635	214414	1000000	72	TGP-SST	205905	529585	1000000	73
CGP-EST	112127	233049	529320	92	CGP-SST	165990	537935	1000000	72
LGP-EST	100285	198551	359006	97	LGP-SST	92080	262425	631395	87

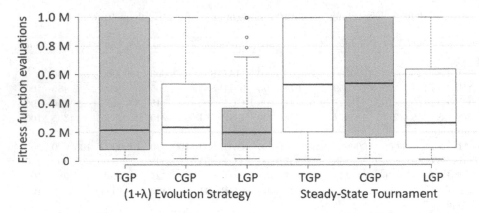

Fig. 2. Comparison of the three genetic programming methods on the task of designing Siegenthaler Boolean functions with 12 inputs, for two different population schemes.

Tables 3, 4, 5 show the importance of crossover for TGP. Making a comparison between the two population schemes, we see that only using mutation increases the median number of evaluations by 2–5 times, as well as severely impacting TGP's reliability, making it the only GP method with less than 100% success rate when designing bent and balanced Boolean functions. Conversely, the results shown in Table 6 illustrate that the use of SST scheme can also have a negative effect and that different population schemes may be ideal for different tasks, even when utilizing the same GP method.

The results provided by CGP show that it indeed performs better with a population scheme that uses mutation as its only genetic operator. The design of 12-input bent functions being the only task where CGP performed better using SST than the EST setup.

The most interesting results, however, are shown in Table 6. LGP which throughout most of the experiments performed in a manner similar to CGP but worse. It manages to be greatly successful with the design of Siegenthaler functions. Though still performing poorly when evolving Siegenthaler functions with a small a number of inputs, LGP managed to cope with additional inputs significantly better than other approaches and over-performed them both when designing Siegenthaller functions with 12 inputs, as is highlighted in Fig. 2. In addition, LGP had the greatest number of successful runs for both population schemes and functions of 10 and 12 inputs.

6 Conclusion and Future Work

In this paper, we have discussed the use of Boolean functions in stream ciphers and examined how suitable various GP methods are for designing them. We evolved functions possessing cryptographically significant properties of nonlinearity, balancedness, correlation immunity, and algebraic degree, with an even number of 6 to 12 inputs. For each GP method, we have implemented two different population schemes, Steady-State Tournament and $(1 + \lambda)$ Evolutionary Strategy, and to provide a fair comparison, we have performed a one-at-a-time parameter optimization to find the most suitable setup for each method and scheme.

Our results have confirmed that genetic algorithms which evolve functions on the truth table level cope poorly with an increased number of function inputs and can not compete with GP approaches. All three examined GP methods have been shown as competitive, with TGP being best suited for bent and resilient functions, CGP for balanced functions, and LGP for resilient functions with algebraic degree approaching the Siegenthaler's inequality.

For CGP and LGP, use of the SST scheme failed to provide better performance than EST, implying that a simple one-point crossover is not sufficiently complex to improve their performance. Our results also do not confirm that LGP is better than CGP at dealing with an increasing number of function inputs when designing bent Booleans functions. However, they do show this to be the case when evolving resilient functions with a high algebraic degree approaching the Siegenthaler's inequality.

To the best of our knowledge, this is the first work to design balanced, resilient and Siegenthaler functions using LGP, and to experiment with multiple population schemes for each of the GP methods while designing cryptographic Boolean functions. Thanks to this we have shown that the ideal choice of population scheme depends not only on the GP method but also on the type of function being evolved.

Still, our work is just one step in the exploration of the diverse and difficult domain of cryptographic Boolean functions. Future works could expand upon it by including other cryptographic criteria like algebraic and fast algebraic immunity, or focus on the evolution of functions with other cryptographical uses, like the design of Boolean functions with high correlation immunity and minimal hamming weight that can provide protection against side channel cryptographic attacks. Lastly, it would be interesting to experiment with other population schemes and examine their influence on each GP method and type of Boolean function evolved.

Acknowledgments. This work was supported by Czech Science Foundation project 19-10137S.

References

1. Vernam, G.S.: Cipher printing telegraph systems: for secret wire and radio telegraphic communications. J. AIEE **45**(2), 109–115 (1926)
2. Goossens, K.: Automated creation and selection of cryptographic primitives. Master's thesis, Katholieke Universiteit Leuven (2005)

3. Picek, S., Marchiori, E., Batina, L., Jakobovic, D.: Combining evolutionary computation and algebraic constructions to find cryptography-relevant Boolean functions. In: Bartz-Beielstein, T., Branke, J., Filipič, B., Smith, J. (eds.) PPSN 2014. LNCS, vol. 8672, pp. 822–831. Springer, Cham (2014). https://doi.org/10.1007/978-3-319-10762-2_81

4. Forrié, R.: The strict avalanche criterion: spectral properties of Boolean functions and an extended definition. In: Goldwasser, S. (ed.) CRYPTO 1988. LNCS, vol. 403, pp. 450–468. Springer, New York (1990). https://doi.org/10.1007/0-387-34799-2_31

5. Fino, B.J., Algazi, V.R.: Unified matrix treatment of the fast Walsh-Hadamard transform. IEEE Trans. Comput. C-25(11), 1142–1146 (1976)

6. Meier, W., Pasalic, E., Carlet, C.: Algebraic attacks and decomposition of Boolean functions. In: Cachin, C., Camenisch, J.L. (eds.) EUROCRYPT 2004. LNCS, vol. 3027, pp. 474–491. Springer, Heidelberg (2004). https://doi.org/10.1007/978-3-540-24676-3_28

7. Wu, H.: Cryptanalysis and design of stream ciphers. A Ph.D. thesis of Katholieke Universiteit Leuven, Belgium (2008)

8. Carlet, C.: Boolean functions for cryptography and error correcting codes. Boolean Models Meth. Math. Comput. Sci. Eng. 2, 257–397 (2010)

9. Armknecht, F.: Algebraic attacks on certain stream ciphers. Ph.D. thesis, University of Rennes (2006)

10. Massey, J.: Shift-register synthesis and BCH decoding. IEEE Trans. Inf. Theory 15(1), 122–127 (1969)

11. Norton, G.H.: The Berlekamp-Massey algorithm via minimal polynomials. arXiv preprint arXiv:1001.1597 (2010)

12. Didier, F.: Attacking the filter generator by finding zero inputs of the filtering function. In: Srinathan, K., Rangan, C.P., Yung, M. (eds.) INDOCRYPT 2007. LNCS, vol. 4859, pp. 404–413. Springer, Heidelberg (2007). https://doi.org/10.1007/978-3-540-77026-8_32

13. Hodžić, S., Wei, Y., Pašalić, E., Bajrić, S.: Optimizing the placement of tap positions. Ph.D. thesis, Univerza na Primorskem, Fakulteta za matematiko, naravoslovje in informacijske tehnologije (2015)

14. Picek, S., Jakobovic, D., Miller, J.F., Batina, L., Cupic, M.: Cryptographic boolean functions: one output, many design criteria. Appl. Soft Comput. 40, 635–653 (2016)

15. Carlet, C., Feng, K.: An infinite class of balanced functions with optimal algebraic immunity, good immunity to fast algebraic attacks and good nonlinearity. In: Pieprzyk, J. (ed.) ASIACRYPT 2008. LNCS, vol. 5350, pp. 425–440. Springer, Heidelberg (2008). https://doi.org/10.1007/978-3-540-89255-7_26

16. Chose, P., Joux, A., Mitton, M.: Fast correlation attacks: an algorithmic point of view. In: Knudsen, L.R. (ed.) EUROCRYPT 2002. LNCS, vol. 2332, pp. 209–221. Springer, Heidelberg (2002). https://doi.org/10.1007/3-540-46035-7_14

17. Kavut, S., Maitra, S., Yücel, M.D.: There exist Boolean functions on n (odd) variables having nonlinearity $> 2^{n-1} - 2^{\frac{n-1}{2}}$ if and only if $n > 7$ (2006)

18. Canteaut, A., Trabbia, M.: Improved fast correlation attacks using parity-check equations of weight 4 and 5. In: Preneel, B. (ed.) EUROCRYPT 2000. LNCS, vol. 1807, pp. 573–588. Springer, Heidelberg (2000). https://doi.org/10.1007/3-540-45539-6_40

19. Braeken, A.: Cryptographic properties of Boolean functions and S-boxes. Ph.D. thesis (2006)

20. Tarannikov, Y., Korolev, P., Botev, A.: Autocorrelation coefficients and correlation immunity of Boolean functions. In: Boyd, C. (ed.) ASIACRYPT 2001. LNCS, vol. 2248, pp. 460–479. Springer, Heidelberg (2001). https://doi.org/10.1007/3-540-45682-1_27

21. Ronjom, S., Helleseth, T.: A new attack on the filter generator. IEEE Trans. Inf. Theory 53(5), 1752–1758 (2007)

22. Courtois, N.T., Meier, W.: Algebraic attacks on stream ciphers with linear feedback. In: Biham, E. (ed.) EUROCRYPT 2003. LNCS, vol. 2656, pp. 345–359. Springer, Heidelberg (2003). https://doi.org/10.1007/3-540-39200-9_21

23. Cramer, N.L.: A representation for the adaptive generation of simple sequential programs. In: Proceedings of the First International Conference on Genetic Algorithms, pp. 183–187 (1985)

24. Kalkreuth, R., Rudolph, G., Droschinsky, A.: A new subgraph crossover for Cartesian genetic programming. In: McDermott, J., Castelli, M., Sekanina, L., Haasdijk, E., García-Sánchez, P. (eds.) EuroGP 2017. LNCS, vol. 10196, pp. 294–310. Springer, Cham (2017). https://doi.org/10.1007/978-3-319-55696-3_19

25. Brameier, M.: On linear genetic programming. Ph.D. thesis, Universitätsbibliothek Technische Universität Dortmund (2004)

26. Millan, W., Clark, A., Dawson, E.: An effective genetic algorithm for finding highly nonlinear boolean functions. In: Han, Y., Okamoto, T., Qing, S. (eds.) ICICS 1997. LNCS, vol. 1334, pp. 149–158. Springer, Heidelberg (1997). https://doi.org/10.1007/BFb0028471

27. Picek, S., Jakobovic, D., Golub, M.: Evolving cryptographically sound Boolean functions. In: Proceedings of the 15th Annual Conference Companion on Genetic and Evolutionary Computation, pp. 191–192. ACM (2013)

28. Hrbacek, R., Dvorak, V.: Bent function synthesis by means of Cartesian genetic programming. In: Bartz-Beielstein, T., Branke, J., Filipič, B., Smith, J. (eds.) PPSN 2014. LNCS, vol. 8672, pp. 414–423. Springer, Cham (2014). https://doi.org/10.1007/978-3-319-10762-2_41

29. Hrbacek, R.: Bent functions synthesis on Intel Xeon Phi coprocessor. In: Hliněný, P., et al. (eds.) MEMICS 2014. LNCS, vol. 8934, pp. 88–99. Springer, Cham (2014). https://doi.org/10.1007/978-3-319-14896-0_8

30. Husa, J., Dobai, R.: Designing bent Boolean functions with parallelized linear genetic programming. In: Proceedings of the Genetic and Evolutionary Computation Conference Companion, pp. 1825–1832. ACM (2017)

31. Picek, S., Jakobovic, D., Miller, J.F., Marchiori, E., Batina, L.: Evolutionary methods for the construction of cryptographic Boolean functions. In: Machado, P., et al. (eds.) EuroGP 2015. LNCS, vol. 9025, pp. 192–204. Springer, Cham (2015). https://doi.org/10.1007/978-3-319-16501-1_16

32. Picek, S., Carlet, C., Jakobovic, D., Miller, J.F., Batina, L.: Correlation immunity of Boolean functions: an evolutionary algorithms perspective. In: Proceedings of the 2015 Annual Conference on Genetic and Evolutionary Computation, pp. 1095–1102. ACM (2015)

33. Picek, S., Carlet, C., Guilley, S., Miller, J.F., Jakobovic, D.: Evolutionary algorithms for Boolean functions in diverse domains of cryptography. Evol. Comput. 24(4), 667–694 (2016)

34. Picek, S., Guilley, S., Carlet, C., Jakobovic, D., Miller, J.F.: Evolutionary approach for finding correlation immune Boolean functions of order t with minimal hamming weight. In: Dediu, A.-H., Magdalena, L., Martín-Vide, C. (eds.) TPNC 2015. LNCS, vol. 9477, pp. 71–82. Springer, Cham (2015). https://doi.org/10.1007/978-3-319-26841-5_6

35. Picek, S., Jakobovic, D.: Evolving algebraic constructions for designing bent Boolean functions. In: Proceedings of the Genetic and Evolutionary Computation Conference 2016, pp. 781–788. ACM (2016)
36. Husa, J., Kalkreuth, R.: A comparative study on crossover in Cartesian genetic programming. In: Castelli, M., Sekanina, L., Zhang, M., Cagnoni, S., García-Sánchez, P. (eds.) EuroGP 2018. LNCS, vol. 10781, pp. 203–219. Springer, Cham (2018). https://doi.org/10.1007/978-3-319-77553-1_13
37. Zhang, W., Pasalic, E.: Improving the lower bound on the maximum nonlinearity of 1-resilient boolean functions and designing functions satisfying all cryptographic criteria. Inf. Sci. **376**, 21–30 (2017)
38. Eiben, A.E., Smith, J.E., et al.: Introduction to Evolutionary Computing, vol. 53. Springer, Heidelberg (2003). https://doi.org/10.1007/978-3-662-05094-1

Evolving AVX512 Parallel C Code Using GP

William B. Langdon[1(✉)] and Ronny Lorenz[2]

[1] Computer Science, CREST, UCL, London WC1E 6BT, UK
W.Langdon@cs.ucl.ac.uk
[2] Institute for Theoretical Chemistry, University of Vienna, 1090 Vienna, Austria
http://www.cs.ucl.ac.uk/staff/W.Langdon/,
https://www.tbi.univie.ac.at/~ronny/cv.html,
http://crest.cs.ucl.ac.uk/, http://www.tbi.univie.ac.at/

Abstract. Using 512 bit Advanced Vector Extensions, previous development history and Intel documentation, BNF grammar based genetic improvement automatically ports RNAfold to AVX, giving up to a 1.77 fold speed up. The evolved code pull request is an accepted GI software maintenance update to bioinformatics package ViennaRNA.

Keywords: RNA secondary structure prediction ·
Genetic programming · GGGP · SIMD parallel computing ·
Software engineering · RCS · SBSE

1 Background: RNA, Genetic Improvement and RNAfold

RNA (like DNA) is a long chain biomolecule whose individual components (bases) have distinct side-binding affinities as well as forming strong links along the chain. Side bindings between elements of the chain cause RNA molecules to fold up in particular ways (known as their secondary structure). Giving them more or less stable shapes. The shapes of RNA molecules strongly influence their chemistry, including gene regulation. The ViennaRNA package [1], particularly RNAfold, is often used to predict secondary structures from RNA base sequence data. We show genetic programming can automatically convert existing RNAfold code to new AVX 512 bit parallel instructions.

Unlike our earlier work with pknotsRG [2] initially there wasn't a parallel version of RNAfold. However in [3] we used grow and graft genetic programming to create a parallel vector (SSE128) based implementation for ViennaRNA release 2.3.0. This was adopted and has been shipped with the ViennaRNA package since release 2.4.7.

At the time we were frustrated in our attempt to go as far as the full 512 bit Advanced Vector Extensions (AVX512) as they were only supported in specialised (GPU like) hardware accelerators, i.e. the Intel Phi cards. (ViennaRNA also includes our GPU version of RNAfold [4].) Since then AVX has been made

© Springer Nature Switzerland AG 2019
L. Sekanina et al. (Eds.): EuroGP 2019, LNCS 11451, pp. 245–261, 2019.
https://doi.org/10.1007/978-3-030-16670-0_16

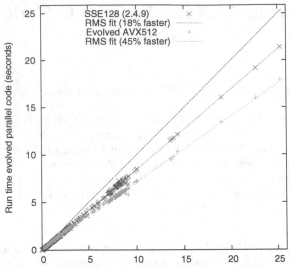

Original runtime ViennaRNA 2.4.9 (noSSE) on RNA STRAND v2.0, Intel Xeon 6126 2.60GHz

Fig. 1. Before and after GI elapsed time of RNAfold on 4663 RNA molecules.

available in certain top end Intel CPU chips and so we renewed our attempt to automatically port RNAfold to AVX512 hardware.

As before we use our grammar based GP system [5]. The BNF grammar is automatically created from our SSE128 code (itself based on ViennaRNA release 2.3.0), the complete revision control system (RCS) history of the manual code we generated for it and the Intel documentation (https://software.intel.com/sites/landingpage/IntrinsicsGuide/ downloaded 26 Jan 2017). At 6122 rules (although only 631 are used) this is perhaps the largest GP grammar.

1.1 RNA_STRAND

We test our evolved version of RNAfold on more than four thousand curated RNA molecules given by RNA_STRAND [7]. All 4666 RNA molecular base sequence were downloaded from http://www.rnasoft.ca/strand/download/.

1.2 GGGP Genetic Improvement System

We use genetic improvement [5,8–10] as part of our Grow and Graft GP (GGGP) [11] approach to optimising program sources. Previously [3] we had profiled RNAfold using GNU gcov. It indicated almost all the execution time was taken by a small fraction of function E_ml_stems_fast in multibranch_loops.c Just four lines of C code inside nested for loops (given in Fig. 2) were responsible, since they were executed billions of times. Previously [3], RNAfold was sped up by porting them to 128 bit SSE parallel Intel instructions. Here genetic programming ports them to 512 bit operations.

```
for (decomp=INF, k=i + 1 + turn; k<=stop; k++, k1j++){
  if((fmi[k] != INF ) && (fm[k1j] != INF)){
    en = fmi[k] + fm[k1j];
    decomp = MIN2(decomp, en);
  }
}
```

Fig. 2. Original RNAfold 2.3.0 code in function E_ml_stems_fast()

1.3   Parallel SSE 128-Bit and AVX 512-Bit Vector Instructions

At the end of the second millennium Intel started to increase the parallelism of its flagship 8080 series of processors. The most successful approach remains to integrate a few CPU cores onto single silicon chips. However at about the same time, Intel extended the instruction set with SIMD (single instruction multiple data) vector operations. Initially these allowed up to four 32 bit operations in parallel but the SSE instruction set has been progressively extended and now some processors support AVX (Advanced Vector Extensions) of 512 bits (e.g. sixteen 32 bit int operations in parallel).

Table 1. GGGP to improve SSE version of RNAfold

Representation	variable list of replacements, deletions and insertions into BNF grammar (6122 rules)
Fitness	Compile (GCC 7.3.1 -O2, -DNDEBUG -march=native -mtune=native) to modified object code, run on 10 000 random test cases. Two objectives: number of tests past and elapsed time (see Sect. 2.10)
Population	5000, panmictic, elite 10, generational
Parameters	Initial population of random single mutants. 50% truncation selection. 50% two point crossover, 50% mutation. No size limit. Stop after 50 generations

During the original manual phase, intermediate versions of the manually written SSE code were held in a revision control system (RCS), see reference [3, Figs. 2 and 3]. The appendix (Fig. 9) shows the initial (SSE 128) seed code. Additionally the whole of the Intel SSE/AVX library was available to genetic programming via the automatically created grammar, as an extended code base [12]. 5694 rules were derived from the SSE/AVX documentation. A further 168 were derived from RCS. Finally 141 came from the manually written SSE code Fig. 9 (i.e. the usual GI source seed code [5]). Of these 68 are fixed and provide the framework within which GI operates on the remaining 73.

The next Sect. 2 deals with setting up the grammar (Sect. 2.1) and type restrictions (Sects. 2.2–2.5) before dealing with evolutionary parameters, such

```
const int end = 1 + stop - k;
int     cnt;
__m128i  inf = _mm_set1_epi32(INF);

for (cnt = 0; cnt < end - 3; cnt += 4) {
  __m128i  a     = _mm_loadu_si128((__m128i *)&fmi_tmp[k + cnt]);
  __m128i  b     = _mm_loadu_si128((__m128i *)&fm[k1j + cnt]);
  __m128i  c     = _mm_add_epi32(a, b);
  __m128i  mask1 = _mm_cmplt_epi32(a, inf);
  __m128i  mask2 = _mm_cmplt_epi32(b, inf);
  __m128i  res   = _mm_or_si128(_mm_and_si128(mask1, c),
                                _mm_andnot_si128(mask1, a));

  res = _mm_or_si128(_mm_and_si128(mask2, res),
                     _mm_andnot_si128(mask2, b));
  const int en = horizontal_min_Vec4i(res);
  decomp = MIN2(decomp, en);
}

for (; cnt < end; cnt++) {
  if ((fmi[k + cnt] != INF) && (fm[k1j + cnt] != INF)) {
    const int en = fmi[k + cnt] + fm[k1j + cnt];
    decomp = MIN2(decomp, en);
  }
}

k   += cnt;
k1j += cnt;
```

Fig. 3. Start point for genetic improvement (ViennaRNA patch 2.4.9). Notice SSE 128 code already added [3], cf. Fig. 2.

```
const int end = 1 + stop - k;
int     cnt;
__m128i  inf = _mm_set1_epi32(INF);

#ifdef AVX512
  for(cnt = 0; cnt < end - (16- 1); cnt += 16)  {
    __m512i  a     = _mm512_loadu_si512((__m512i*)&fmi[k + cnt]);
    __m512i  b     = _mm512_loadu_si512((__m512i*)&fm[k1j + cnt]);
    __m512i  c     = _mm512_add_epi32(a, b);
    __m512i  min1  = _mm512_shuffle_epi32(c, _MM_SHUFFLE(0,0,3,2));
    __m512i  min2  = c;
    __m512i  min3  = _mm512_shuffle_epi32(min2, _MM_SHUFFLE(0,0,0,1));
    __m512i  min4  = _mm512_min_epi32(min2,min3);
    en = _mm512_reduce_min_epi32(min4);
    decomp = MIN2(decomp, en);
  }
#endif /*AVX512*/
  //second for loop used by both 128 bit SSE and 512 bit AVX code
  for (; cnt < end; cnt++) {
    if ((fmi[k + cnt] != INF) && (fm[k1j + cnt] != INF)) {
      const int en = fmi[k + cnt] + fm[k1j + cnt];
      decomp = MIN2(decomp, en);
    }
  }

  k   += cnt;
  k1j += cnt;
```

Fig. 4. AVX 512 code automatically evolved from Fig. 3

as population size (Sect. 2.6, see also Table 1), creating the initial population (Sect. 2.7) crossover and mutation (Sect. 2.8) and fitness and selection (Sects. 2.9–2.12). Some sections, e.g. Sect. 2.5, are very detailed and included for completeness.

Section 3 gives the results. Whilst Sect. 4 discusses, although it was not needed here, if a Pareto approach might been better and suggests that perhaps, despite the noisy fitness function, elitism is not in fact necessary. Finally we conclude (Sect. 5) that the evolved AVX 512 code is more than six times faster than the sequential code.

2   The Grammar Based Genetic Programming System

For our genetic programming [13–15] system we used GISMOE [5,16–18]. GIS-MOE (Fig. 5) creates a BNF grammar, which represents the original program's source code and legitimate changes to it. It then creates and evolves a population of changes to the BNF. Each modified BNF is expanded to give a modified (i.e. mutated) C source file, which is then compiled (Sect. 2.9) and tested (Sect. 2.10).

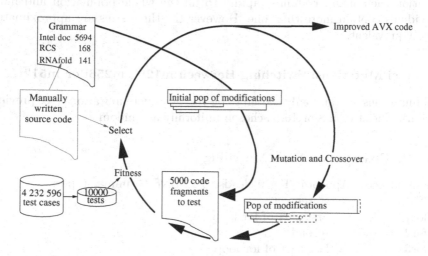

Fig. 5. Genetic improvement evolutionary cycle starts with the original C code (red, see appendix Fig. 9). The grammar tries to ensure many mutants compile, run, and terminate in ≤2s. Fitness is given by run time and comparing with the original code's answer. (Color figure online)

2.1   Representation: Variable Length Genome

Each variable length linear GP individual is an ordered list of changes to the original BNF grammar. Each generation, mutation can append a new change to the list or two point crossover can create a new list from randomly chosen parts of two parent lists [19, p. 18]. (Section 3.2 contains an example individual.)

2.2   BNF Generic Vector Types veci (m128i, m256i or m512i)

Since we want evolution to have access to the new vector instructions (i.e. 256 and 512 bit) but these are not used in the existing code, we automatically convert BNF grammar rules for existing 128 bit SSE code to a generic vector type "<veci_". (See also next section.) This can be either 4 int, 8 int or 16 int. There are three special meta mutations vecsize=4, vecsize=8 and vecsize=16, which convert veci in BNF rules to m128i, m256i or m512i types. An individual can contain multiple vecsize=meta-mutations. Variable length linear GP individuals are interpreted from left to right. Therefore veci is converted to whichever mnnni type is active. (By default the original code, i.e. m128i, is active.)

Everyone in the initial population must be unique, therefore there are initially only three individuals contain vecsize=metamutations. However during evolution their number climbs rapidly to fill the whole population and many individuals contain more than one. However the three sizes are approximately equally prevalent.

2.3   veci Mutations: Switching Between m128i, m256i or m512i

Half mutations are randomly chosen to be vecsize = meta mutations (see previous section). The size, 4, 8 or 16, is chosen uniformly at random.

2.4   BNF Grammar Type Matching

The input code (Appendix Fig. 9) leads to 12 BNF Grammar types:

39	""	statement
2	for1	first part of for loop
2	for2	second part
2	for3	last part of for loop
2	inti	int
1	int(veci)	int function with generic input
8	const03	0, 1, 2 or 3
2	unsigned-char(const03,const03,const03,const03)	function with 4 inputs
9	veci	generic vector
2	veci(veci-const*)	function pointer input, return vector
2	veci(veci,int)	function with vector and int arguments
3	veci(veci,veci)	function with two vector inputs

The first four types have been repeatedly used by GISMOE [5,20]. There are 39 simple lines (i.e. type ""). These can be deleted, replaced by another line, swapped or inserted before another line. All genetic operations are between BNF grammar rules of the same type. The two for loops in the input code (appendix Fig. 9) each give rise to three rules (of types: for1, for2 and for3).

2.5   Vector Type Matching Rules

The remaining types (i.e. "inti" to the end of the table in Sect. 2.4) are new and relate to expressions rather than complete lines of code. They can be deleted, replaced and exchanged. Rather than leave a hole in the source code, each type has a delete operation. For example: deleting a constant actually causes it to be replaced by zero, deleting a vector causes it to be replaced by a default vector "a", deleting a pointer to vector means replacing it with a pointer to "a" and deleting functions causes them to be replaced by their first argument. Notice 5 types include "veci" and so expand to 15 possibilities, making 22 types in total.

The Intel SIMD intrinsics documentation defines 5694 library functions. The documentation includes their input arguments' types and return type. This gives 1337 types. However to be used the types must match rules given by the input code (Appendix Fig. 9). There are several aspects of matching. Firstly, the input code uses generic veci types (Sect. 2.2 above). Secondly, the Intel intrinsics use multiple types to represent pointers and integers. For example, the type int(veci) matches functions which return int, mmask8, mmask16, mmask32 or unsigned-int and (depending on vecsize, Sect. 2.2) takes vector inputs of type m128i, m256i or m512i. The possible function type matches for this RNAfold experiment are given in Table 2.

When a new mutant code change is made, the Intel intrinsics whose return type matches the current vecsize are eligible and one of them (depending upon mutant type) maybe chosen. However if there is a type error during fitness evaluation when the mutated C code is generated, there is a fixup process which selects the Intel intrinsic of the right type whose name matches as closely as possible. Also there are a few veci(veci,int) and veci(veci,veci) types where veci takes different values. Since vecsize can only take one value at a time, these are resolved by the fixup process. E.g., m128i(m256i,int) (3rd box in Table 2) may be replaced by one of the m128i(m128i,int), m256i(m256i,int), m512i(m512i,const-int), etc. functions. Which one depends upon the current vecsize but is fixed so that the mapping from genotype to phenotype is deterministic.

In addition to the 502 intrinsics in Table 2, during evolution (Sect. 3), the fixup process substitutes other rules from the input (17), the history (10), an additional 5 Intel intrinsics (not in the table) or either stub. (Two stub functions, _mm_cvtsi256_si32 _mm_cvtsi512_si32, are missing from the Intel intrinsics documentation. They were defined by hand in C code to correspond to _mm_cvtsi128_si32.)

2.6   Population Size

In order to give a reasonable chance of including most of the eligible Intel SIMD intrinsics, the population size was increased dramatically to 5000 mutants. The initial generation (Sect. 3) used 396 intrinsics directly.

2.7   Initial Population

The initial population of 5000 unique individuals is created using mutation.

Table 2. Each box of three columns lists the number of Intel intrinsics which match the generic type above the table. The columns are (left to right) for 128, 256 and 512 bit vectors. I.e. 4, 8 and 16 int parallel operations.

int(veci)

int(m128i)	3	int(m256i)	2	int(m512i)	7
mmask8(m128i)	3	mmask8(m256i)	2	mmask8(m512i)	1
mmask16(m128i)	1	mmask16(m256i)	1	mmask16(m512i)	1
mmask32(m128i)	0	mmask32(m256i)	1	mmask32(m512i)	1
unsigned(m128i)	0	unsigned(m256i)	0	unsigned(m512i)	2

veci(veci-const*)

m128i(m128i*)	1	m256i(m256i*)	0	m512i(m512i*)	0
m128i(m128i-const*)	4	m256i(m256i-const*)	4	m512i(m512i-const*)	0
m128i(void-const*)	6	m256i(void-const*)	0	m512i(void-const*)	5

veci(veci,int)

m128i(m128i,int)	16	m256i(m256i,int)	8	m512i(m512i,int)	6
m128i(m256i,int)	2	m256i(m128i,int)	0	m512i(m128i,int)	0
m128i(m512i,int)	2	m256i(m512i,int)	2	m512i(m256i,int)	0
m128i(m128i,const-int)	4	m256i(m128i,const-int)	0	m512i(m128i,const-int)	0
m128i(m256i,const-int)	2	m256i(m256i,const-int)	13	m512i(m256i,const-int)	0
m128i(m512i,const-int)	0	m256i(m512i,const-int)	0	m512i(m512i,const-int)	3
m128i(m128i,unsigned)	1	m256i(m256i,unsigned)	1	m512i(m512i,unsigned)	9
				m512i(m512i,MM-*)	4

veci(veci,veci)

m128i(m128i,m128i)	136	m256i(m256i,m256i)	118	m512i(m512i,m512i)	110
		m256i(m128i,m128i)	2	m512i(m512i,m128i)	9
		m256i(m256i,m128i)	9		

Total usable intrinsics $502 + 5$ (see end of Sect. 2.5) $= 507$

All these available intrinsic functions are used by evolution.

2.8   Genetic Operations: Mutation and Crossover

Half the new members of each generation are created by mutation and half by two point crossover. Excluding the small elite (Sect. 2.12), every member of the generation is unique. Both mutation and crossover attempt to avoid repeating the same changes by removing such duplications from the new child's genome before applying unique requirements. Similarly both apply limited scoping checks to the genome [11,21]. If any of these checks fail, the change is discarded and a new random change is attempted.

As mentioned above (Sect. 2.3) half the time mutation attempts to set vec-size. (However the uniqueness requirement limits the extent to which vecsize floods the population.) If chosen, vecsize is inserted at a random point in the individual's genome. Whereas other mutations are appended to it.

If not a vecsize mutation, one of the 73 variable BNF grammar rules from the input code is chosen at random to be the target of a change. There are four changes: delete, replace, insert and swap. Any grammar rule can be deleted but for replacement, insert and swap, the replacement grammar rule must be of the same type.

2.9   Compiling Gcc -O2 -DNDEBUG -march=native -mtune=native

Both reference (see Sect. 2.11) and evolved code were compiled with the same compiler (GCC 7.3.1) and options (-O2 -DNDEBUG -fmax-errors=1 and -march=native -mtune=native). Notice we use the same default optimisations (-O2 -DNDEBUG) as the ViennaRNA release kit. The gcc command line option -DNDEBUG is used to disable some runtime checks in the supporting code. It does not directly affect the evolved code. Since any compilation error is regarded as fatal, -fmax-errors=1 is used to terminate failed compilations as quickly and tersely as possible.

By default the compiler will not generate either SSE or AVX instructions. Therefore, -march=native -mtune=native gcc command line options are used to generate optimised SSE and AVX instructions.

As before [3], we prevent some semantically equivalent mutants by insisting that the compiled object code is not identical to that of the un-mutated code.

2.10   Fitness Function: Run 10 000 Times, Elapsed Time and Accuracy

To get a realistic fitness function for training the GI, we started by profiling the RNAfold code on a real RNA molecule. This logged all 4 232 596 calls to the modular decomposition code in order. Each call was converted into a test case, giving the inputs to the code and the required output (i.e. the test oracle). Due to the way RNAfold uses dynamic programming, these test cases are also roughly ordered by difficulty and run time.

To both decrease the chance of over fitting and to reduced run time only a fraction of the test cases are used. To give a good spread of test case difficulty and runtime the 4 232 596 are divided into five equal sequences. Each generation a sequence of 2000 tests is chosen uniformly at random from each group. Making a total of 10 000 test cases.

Assuming the mutant compiled ok, it is run and the number of tests passed and time taken, for both the original code and the mutant code, is recorded. The whole process is subject to a total CPU limit of two seconds. (Normally it will take about 0.2 s.)

2.11   Selection for Speed and Correctness

As is often the case, we have two aspects of fitness. (1) the code should be accurate. (2) it should be fast. These are decided by testing (see previous section). Instead of combing these in a fixed way our selection is similar to VEGA [22]. (See also Sect. 4.) All the individuals which completed the tests (i.e. did not abort during testing) and passed at least one test are sorted into two lists by speed and accuracy. One list is sorted first by number of tests passed (ties are resolved by relative elapsed time). The second is sorted by relative elapsed time. Ties are unlikely but for symmetry are resolved by number of tests passed. (To compensate for short term fluctuations in computer speed we use the ratio of run time of the mutant and that of the reference code measured in the same exe image within a few milliseconds of each other. Indeed to obtain stable results both reference and mutated code are run eleven times and the first quartile, i.e. 3rd fastest, run time is used.) Mutants are taken alternatively from the two lists. When a mutant has already been selected, because it came earlier in the other list, it is not selected again. In effect that list loses a turn and selection goes immediately to the other list.

As mentioned in Sect. 2.8, two children are created in the next generation per selected parent. However if less than half the population are selected to be parents the next generation is made up to the full population size by creating random individuals.

It seems this selection does reduced the tendency of the population to converge but also the population seems to diverge into two separate camps (as reported for VEGA [23]), see Fig. 7.

2.12   10 Member Elite

Up to the first ten members of the list sorted by tests passed (see previous section) are automatically carried unchanged directly into the next generation [24, p. 101]. Normally they will also be parents of children (created with additional mutations or by crossover).

Since it is easy to retain the best of each generation, we had not been overly worried by evolution's tendency to lose (even the best) individuals from the population. Nonetheless, due to the noisy fitness function, we tried elitism here and it did indeed stabilise the best (see Figs. 6 and 7). Nonetheless having the best of run individual in the last generation does not seem a big gain and in practise we were still free to chose the best mutant from any generation (Sect. 3.2).

3   Results

The next section describes the evolutionary processes, before Sect. 3.2 describes why we chose the best mutant from generation 34. Section 3.3 explains how its genome translates into coding changes. Whilst Sect. 3.4 shows it gives a six fold speed up. Then Sect. 3.5 measures it in situ, i.e. within RNAfold, and shows it speeds up the whole of RNAfold by 45% on real RNA molecules.

3.1   Evolving the Population

Several hundred individuals which pass all their tests and are more than 10% faster evolved (Fig. 6 red ×). But most of population evolve to be either much faster and fail many tests or pass all their tests but give no speed up Fig. 7. Figure 7 highlights new code which passes all its tests and is at least 40% faster (×) or is more than 60% faster (+).

During evolution 13% of mutants fail to compile, 0.3% compile ok but their object code is identical to the seed code's (end of Sect. 2.9), 1.5% fail at runtime (e.g. segfault or CPU time limit exceeded) and 85% run all ten thousand tests.

3.2   Choosing the Winner

We choose the best individual in generation 34 since it was the first to be more than 60% faster (actually 69.9%) and pass all 10 000 of the tests used in its generation. This evolved individual contains 8 mutations:

vecsize=16
<_modular_decomposition.c_100>x<_modular_decomposition.c_77>
<veci_2modular_decomposition.c_110>
<int(veci)_1modular_decomposition.c_116><int(m512i)_IntrinsicsGuide.txt_8506>
<veci(veci,int)_1modular_decomposition.c_112><m256i(m512i,int)_IntrinsicsGuide.txt_4614>
<_modular_decomposition.c_97>x<_modular_decomposition.c_84>
<const03_5modular_decomposition.c_114>
<veci(veci,veci)_1modular_decomposition.c_113>

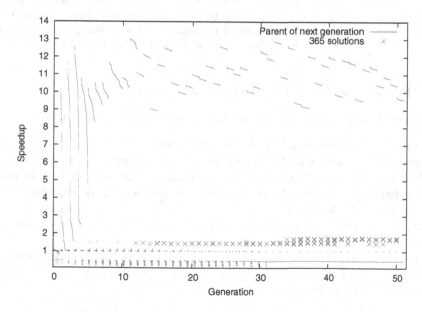

Fig. 6. Evolution of breeding population with elitism. Red × evolved individual passes all tests and speed up exceeds 10%. With our two criteria selection (Sect. 2.11, approximately half each breeding population is faster than the original code but fails one or more tests (see also Fig. 7). (Color figure online)

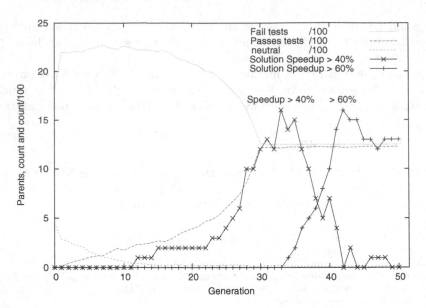

Fig. 7. Evolution of breeding population (2500). Solutions black lines (×, +). Initially most parents fail tests but are faster (solid purple), until gen 30. When the parents which pass all the tests but give no speed up (blue dashed) has grown to almost half (1216). Initially 448 parents (red dashed) pass all 10 000 tests but give no speed up. Gen 50 ≈50% are >10 fold faster but fail many tests (Fig. 6). (Color figure online)

3.3 Explaining the Wining Evolved Program

The evolved individual (8 mutations) can be reduced to three, since:

- Swapping lines 100 and 77 makes no difference.
- Similarly swapping lines 97 and 84 makes no difference.
- Deleting rule <veci_2modular_decomposition.c_110> again makes no difference (since delete substitutes a default variable for the deleted one, which turns out to be identical).
- Similarly <const03_5modular_decomposition.c_114> replaces a zero with another zero.
- The mutation which changes the right hand side of line 112 also makes no difference since the variable it writes to (min1) is no longer used. (Originally it was set on line 112 and used on line 113.)

The three remaining essential changes that are left are: vecsize=16 <int(veci)_1modular_decomposition.c_116><int(m512i)_IntrinsicsGuide.txt_8506> <veci(veci,veci)_1modular_decomposition.c_113>

- vecsize=16 converts all generic SSE code to 512 bit code.
- Line 8506 of the Intel documentation IntrinsicsGuide.txt is _mm512_reduce_ min_epi32, so <int(veci)_1modular_decomposition.c_116><int(m512i)_Intri nsicsGuide.txt_8506> causes en = _mm_cvtsi128_si32(min4) to be replaced with en = _mm512_reduce_min_epi32(min4). This is a key step, since it causes the output of the code (en) to be set to the minimum of sixteen (32 bit) int values held in 512 bit vector min4.
- The remaining mutation, increases efficiency by allowing the optimising compiler (note -O2) to avoid calculating a result by spotting it is being written into a variable which is not used.

Other efficiency gains might be possible by removing other unneeded operations. (Conceivable the compiler has already splotted them without needing explicit changes to the source code.) As is usual in GI practise [5], only the critical changes were retained. The evolved code gives exactly the same answers as the original sequential code.

3.4   Improved Performance Inside RNAfold

To estimate the total improvement in performance we ran RNAfold on a long real sequence eleven times: for no SSE, for the 2.4.9 SSE128 released code and for the new evolved AVX512 code. To allow a regression plot (see Fig. 8) each of these were repeated with the critical code repeated from one to ten times. The gnuplot (version 4.6.2) fit function was used to calculate the RMS least errors linear fit (Fig. 8).

In summary, on long sequences, the SSE128 code in release 2.4.9 is 3.428 ± 0.002 times faster than the nonSSE code and the newly evolved AVX512 code is 1.774 ± 0.003 faster than the released SSE128 code. I.e. 6.083 ± 0.002 times faster than the non SSE release 2.4.9 code.

3.5   Performance on Hold Out Data

The new mutant AVX 512 code was inserted into the current release (2.4.9) of RNAfold by hand and tested on the whole of RNA_STRAND (see Fig. 1). In all 4 666 cases it gives identical results. Even on the five corrupt fasta strings in RNA_STRAND v2.0 where RNAfold fails, the evolved and the released code give the same error messages. It is on average 45% faster. (Remember this includes the whole code, not just the dynamic programming inner loop.)

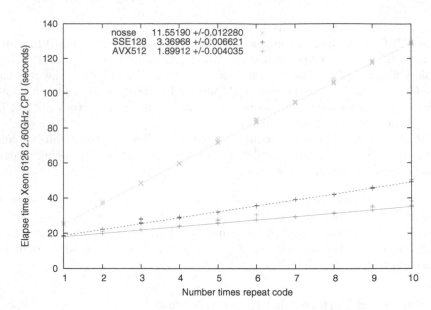

Fig. 8. Estimating true runtime with linear regression. Running multiple times (x-axis) allows the individual elapsed time to be estimated from the gradient. CRW_00528 (4382 bases) using newly evolved AVX512 (lowest +) code, the release 2.4.9 code configured with –enable-sse (+) and without (nosse × top).

4   Discussion: Population Convergence

The (VEGA [22] like) selection by two individual objectives (Sect. 2.11) avoids arbitrary weighting of the objectives but it appears (as suggested by Goldberg [23]) that it leads to the population dividing between the two objectives, so some programs are very fast but pass few tests and others pass all their tests but give little speed-up (Fig. 7). Whereas an approach like NSGA II [25] could potentially lead to more intermediate mutants (i.e. give a speed up but fail some tests) [26,27]. It may be having the larger population allows sufficient diversity in the two sub-populations to allow evolution to progress without the need for more refined selection techniques.

It appears that the use of an elite group ([24, p. 101] Sect. 2.12) does stably allow its preservation from one generation to the next. However the number of good fast mutants does not grow exponentially but quickly stabilises at three or four more than the size of the elite. As mentioned in Sect. 2.12, it may be these dozen or so high fitness individuals are not essential to allow evolution to progress.

5   Conclusions

Using a standard computer under a standard operating system (Linux) without specialised customisation to either, we have demonstrated that evolution can optimise a critical function written in C. Automatically creating AVX instructions, it gives almost a doubling in speed (1.71 fold) on top of hand written SSE instructions (6.1× the sequential code).

The new code has been included in ViennaRNA since release 2.4.11.

Acknowledgements. Funded by EPSRC grant EP/M025853/1.

GP code will be available via the author's home page.

Appendix

```
int
modular_decomposition(const int i, const int ij, const int j, const int turn, const int fmi[(2913+1)], const int fm[5000000]) {
  int k = 0;
  int kij = 0;
  int stop = 0;
  int end = 0;
  int en = 0;
  int decomp = INF;
  k += i;
  k += turn;
  k += 1;
  k += 1;
  k += -1;
  k += 2;
  k += -2;
  k += 3;
  k += -3;
  kij += ij;
  kij += turn;
  kij += 2;
  kij += 1;
  kij += -1;
  kij += 2;
  kij += -2;
  kij += 3;
  kij += -3;
  stop += j;
  stop += -turn;
  stop += -2;
  stop += 1;
  stop += -1;
  stop += 2;
  stop += -2;
  stop += 3;
  stop += -3;
  {
  end += 1;
  end += stop;
  end += -k;
  end += 1;
  end += -1;
  end += 2;
  end += -2;
  end += 3;
  end += -3;
  int i;
  for(i=0;i<end-3;i+=4) {
    __m128i a = _mm_loadu_si128((__m128i*)&fmi[k +i]);
    __m128i b = _mm_loadu_si128((__m128i*)&fm[kij+i]);
    __m128i c = _mm_add_epi32(a,b);
    __m128i min1 = _mm_shuffle_epi32(c, _MM_SHUFFLE(0,0,3,2));
    __m128i min2 = _mm_min_epi32(c,min1);
    __m128i min3 = _mm_shuffle_epi32(min2, _MM_SHUFFLE(0,0,0,1));
    __m128i min4 = _mm_min_epi32(min2,min3);
    en = _mm_cvtsi128_si32(min4);
    decomp = MIN2(decomp, en);
  }
  for(;i<end;i++) {
    en = fmi[k +i]+fm[kij+i];
    decomp = MIN2(decomp, en);
  }
  }
  return decomp;
}
```

Fig. 9. Starting point for evolution. C code derived from ViennaRNA (Figs. 2 and 3).Pairs of lines of code were manually added before the for loops to provide feed stock for evolution [2] but appear not to have been helpful. Instead successful changes (Fig. 4) come from the for loops themsleves.

References

1. Lorenz, R., et al.: ViennaRNA Package 2.0. AMB 6(1) (2011). https://doi.org/10. 1186/1748-7188-6-26
2. Langdon, W.B., Harman, M.: Grow and graft a better CUDA pknotsRG for RNA pseudoknot free energy calculation. In: Langdon, W.B., et al. (eds.) Genetic Improvement 2015 Workshop, pp. 805–810. ACM, Madrid, 11–15 July 2015. https://doi.org/10.1145/2739482.2768418
3. Langdon, W.B., Lorenz, R.: Improving SSE parallel code with grow and graft genetic programming. In: Petke, J., et al. (eds.) GI-2017, pp. 1537–1538. ACM, Berlin, 15–19 July 2017. https://doi.org/10.1145/3067695.3082524
4. Langdon, W.B., Lorenz, R.: CUDA RNAfold. Technical report RN/18/02, Computer Science, University College, London, Gower Street, London, 27 March 2018. https://doi.org/10.1101/298885
5. Langdon, W.B., Harman, M.: Optimising existing software with genetic programming. IEEE Trans. EC **19**(1), 118–135 (2015). https://doi.org/10.1109/TEVC. 2013.2281544
6. Langdon, W.B., Petke, J., Lorenz, R.: Evolving better RNAfold structure prediction. In: Castelli, M., Sekanina, L., Zhang, M., Cagnoni, S., García-Sánchez, P. (eds.) EuroGP 2018. LNCS, vol. 10781, pp. 220–236. Springer, Cham (2018). https://doi.org/10.1007/978-3-319-77553-1_14
7. Andronescu, M., Bereg, V., Hoos, H.H., Condon, A.: RNA STRAND: the RNA secondary structure and statistical analysis database. BMC Bioinform. **9**(1), 340 (2008). https://doi.org/10.1186/1471-2105-9-340
8. Langdon, W.B.: Genetic improvement of programs. In: Matousek, R. (ed.) 18th International Conference on Soft Computing, MENDEL 2012. Brno University of Technology, Brno, 27–29 June 2012. Invited Keynote http://www.cs.ucl.ac.uk/ staff/W.Langdon/ftp/papers/Langdon_2012_mendel.pdf
9. Petke, J., Haraldsson, S.O., Harman, M., Langdon, W.B., White, D.R., Woodward, J.R.: Genetic improvement of software: a comprehensive survey. IEEE Trans. EC **22**(3), 415–432 (2018). https://doi.org/10.1109/TEVC.2017.2693219
10. Langdon, W.B.: Genetically improved software. In: Gandomi, A.H., Alavi, A.H., Ryan, C. (eds.) Handbook of Genetic Programming Applications, pp. 181–220. Springer, Cham (2015). https://doi.org/10.1007/978-3-319-20883-1_8
11. Langdon, W.B., Lam, Brian Y.H., Petke, J., Harman, M.: Improving CUDA DNA analysis software with genetic programming. In: Silva, S., et al. (eds.) GECCO 2015, pp. 1063–1070. ACM, Madrid, 11–15 July 2015. https://doi.org/10.1145/ 2739480.2754652
12. Petke, J., Harman, M., Langdon, W.B., Weimer, W.: Specialising software for different downstream applications using genetic improvement and code transplantation. IEEE Trans. SE **44**(6), 574–594 (2018). https://doi.org/10.1109/TSE.2017. 2702606
13. Koza, J.R.: Genetic Programming: On the Programming of Computers by Natural Selection. MIT Press, Cambridge (1992)
14. Banzhaf, W., Nordin, P., Keller, R.E., Francone, F.D.: Genetic Programming - An Introduction. Morgan Kaufmann, San Francisco (1998)
15. Poli, R., Langdon, W.B., McPhee, N.F.: A field guide to genetic programming (2008), (With Contributions by J.R. Koza). Published via http://lulu.com and freely available at http://www.gp-field-guide.org.uk

16. Langdon, W.B.: Genetic improvement of software for multiple objectives. In: Barros, M., Labiche, Y. (eds.) SSBSE 2015. LNCS, vol. 9275, pp. 12–28. Springer, Cham (2015). https://doi.org/10.1007/978-3-319-22183-0_2

17. Langdon, W.B.: Genetic improvement GISMOE blue software tool demo. Technical report RN/18/06, University College, London, 22 September 2018. http://www.cs.ucl.ac.uk/fileadmin/user_upload/blue.pdf

18. Lopez-Lopez, V.R., Trujillo, L., Legrand, P.: Novelty search for software improvement of a SLAM system. In: Alexander, B., et al. (eds.) 5th Edition of GI @ GECCO 2018, pp. 1598–1605. ACM, Kyoto, 15–19 July 2018. https://doi.org/10.1145/3205651.3208237

19. Langdon, W.B., Lam, B.Y.H., Modat, M., Petke, J., Harman, M.: Genetic improvement of GPU software. Genet. Program. Evol. Mach. **18**(1), 5–44 (2017). https://doi.org/10.1007/s10710-016-9273-9

20. Langdon, W.B.: Evolving better RNAfold C source code. Technical report RN/17/08, University College, London (2017). https://doi.org/10.1101/201640

21. Langdon, W.B., Lam, B.Y.H.: Genetically improved BarraCUDA. BioData Min. **20**(28) (2017). https://doi.org/10.1186/s13040-017-0149-1

22. Schaffer, J.D.: Multiple objective optimization with vector evaluated genetic algorithms. In: Grefenstette, J.J. (ed.) Proceedings of an International Conference on Genetic Algorithms and the Applications, pp. 93–100. Carnegie-Mellon University, Pittsburgh, 24–26 July 1985. http://www.cs.ucl.ac.uk/staff/W.Langdon/ftp/papers/icga1985/icga85_schaffer.pdf

23. Goldberg, D.E.: Genetic Algorithms in Search Optimization and Machine Learning. Addison-Wesley, Boston (1989)

24. De Jong, K.A.: An analysis of the behavior of a class of genetic adaptive systems. Ph.D. thesis, Computer and Communications Sciences, Michigan, USA (1975)

25. Deb, K., Pratap, A., Agarwal, S., Meyarivan, T.: A fast and elitist multiobjective genetic algorithm: NSGA-II. IEEE Trans. EC **6**(2), 182–197 (2002). https://doi.org/10.1109/4235.996017

26. Langdon, W.B.: Genetic Programming and Data Structures. Kluwer, Boston (1998). https://doi.org/10.1007/978-1-4615-5731-9

27. Knieper, T., Defo, B., Kaufmann, P., Platzner, M.: On robust evolution of digital hardware. In: Hinchey, M., Pagnoni, A., Rammig, F.J., Schmeck, H. (eds.) BICC 2008. ITIFIP, vol. 268, pp. 213–222. Springer, Boston (2008). https://doi.org/10.1007/978-0-387-09655-1_19

Hyper-bent Boolean Functions
and Evolutionary Algorithms

Luca Mariot[1]([⊠]), Domagoj Jakobovic[2], Alberto Leporati[1], and Stjepan Picek[3]

[1] DISCo, Università degli Studi di Milano-Bicocca,
Viale Sarca 336/14, 20126 Milan, Italy
{luca.mariot,alberto.leporati}@unimib.it
[2] Faculty of Electrical Engineering and Computing, University of Zagreb,
Unska 3, Zagreb, Croatia
domagoj.jakobovic@fer.hr
[3] Cyber Security Research Group, Delft University of Technology,
Mekelweg 2, Delft, The Netherlands
S.Picek@tudelft.nl

Abstract. Bent Boolean functions play an important role in the design of secure symmetric ciphers, since they achieve the maximum distance from affine functions allowed by Parseval's relation. Hyper-bent functions, in turn, are those bent functions which additionally reach maximum distance from all bijective monomial functions, and provide further security towards approximation attacks. Being characterized by a stricter definition, hyper-bent functions are rarer than bent functions, and much more difficult to construct. In this paper, we employ several evolutionary algorithms in order to evolve hyper-bent Boolean functions of various sizes. Our results show that hyper-bent functions are extremely difficult to evolve, since we manage to find such functions only for the smallest investigated size. Interestingly, we are able to identify this difficulty as not lying in the evolution of hyper-bent functions itself, but rather in evolving some of their components, i.e. bent functions. Finally, we present an additional parameter to evaluate the performance of evolutionary algorithms when evolving Boolean functions: the diversity of the obtained solutions.

Keywords: Bent functions · Hyper-bent functions ·
Genetic programming · Genetic algorithms · Evolution strategies

1 Introduction

Boolean functions are mathematical objects with numerous applications in cryptography, coding theory, and sequences. As such, they received a great deal of attention by the research community in the last decades. *Bent* Boolean functions, which exist only for even numbers of input variables, are those functions that have maximal nonlinearity, that is, they have the highest possible distance

© Springer Nature Switzerland AG 2019
L. Sekanina et al. (Eds.): EuroGP 2019, LNCS 11451, pp. 262–277, 2019.
https://doi.org/10.1007/978-3-030-16670-0_17

from the set of *affine functions*. For their construction, one can employ a number of different *primary constructions* (where bent functions are generated from scratch) or *secondary constructions* (where one uses already known bent functions to construct larger ones). As an alternative method one can use heuristics, among which *evolutionary algorithms* proved to be especially adept in the last years. In fact, the sheer amount of successful results obtained with evolutionary algorithms makes the evolution of bent functions almost an easy problem [1–3]. Naturally, this is a somewhat oversimplified claim, since we can always aim to evolve Boolean functions of a size large enough that the process will be unfeasible from the computational perspective. There exist a sub-class of bent functions, namely *hyper-bent* functions, that have even stronger properties and are rarer than bent functions. Indeed, Hyper-bent functions are not only as far as possible from all affine functions, but also from all coordinate functions of all *bijective monomials*. Consequently, they can provide a good source of nonlinearity when designing block ciphers [4]. Unfortunately, since hyper-bent Boolean functions are rarer than bent Boolean functions, it could be that such functions are also more difficult to generate than bent functions. Still, the authors of [4] proved that hyper-bent functions exist for every even n, as general bent functions.

In this paper, we examine whether evolutionary algorithms can be a suitable technique for constructing hyper-bent functions, since such techniques proved to be very powerful when considering the generation of bent functions. More precisely, we consider some well-known techniques like genetic algorithms (GA), genetic programming (GP), and evolution strategy (ES) that proved to be able to evolve bent functions for a number of different sizes. We pose the following research questions. Can evolutionary algorithms be used to obtain hyper-bent functions in various sizes? If so, we are additionally interested in what is the richness of the solution set. More precisely, a common argument for using heuristics is that it allows us to obtain a number of different solutions which is sometimes not possible with algebraic constructions. Consequently, we will not only be interested in obtaining hyper-bent functions but also, but also in examining the number of different hyper-bent functions we are able to construct. In order to provide answers to the defined questions, we examine which of the considered evolutionary techniques achieves the best results. To the best of our knowledge, this is the first time evolutionary algorithms are considered for the evolution of hyper-bent functions. The obtained results show us that this problem is much more difficult for evolutionary algorithms than the evolution of bent functions, since the latter has already been investigated in the literature with the same techniques described in this paper. Hence, this problem could also represent a good benchmark for evaluating the performance of heuristics.

This paper is organized as follows. In Sect. 2, we briefly introduce bent and hyper-bent Boolean functions along with their relevant properties. Section 3 gives an overview of related work. Section 4 discusses the evolutionary algorithms we consider as well as the fitness functions we use in the experiments. In Sect. 5, we present results from evolutionary algorithms for a number of relevant Boolean function sizes. Additionally, we provide a review of the obtained results and we discuss possible future research directions. Finally, in Sect. 6, we summarize the main contributions of this work and conclude the paper.

2   Background

We refer the reader to [5] for a thorough introduction to the notions about Boolean functions discussed in this section.

In this work, we consider the set $\{0,1\}$ as the *finite field* with two elements, equipped with the XOR $\oplus$ and AND $\cdot$ as field operations. Given a positive integer $n \in \mathbb{N}$, the set $\{0,1\}^n$ of all binary strings of length n, which is composed of 2^n elements, is denoted as $\mathbb{F}_2^n$ and it is regarded as a vector space over $\mathbb{F}_2$. Additionally, to introduce the definition of hyper-bent functions, we will also consider $\{0,1\}^n$ as the finite field $\mathbb{F}_{2^n}$. The elements of $\mathbb{F}_{2^n}$ can be interpreted as *polynomials* in $\mathbb{F}_2[x]$ modulo an irreducible polynomial $p(x)$ of degree n, i.e. as elements of the quotient ring $\mathbb{F}_2[x]/p(x)$. In particular, the components of an n-bit string identify the coefficients of the associated polynomial in $\mathbb{F}_{2^n}$. Since $\mathbb{F}_{2^n}$ is isomorphic to $\mathbb{F}_2^n$, it can also be considered as a $\mathbb{F}_2$-vector space [6].

A *Boolean function* of n variables is a map $f : \mathbb{F}_2^n \rightarrow \mathbb{F}_2$, and it can be uniquely represented by its *truth table* (TT), which is a vector $\Omega_f \in \mathbb{F}_2^n$ of length 2^n that specifies the output value $f(x)$ for each possible input vector $x \in \mathbb{F}_2^n$. Usually, it is assumed that the function values in Ω_f are lexicographically ordered with respect to their inputs.

The *dot product* of two vectors $a, b \in \mathbb{F}_2^n$ is defined as $a \cdot b = a_1 b_1 \oplus \cdots \oplus a_n b_n$, and it satisfies the axioms of *inner product* over the vector space $\mathbb{F}_2^n$. The *absolute trace* of an element x in the finite field $\mathbb{F}_{2^n}$ equals:

$$Tr(x) = x + x^2 + \ldots + x^{2^{n-1}}, \tag{1}$$

The trace $Tr(ax)$ of $a, x \in \mathbb{F}_{2^n}$ is also an inner product over $\mathbb{F}_{2^n}$, where ax represents the field multiplication of the elements a and x (that is, polynomial multiplication). A Boolean function L_a is called *linear* if it is defined by an inner product. Thus, in the case of the vector space $\mathbb{F}_2^n$, $a \in \mathbb{F}_2^n$ is an n-bit vector and $L_a : \mathbb{F}_2^n \rightarrow \mathbb{F}_2$ is defined as $L_a(x) = a \cdot x$ for all $x \in \mathbb{F}_2^n$. Likewise, for the finite field case $a \in \mathbb{F}_{2^n}$ is a polynomial in $\mathbb{F}_2[x]/p(x)$ and $L_a : \mathbb{F}_{2^n} \rightarrow \mathbb{F}_2$ is defined as $L_a(x) = Tr(ax)$. Linear functions which also sum a constant $a_0 \in \mathbb{F}_2$ to the inner product are called *affine*.

The *Walsh-Hadamard transform* is another unique representation for Boolean functions. Formally, given $f : \mathbb{F}_2^n \rightarrow \mathbb{F}_2$, its Walsh-Hadamard transform $W_f : \mathbb{F}_2^n \rightarrow \mathbb{Z}$ is defined for all $a \in \mathbb{F}_2^n$ as:

$$W_f(a) = \sum_{x \in \mathbb{F}_2^n} (-1)^{f(x) \oplus a \cdot x}. \tag{2}$$

In other words, the coefficient $W_f(a)$ measures the correlation between f and the linear function $a \cdot x$. The maximum absolute value $|W_f(a)|$ of W_f over all $a \in \mathbb{F}_2^n$ is also called the *spectral radius* of f. *Parseval's relation* states that the sum of the squared Walsh coefficients is constant for every Boolean function $f : \mathbb{F}_2^n \to \mathbb{F}_2$:

$$\sum_{a \in \mathbb{F}_2^n} W_f(a)^2 = 2^{2n}. \tag{3}$$

The *nonlinearity* of a Boolean function $f : \mathbb{F}_2^n \to \mathbb{F}_2$ is defined as the minimum Hamming distance between the truth table of f and the set of affine functions, and it can be expressed as [5]:

$$Nl_f = 2^{n-1} - \frac{1}{2} \max_{a \in \mathbb{F}_2^n} |W_f(a)|, \tag{4}$$

from which it follows that the lower the spectral radius of a Boolean function is, the higher its nonlinearity will be. By Parseval's relation, one can see that the the spectral radius is minimum if and only if all Walsh coefficients equal $\pm 2^{\frac{n}{2}}$. Functions satisfying this property are called *bent*, and they achieve the maximum value of nonlinearity $2^{n-1} - 2^{\frac{n}{2}-1}$. Since the Walsh-Hadamard coefficients must be integer numbers, it follows that bent functions exist only when the number of variables n is even.

Remark that the notion of bent function is independent of the underlying inner product used in the Walsh-Hadamard transform [5]. Hence, one could substitute the dot product $a \cdot x$ in Eq. 2 with the trace $Tr(ax)$ and obtain the same set of bent functions. This is important to introduce the notion of hyper-bent functions, the main object we are interested in this paper. In fact, hyper-bent functions are characterized through the so-called *extended Walsh-Hadamard transform*, which is defined over the finite field $\mathbb{F}_{2^n}$ as follows:

$$W_f(a, i) = \sum_{x \in \mathbb{F}_2^n} (-1)^{f(x) \oplus Tr(ax^i)}, \tag{5}$$

where i is coprime with $2^n - 1$, i.e., $gcd(i, 2^n - 1) = 1$. Hence, in the extended transform we are computing several spectra of Walsh coefficients. The reason to consider the linear functions defined by $Tr(ax^i)$ is that x^i represents a *bijective monomial*, since i is coprime with $2^n - 1$. By considering only $i = 1$, one gets the usual Walsh-Hadamard transform.

A function f which is bent with respect to the extended transform for all i coprime with $2^n - 1$ is called *hyper-bent*. Thus, hyper-bent functions have the highest Hamming distance from all affine functions defined by bijective monomials. Notice that the number of indices i coprime with $2^n - 1$ is determined by *Euler's totient function* $\phi(2^n - 1)$, which grows as $O(2^n)$. Thus, the number of indices against which the bent property must be checked with the extended transform is exponential in the number of variables n, and it becomes computationally expensive already for small values of n.

Table 1 reports the number of Boolean functions $\mathcal{B}_n$ of n variables, the number of exponents $\mathcal{I}_n$ coprime with $2^n - 1$, the number of bent $\mathcal{A}_n$ and hyper-bent $\mathcal{H}_n$ Boolean functions, and their nonlinearity Nl_f for sizes $n = 4, 6, 8$. The values for $\mathcal{A}_n$ and $\mathcal{H}_n$ have been taken from [7].

Table 1. The number and nonlinearity of Boolean functions for various input sizes n.

n	4	6	8
$\mathcal{B}_n$	2^{16}	2^{64}	2^{256}
$\mathcal{I}_n$	8	36	128
$\mathcal{A}_n$	896	$\approx 2^{32.3}$	$\approx 2^{106.3}$
$\mathcal{H}_n$	56	252	48 620
Nl_f	6	28	120

Observe that for $n = 4$ variables the resulting space of Boolean functions can be completely enumerated, since it is composed of only $2^{16} = 65\,536$ elements.

3   Related Work

Golomb and Gong proposed hyper-bent functions as components of Substitution Boxes to ensure the security of cryptographic algorithms [8]. Charpin and Gong investigate how to classify hyper-bent functions [9]. Carlet and Gaborit show that hyper-bent functions known at the time belong to the $\mathcal{PS}_{ap}^{\#}$ class. Next, they show how such functions can be obtained from certain codewords of extended cyclic code with small dimension and they enumerate hyper-bent functions up to $n = 10$ [7].

As already mentioned, there is a significant corpus of papers dealing with the heuristic generation of bent Boolean functions. At the same time, there are no works, to the best of our knowledge, considering the heuristic generation of hyper-bent functions. Consequently, here we remind the readers on relevant results in the heuristic generation of bent Boolean functions as well as the relevant theoretical results.

As far as we are aware, the first paper that uses evolutionary algorithms with a goal of evolving cryptographic Boolean functions dates back to 1997. There, Millan et al. used genetic algorithms to evolve Boolean functions with high nonlinearity [10]. Millan, Clark, and Dawson experimented with GA to evolve Boolean functions that have high nonlinearity [11]. They used a combination of GAs and hill climbing together with a resetting step in order to find Boolean functions up to 12 inputs with high nonlinearity. Millan, Fuller, and Dawson proposed an adaptive strategy for a local search algorithm for the generation of Boolean functions with high nonlinearity [12]. Additionally, they introduced the notion of the graph of affine equivalence classes of Boolean functions.

Picek, Jakobovic, and Golub used GA and GP to find Boolean functions that have several optimal cryptographic properties [13]. To the best of our knowledge, this is the first application of GP for evolving cryptographic Boolean functions. Hrbacek and Dvorak used CGP in order to evolve bent Boolean functions of sizes up to 16 inputs [1]. The authors experimented with several configurations of algorithms in order to speed up the evolution process where they did not limit the number of generations in their search. With such an approach, they succeeded in finding bent function in each run for sizes between 6 and 16 inputs. Mariot and Leporati designed a discrete particle swarm optimizer to search for balanced Boolean functions with high nonlinearity [14]. The same authors in [15] used GA where the genotype consists of the Walsh-Hadamard values in order to evolve semi-bent Boolean functions by spectral inversion. An analysis of the efficiency of several evolutionary algorithms when evolving Boolean functions satisfying different cryptographic criteria is given in [16]. Picek and Jakobovic experimented with GP with a goal of evolving algebraic constructions, which are then used to construct bent functions [2]. There, the authors are able to find bent Boolean functions for sizes up to 24 inputs. Picek, Sisejkovic, and Jakobovic experimented with several immunological algorithms to construct bent or balanced, highly nonlinear Boolean functions [3].

4 Experimental Setup

In this section, we describe the solution representations, the search algorithms, and the fitness functions we use in our experiments.

4.1 Truth Table Representation

The most intuitive encoding for a Boolean function is in the form of a truth table. In this case, individuals are represented as strings of bits which present truth tables of Boolean functions. The string length is determined by the number of Boolean variables n and equals 2^n. Using this encoding, we experiment with two search algorithms: a genetic algorithm and an evolution strategy.

For GA, we employ a 3-tournament selection which serves to eliminate the worst individual among three randomly selected ones. After the elimination, a new individual is produced using the crossover operator applied on the remaining two. The new individual immediately undergoes mutation subject to a given individual mutation rate.

The crossover operators are one-point and uniform crossover, performed uniformly at random for each new offspring. The mutation operator is selected uniformly at random between a simple mutation, where a single bit is inverted, and a mixed mutation, which randomly shuffles the bits in a randomly selected subset. The mutation probability is used to select whether an individual would be mutated or not, and the mutation operator is executed only once on a given individual. For example, if the mutation probability is 0.7, then on average 7 out of every 10 new individuals will be mutated and one mutation will be performed on each of those individuals. The selected population size equals 100 individuals, whereas the individual mutation probability is 0.3.

When experimenting with evolution strategy (ES), we use $(\mu + \lambda)$-ES. In this algorithm, in each generation, parents compete with offspring and from their joint set μ fittest individuals are kept. In our experiments, offspring population size λ has a value equal to 5 and parent population size μ has a value of 100.

Although it is not standard in the ES literature to have such a big population, we adopted this size since some works (see e.g. [16]) showed that it brings good results when evolving cryptographic Boolean functions. For further information on ES, we refer interested readers to [17–19].

4.2   Tree Representation

Tree encoding is commonly related to genetic programming (GP) in which the data structures that undergo optimization are executable expressions) [17]. Each individual in a GP population represents a computable expression, whose most common form are symbolic expressions corresponding to parse trees. A tree can represent a mathematical expression, a rule set or a decision tree. The building elements in a tree-based GP are functions (inner nodes) and terminals (leaves, problem variables).

As opposed to truth table encoding, the other option we consider is to use a symbolic representation of a Boolean function. This is performed in a way such that genetic programming can be used to evolve a Boolean function in the form of a syntactic tree. Here, the terminal set consists of the n input Boolean variables, denoted $\{v_0, \ldots, v_n\}$. The function set (i.e., the set of inner nodes of a tree) should consist of appropriate Boolean operators that allow the definition of any function with n inputs.

The function set for genetic programming in all the experiments consists of Boolean functions OR, XOR, AND (taking two arguments), NOT (one argument), and IF. The function IF takes three arguments and returns the second argument if the first one evaluates to 'true', and the third one otherwise.

In our experiments, GP uses the same steady-state tournament selection algorithm of GA. The variation operators are simple tree crossover, uniform crossover, size fair, one-point, and context preserving crossover (selected at random), and subtree mutation [20]. All our experiments suggest that having a maximum tree depth equal to the number of Boolean variables is sufficient (i.e., tree depth equals n). The initial population is created at random and the population size equals 500.

4.3   Boolean Construction Representation

Finally, we experiment with a concept that stems from Boolean algebraic construction methods. In this setting, we try to construct a Boolean function of $n+2$ variables using previously obtained Boolean functions of n variables. The process can be described as: first, the optimization problem is solved for a Boolean function size which allows the solutions to be found without much effort. In many cases, we can always start with the number of Boolean variables that produces the search space that can be scanned by an exhaustive search. For this problem, we start with $n = 4$, where we can enumerate all possible solutions, e.g. 56 hyper-bent Boolean functions of size 4.

The second step is to use GP to evolve Boolean functions of size $n + 2$; in this case, the terminals are not the $n + 2$ Boolean variables; rather, the terminal set includes *four* predefined Boolean functions with n variables which were previously obtained. These terminals are denoted with f_0, f_1, f_2, and f_3 (input functions). Additionally, the terminal set includes two independent Boolean variables, v_0 and v_1 (since we are constructing functions in size $n+2$). The choice of having four input functions is inspired by the Rothaus' construction [21], where three bent functions such that their exclusive OR is again bent (consequently, we can consider this as having four functions) are used to construct new bent functions.

Finally, each resulting construction (a GP expression) that includes input functions and additional two variables, represents a new Boolean function with $n+2$ variables. This construction is based on truth table representations of input functions, which are the size of 2^n, and the extension with two Boolean variables which, for every combination of their values, extend the resulting truth table to the size of 2^{n+2}.

With the goal of obtaining hyper-bent functions of size $n+2$, the input functions are presumed to be hyper-bent themselves. Since 56 hyper-bent functions of size 4 are available, we define 14 sets with four input functions in each set, and every construction is evaluated using each of the 14 input sets. As we are interested only in optimizing the objective value (and not in finding a general construction method), we assign the fitness function of a candidate construction as the best objective value among all possible 14 resulting functions.

Note that this process can be self-sustained: if the resulting construction produces the desired solutions, then those solutions can be used as input functions for the next construction stage for size $n + 4$. However, in each stage we need to add only two additional Boolean variables; this greatly reduces the search size and computational effort, as compared to the common approach where the solution size is increased with increasing number of Boolean variables. Apart from the terminals set, all the other GP parameters, as well as the function set, remain the same.

4.4   Fitness Functions

The first fitness function aims to find hyper-bent functions by considering the nonlinearity of the bent function for each i (see Eq. (5)). The aim is to minimize the following expression:

$$fitness_1 = \sum_{i=1, gcd(i, 2^n - 1)=1}^{2^n - 1} \left(2^{n-1} - 2^{\frac{n}{2} - 1} - Nl_{f_i} \right). \tag{6}$$

Note, we simply subtract the nonlinearity value of the obtained function from the nonlinearity of the bent function (see Sect. 2). This fitness function we also use when trying to obtain hyper-bent functions via secondary constructions.

In the second fitness function, we consider the whole Walsh-Hadamard spectrum and we penalize proportionally to the number of the Walsh-Hadamard coefficients differing from $2^{\frac{n}{2}}$ (see Eq. (3) and conclusions stemming from it). Again, the goal is to minimize the following expression:

$$fitness_2 = \sum_{i=1,gcd(i,2^n-1)=1}^{2^n-1} T_i, \tag{7}$$

where T_i equals the number of times the Walsh-Hadamard coefficients differ from $2^{\frac{n}{2}}$ for each i.

Finally, we define the third fitness function which tries to increase the number of cases in the above sum (different values of i) where the obtained function is bent. This factor is multiplied with an arbitrary constant to make it a primary criterion; the secondary criterion is then simply the relative difference from maximum nonlinearity for all the cases where a bent function is not found. The third objective function is defined as a minimization of the following:

$$fitness_3 = 100 \times S_i + \sum_{i=1,gcd(i,2^n-1)=1}^{2^n-1} \left(1 - \frac{Nl_{f_i}}{2^{n-1} - 2^{\frac{n}{2}-1}} \right), \tag{8}$$

where S_i equals the number of different values of i in the above sum where the Nl_{f_i} was not equal to maximum nonlinearity.

5   Experiments

In this section, we first give results for GA, GP, and ES. Afterward, we discuss the difficulty of this problem, the diversity of solutions, and finally, possible future works.

5.1   Results

We present the results of three evolutionary algorithms (GA, ES, and GP) based on the function size and fitness functions defined in the previous section. Regardless of the representation and the search algorithm, every experiment is repeated 30 times and the stopping condition is the number of fitness evaluations, which is set to 500 000.

In the case of Boolean functions of 4 variables, all tested methods easily converge to fitness value of zero, i.e., we find a hyper-bent function in every run. This is expected since the search size is relatively small and this presents no problem regardless of the solution encoding. Consequently, we do not present results for $n = 4$.

For size $n = 6$, the results for all the three algorithms and three variants of fitness functions are shown in Table 2. Since we are using different fitness functions that are not comparable, we divide the results in two groups: the left side of the table shows the statistics of obtained solutions using the objective function for which they were evolved (i.e. a separate set for each one of the three fitness functions). The right half of the table shows the same solutions evaluated on the basis of the first fitness function, so the efficiency of different fitness functions can be assessed.

In terms of objective functions, we can see that the first and the third fitness function provided solutions of the same quality, while the second fitness was inferior to the other two. At the same time, the tree encoding used with GP was convincingly better than the truth table representation, regardless of the search algorithm. Unfortunately, no single algorithm or objective function was able to provide a hyper-bent Boolean function of 6 variables, which would correspond to the fitness value of zero. When considering the obtained results, we can see that GP performs the best. More precisely, for all three fitness functions, it outperforms GA and ES significantly. At the same time, the behavior for GA and ES is similar but still slight advantage goes to GA.

Table 2. Optimization results for Boolean functions of 6 variables

	Optimized objective				Evaluated by $fitness_1$			
$fitness_1$	min	median	avg	max				
ES	108	132	129.64	144				
GA	108	132	135.03	156				
GP	72	120	113.73	156				
$fitness_2$	min	median	avg	max	min	median	avg	max
ES	1 260	1 365	1 361.8	1 422	156	216	217.69	276
GA	1 206	1 380	1 375.9	1 458	144	216	215.51	288
GP	648	864	881.4	1 272	72	204	207.87	432
$fitness_3$	min	median	avg	max	min	median	avg	max
ES	3 602.7	3 603.8	3 603.6	3 604.3	108	132	128.55	144
GA	3 005.3	3 604.1	3 596.5	3 604.3	108	144	137.77	156
GP	1 802.1	3 003.6	2 570.7	3 604.3	72	120	116	144

As for the Boolean construction method (see Sect. 4.3), it has also proven unsuccessful; this approach is tested with GP using groups of different previously obtained hyper-bent functions of 4 variables as input functions. In every GP run, the obtained constructed function converged to the objective value of 120 according to the first fitness function. Since the results do not show any deviation, we omit those from the table. Interestingly, our experiments show that the well-known Rothaus construction cannot be used to construct hyper-bent functions despite the fact that it can construct bent functions. To the best of our knowledge, this information is new (albeit, not unexpected). Additionally, since we have been unable to obtain hyper-bent functions for $n = 6$, we were unable to apply this construction method for larger sizes.

Finally, Table 3 presents the results for optimizing Boolean functions of 8 variables, in the same form as Table 2. From the results obtained for $n = 8$, we can see that for $fitness_1$, GP is now actually the worst performing algorithm. For the second scenario, $fitness_2$, we observe GP has the best value but as evaluated relative to $fitness_1$, GP is still the worst performing algorithm. Only for $fitness_3$ GP is the best choice. Similar as for $n = 6$, the performance of GA and ES is similar with a small advantage for GA (except for $fitness_2$ as evaluated relative to $fitness_1$).

Table 3. Optimization results for Boolean functions of 8 variables

	Optimized objective				Evaluated by $fitness_1$			
$fitness_1$	min	median	avg	max				
ES	1 472	1 504	1 501.7	1 536				
GA	1 440	1 520	1 527.6	1 616				
GP	1 536	1 632	1 630	1 696				
$fitness_2$	min	median	avg	max	min	median	avg	max
ES	27 704	27 824	27 818	27 944	1 920	2 096	2 107	2 288
GA	27 568	27 808	27 793	28 016	1 984	2 080	2 095.5	2 272
GP	17 456	17 456	17 456	17 456	3 456	3 456	3 456	3 456
$fitness_3$	min	median	avg	max	min	median	avg	max
ES	254.04	254.06	254.05	254.06	2 016	2 088	2 149.3	2 336
GA	254.04	254.06	254.05	254.06	1 936	1 984	2 008	2 160
GP	248.06	248.06	248.07	248.1	1 792	1 872	1 938.6	2 176

In Figs. 1a until 1e, we display results for $n = 6$, all algorithms and fitness functions 1, 2, and 3, respectively. We display the average values as averaged over 30 runs. When considering $fitness_1$, we see that GP obtains the best results but also the worst average, which indicates that the results are not always stable and more evaluations could be needed. Evolution strategy, on the other hand, performs the worst while having all the values approximately the same. The second fitness allows GP to become more stable and we can actually see that even the worst values for GP are better than median values for GA and ES. Interestingly, we also observe that GA has better general performance than ES but worse outlier solutions. Finally, for $fitness_3$, we see a significant difference in the performance between GP on one side and GA and ES on the other side. More precisely, GA and ES perform similarly and much worse than GP where only the best GA outliers reach the worst performance for GP.

Figures 1b until 1f give results for case when $n = 8$. As already noted, for $fitness_1$ we surprisingly see that GP performs the worst. What is more, both GA and ES exhibit significantly better performance where GA is the best. In the second scenario ($fitness_2$), the situation changes completely and now GP is by far the best. At the same time, the difference between GA and ES is quite small. Finally, for $fitness_3$, GP is again the best with (almost) no differences between GA and ES.

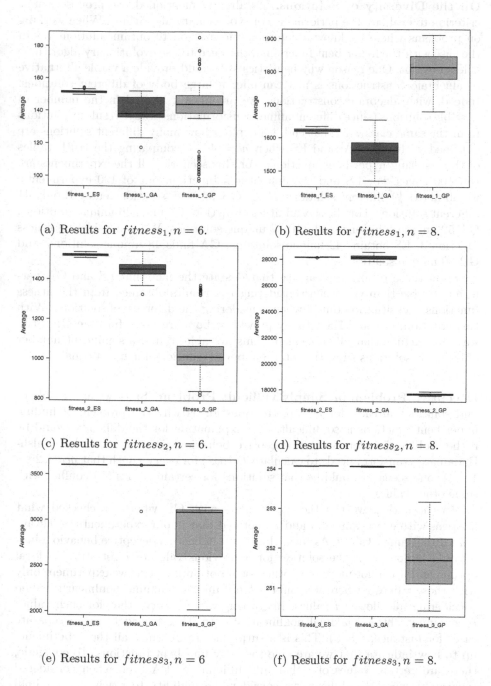

(a) Results for $fitness_1, n = 6$.

(b) Results for $fitness_1, n = 8$.

(c) Results for $fitness_2, n = 6$.

(d) Results for $fitness_2, n = 8$.

(e) Results for $fitness_3, n = 6$

(f) Results for $fitness_3, n = 8$.

Fig. 1. Results for all tested algorithms with $n = 6, 8$.

On the Diversity of Solutions. As already mentioned, we propose a new criterion to evaluate the performance of evolutionary algorithms. When working on problems where we know a deterministic method to obtain solutions (as in the case with the hyper-bent functions), the need to use evolutionary algorithms is less obvious. One reason why heuristics still could provide a viable alternative to algebraic constructions is if it can offer a large body of different solutions. Indeed, with algebraic constructions, we are always limited in the number of possible solutions (since different affine transformations always result in solutions from the same class). To that end, we explore how many different solutions are obtained with GP, GA, and ES when $n = 4$, by comparing the truth tables of the resulting hyper-bent functions. For this scenario, all the experiments are run 100 times. For ES and the first fitness function, out of 100 experiments, we obtained 49 different solutions. Next, GA finds 45 solutions and GP only 41 different solutions. For the second fitness function, ES finds 46 unique solutions, GA 52 unique solutions, and, GP 37 unique solutions. When considering fitness function 3, ES obtains 42 unique solutions, GA finds 48 unique solutions, and GP 30 unique solutions.

From these results, we can see that despite the fact that ES and GA performed worse than GP when considering the results obtained from the fitness functions, the situation differs when considering the diversity of solutions. With this criterion, GA and ES actually proved to be more powerful than GP. Still, we can conclude that all three algorithms are able to find a significant number of different solutions since the total number of solutions for $n = 4$ equals 56.

Deceptive Problem or Simply Difficult Problem. Since we are unable to find solutions already for $n = 6$, the question is why the problem of finding hyper-bent functions is so difficult. One explanation for the difficulty would be if the problem has some sort of deceptive behavior: since the extended Walsh-Hadamard transform needs to calculate $Tr(ax^i)$ for every i such that $gcd(i, 2^n - 1) = 1$, one scenario would be that solutions for certain values of i conflict with some other values.

To check this, we run the experiments with GP where we checked what happens with every value of i and we noticed that in our experiments we always fail for the same i values. As such, this could indicate a deceptive behavior since it could happen that the solution for a previous value of i got stuck in local optima and it cannot move to another local optimum. Next, we experiment only with those values i where we could not obtain the maximal nonlinearity when considering all allowed i values. Interestingly, we observe that for such values i we cannot obtain maximal nonlinearity even if we consider it as a separate case (for instance, $i = 5$). This is a surprising result since all the experiments up to now indicate that we are able to easily find bent functions. To conclude, there are certain values of i that are difficult for EA even when considered separately, where by difficult we consider the inability to reach the maximal nonlinearity. Thus, when considering all possible i values, it is impossible to find hyper-bent functions. From these observations, we conclude that the problem is not deceptive but is "simply" difficult in certain components. Currently, we leave it as an open question what does make certain values of i so difficult.

5.2   Future Work

There are several directions along which the present work can be extended. A first idea to improve the performance of GA would be to design specific crossover and mutation operators which reduce the search space of candidate solutions. In fact, as discussed in Sect. 2, the Walsh-Hadamard coefficients of a bent functions of n variables are all equal to $\pm 2^{\frac{n}{2}}$. Since the Walsh-Hadamard coefficient $W_f(\underline{0})$ of the null vector is the deviation of the function from being balanced, it follows that the truth table of any bent function is composed of $hwt_f = 2^{n-1} \pm 2^{\frac{n}{2}-1}$ ones. This suggests evolving through GA bitstrings with the fixed Hamming weight. A similar strategy was initially proposed by Millan et al. [11] to evolve balanced Boolean functions, where the authors proposed a crossover operator that used counters to keep track of the multiplicities of zeros and ones in the offspring. More recently, the same approach has been used by Mariot and Leporati [15] to evolve plateaued functions by spectral inversion, and by Mariot et al. [22] to evolve binary orthogonal arrays.

In our setting, a possible idea would be to adapt Millan et al.'s counter-based crossover operator to the case of bent functions, with the number of ones in the offspring constrained to be equal to either $2^{n-1} - 2^{\frac{n}{2}-1}$ or $2^{n-1} + 2^{\frac{n}{2}-1}$. Subsequently, it would be interesting to assess if the performances of our GA improve in generating hyper-bent functions.

On a more general level, a computational bottleneck is the calculation of the extended Walsh-Hadamard spectrum. The naive implementation of the Walsh-Hadamard transform has a complexity of $\mathcal{O}(2^{2n})$ where n is the Boolean function size. This can be improved by using the butterfly algorithm where the complexity decreases only to $\mathcal{O}(n2^n)$. Unfortunately, there does not seem to be an easy way to use the butterfly algorithm when calculating the extended Walsh-Hadamard transform. Consequently, the exponential rise in the complexity makes the calculation already extremely difficult for $n \geq 8$, especially coupled with a high number of evaluations occurring in the heuristic approaches. In future work, we plan to explore how to implement more efficiently the extended Walsh-Hadamard spectrum. Alternatively, since it is known that hyper-bent functions have algebraic degree equal to $n/2$ [7], one could try to include this information in the fitness function to speed up the search.

6   Conclusions

In this paper, we consider the evolution of hyper-bent functions, i.e., functions that are bent up to a primitive root change. Hyper-bent functions have real-world applications and are extremely rare objects, which makes them an interesting target for evolutionary algorithms. Our results indicate this problem to be of extreme difficulty and even out of reach for evolutionary algorithms. Indeed, we are able to find hyper-bent functions only for $n = 4$, which is the dimension where also exhaustive search is easily conducted.

Despite the failure in finding hyper-bent functions for $n > 4$, we can still discuss the performance of tested algorithms where we see that GP behaves the best. This is probably not surprising since GP also showed excellent results when evolving bent functions. When considering the diversity of obtained solutions, we observe that GA is the best but all algorithms show very good results: on average every second solution (or third for GP when considering $fitness_3$) is a new one. Finally, our experiments indicate that the difficulty of evolving hyper-bent functions stems from the fact that our algorithms are not able to find certain bent components of hyper-bent functions. It remains to be explored why is that and how can we overcome this obstacle. We note that our results open the problem of evolving hyper-bent functions as a strong benchmark when evaluating the performance of evolutionary algorithms.

Acknowledgments. The authors wish to thank the anonymous referees for their useful comments on improving the presentation quality of the paper. This work has been supported in part by Croatian Science Foundation under the project IP-2014-09-4882. In addition, this work was supported in part by the Research Council KU Leuven (C16/15/058) and IOF project EDA-DSE (HB/13/020).

References

1. Hrbacek, R., Dvorak, V.: Bent function synthesis by means of cartesian genetic programming. In: Bartz-Beielstein, T., Branke, J., Filipič, B., Smith, J. (eds.) PPSN 2014. LNCS, vol. 8672, pp. 414–423. Springer, Cham (2014). https://doi.org/10.1007/978-3-319-10762-2_41

2. Picek, S., Jakobovic, D.: Evolving algebraic constructions for designing bent Boolean functions. In: Proceedings of the 2016 on Genetic and Evolutionary Computation Conference, Denver, CO, USA, 20–24 July 2016, pp. 781–788 (2016)

3. Picek, S., Sisejkovic, D., Jakobovic, D.: Immunological algorithms paradigm for construction of Boolean functions with good cryptographic properties. Eng. Appl. Artif. Intell. **62**, 320–330 (2016)

4. Youssef, A.M., Gong, G.: Hyper-bent functions. In: Pfitzmann, B. (ed.) EURO-CRYPT 2001. LNCS, vol. 2045, pp. 406–419. Springer, Heidelberg (2001). https://doi.org/10.1007/3-540-44987-6_25

5. Carlet, C.: Boolean functions for cryptography and error correcting codes. In: Crama, Y., Hammer, P.L. (eds.) Boolean Models and Methods in Mathematics, Computer Science, and Engineering, pp. 257–397. Cambridge University Press, Cambridge (2010)

6. Lidl, R., Niederreiter, H.: Introduction to Finite Fields and Their Applications. Cambridge University Press, Cambridge (1994)

7. Carlet, C., Gaborit, P.: Hyper-bent functions and cyclic codes. J. Comb. Theory Ser. A **113**(3), 466–482 (2006)

8. Gong, G., Golomb, S.W.: Transform domain analysis of DES. IEEE Trans. Inf. Theory **45**(6), 2065–2073 (1999)

9. Charpin, P., Gong, G.: Hyperbent functions, Kloosterman sums, and Dickson polynomials. IEEE Trans. Inf. Theory **54**(9), 4230–4238 (2008)

10. Millan, W., Clark, A., Dawson, E.: An effective genetic algorithm for finding highly nonlinear Boolean functions. In: Proceedings of the First International Conference on Information and Communication Security, ICICS 1997, pp. 149–158 (1997)
11. Millan, W., Clark, A., Dawson, E.: Heuristic design of cryptographically strong balanced Boolean functions. In: Nyberg, K. (ed.) EUROCRYPT 1998. LNCS, vol. 1403, pp. 489–499. Springer, Heidelberg (1998). https://doi.org/10.1007/BFb0054148
12. Millan, W., Fuller, J., Dawson, E.: New concepts in evolutionary search for Boolean functions in cryptology. Comput. Intell. **20**(3), 463–474 (2004)
13. Picek, S., Jakobovic, D., Golub, M.: Evolving cryptographically sound Boolean functions. In: Proceedings of the 15th Annual Conference Companion on Genetic and Evolutionary Computation, GECCO 2013 Companion, pp. 191–192 (2013)
14. Mariot, L., Leporati, A.: Heuristic search by particle swarm optimization of Boolean functions for cryptographic applications. In: GECCO (Companion), pp. 1425–1426. ACM (2015)
15. Mariot, L., Leporati, A.: A genetic algorithm for evolving plateaued cryptographic Boolean functions. In: Dediu, A.-H., Magdalena, L., Martín-Vide, C. (eds.) TPNC 2015. LNCS, vol. 9477, pp. 33–45. Springer, Cham (2015). https://doi.org/10.1007/978-3-319-26841-5_3
16. Picek, S., Jakobovic, D., Miller, J.F., Batina, L., Cupic, M.: Cryptographic Boolean functions: one output, many design criteria. Appl. Soft Comput. **40**, 635–653 (2016)
17. Bäck, T., Fogel, D., Michalewicz, Z. (eds.): Evolutionary Computation 1: Basic Algorithms and Operators. Institute of Physics Publishing, Bristol (2000)
18. Rozenberg, G., Bäck, T., Kok, J.N.: Handbook of Natural Computing. Springer, Heidelberg (2011). https://doi.org/10.1007/978-3-540-92910-9
19. Beyer, H.G., Schwefel, H.P.: Evolution strategies a comprehensive introduction. Nat. Comput. **1**(1), 3–52 (2002)
20. Poli, R., Langdon, W.B., McPhee, N.F.: A field guide to genetic programming (2008). http://lulu.com, http://www.gp-field-guide.org.uk
21. Dillon, J.F.: Elementary Hadamard difference sets. Ph.D. thesis, University of Maryland (1974)
22. Mariot, L., Picek, S., Jakobovic, D., Leporati, A.: Evolutionary search of binary orthogonal arrays. In: Auger, A., Fonseca, C.M., Lourenço, N., Machado, P., Paquete, L., Whitley, D. (eds.) PPSN 2018, Part I. LNCS, vol. 11101, pp. 121–133. Springer, Cham (2018). https://doi.org/10.1007/978-3-319-99253-2_10

Learning Class Disjointness Axioms Using Grammatical Evolution

Thu Huong Nguyen$^{(\boxtimes)}$ (iD) and Andrea G. B. Tettamanzi (iD)

Université Côte d'Azur, CNRS, Inria, I3S, Nice, France
{thu-huong.nguyen,andrea.tettamanzi}@univ-cotedazur.fr

Abstract. Today, with the development of the Semantic Web, Linked Open Data (LOD), expressed using the Resource Description Framework (RDF), has reached the status of "big data" and can be considered as a giant data resource from which knowledge can be discovered. The process of learning knowledge defined in terms of OWL 2 axioms from the RDF datasets can be viewed as a special case of knowledge discovery from data or "data mining", which can be called "*RDF mining*". The approaches to automated generation of the axioms from recorded RDF facts on the Web may be regarded as a case of inductive reasoning and ontology learning. The instances, represented by RDF triples, play the role of specific observations, from which axioms can be extracted by generalization. Based on the insight that discovering new knowledge is essentially an evolutionary process, whereby hypotheses are generated by some heuristic mechanism and then tested against the available evidence, so that only the best hypotheses survive, we propose the use of Grammatical Evolution, one type of evolutionary algorithm, for mining disjointness OWL 2 axioms from an RDF data repository such as DBpedia. For the evaluation of candidate axioms against the DBpedia dataset, we adopt an approach based on possibility theory.

Keywords: Ontology learning · OWL 2 axiom · Grammatical evolution

1 Introduction

The manual acquisition of formal conceptualizations within domains of knowledge, i.e. ontologies [1] is an expensive and time-consuming task because of the requirement of involving domain specialists and knowledge engineers. This is known as the "*knowledge acquisition bottleneck*". Ontology learning, which comprises the set of methods and techniques used for building an ontology from scratch, enriching, or adapting an existing ontology in a semi-automatic fashion, using several knowledge and information sources [2,3], is a potential approach to overcome this obstacle. An overall classification of ontology learning methods can be found in [3–5]. Ontology learning may be viewed as a special case of knowledge discovery from data (KDD) or data mining, where the data are in a

© Springer Nature Switzerland AG 2019
L. Sekanina et al. (Eds.): EuroGP 2019, LNCS 11451, pp. 278–294, 2019.
https://doi.org/10.1007/978-3-030-16670-0_18

special format and knowledge can consist of concepts, relations, or axioms from a domain-specific application.

Linked Open Data (LOD) being Linked Data[1] published in the form of an Open Data Source can be considered as a giant real-world knowledge base. Such a huge knowledge base opens up exciting opportunities for learning new knowledge in the context of an open world. Based on URIs, HTTP, and RDF, Linked Data is a recommended best practice for exposing, sharing, and connecting pieces of data, information, and knowledge on the Semantic Web. Some approaches to ontology learning from linked data can be found in [6–8]. The advantages of LOD with respect to learning described in [8] is that it is publicly available, highly structured, relational, and large compared with other resources. Ontology learning on the Semantic Web involves handling the enormous and diverse amount of data in the Web and thus enhancing existing approaches for knowledge acquisition instead of only focusing on mostly small and uniform data collections.

In ontology learning, one of the critical tasks is to increase the expressiveness and semantic richness of a knowledge base (KB), which is called *ontology enrichment*. Meanwhile, exploiting ontological axioms in the form of logical assertions to be added to an existing ontology can provide some tight constraints to it or support the inference of implicit information. Adding axioms to a KB can yield several benefits, as indicated in [9]. In particular, class disjointness axioms are useful for checking the logical consistency and detecting undesired usage patterns or incorrect assertions. As for the definition of disjointness [10], two classes are disjoint if they do not possess any common individual according to their intended interpretation, i.e., the intersection of these classes is empty in a particular KB.

A simple example can demonstrate the potential advantages obtained by the addition of this kind of axioms to an ontology. A knowledge base defining terms of classes like *Person, City* and asserting that *Sydney* is both a *Person* and a *City* would be logically consistent, without any errors being recognized by a reasoner. However, if a constraint of disjointness between classes *Person* and *City* is added, the reasoner will be able to reveal an error in the modeling of such a knowledge base. As a consequence, logical inconsistencies of facts can be detected and excluded—thus enhancing the quality of ontologies.

As a matter of fact, very few DisjointClasses axioms are currently found in existing ontologies. For example, in the DBpedia ontology, the query SELECT ?x ?y { ?x owl:disjointWith ?y } executed on November 11, 2018 returned only 25 solutions, whereas the realistic number of class disjointness axioms generated from hundreds of classes in DBpedia (432 classes in DBpedia 2015-04, 753 classes in DBpedia 2016-04) is expected to be much more (in the thousands). Hence, learning implicit knowledge in terms of axioms from a LOD repository in the context of the Semantic Web has been the object of research in several different approaches. Recent methods [11,12] apply top-down or intensional approaches to learning disjointness which rely on schema-level information, i.e., logical and lexical decriptions of the classes. The contributions based on bottom-

---

[1] http://linkeddata.org/.

up or extensional approaches [9,10], on the other hand, require the instances in the dataset to induce instance-driven patterns to suggest axioms, e.g., disjointness class axioms.

Along the lines of extensional (i.e. instance-based) methods, we propose an evolutionary approach, based on grammatical evolution, for mining implicit axioms from RDF datasets. The goal is to derive potential class disjointness axioms of more complex types, i.e., defined with the help of relational operators of intersection and union; in other words, axioms like $Dis(C_1, C_2)$, where C_1 and C_2 are complex class expressions including $\sqcap$ and $\sqcup$ operators. Also, an evaluation method based on possibility theory is adopted to assess the certainty level of induced axioms.

The rest of the paper is organized as follows: some related works are described briefly in Sect. 2. In Sect. 3, some background is provided. OWL 2 classes disjointness axioms discovery with a GE approach is presented in Sect. 4. An axiom evaluation method based on possibility theory is also presented in this section. Section 5 provides experimental evaluation and comparison. Conclusions and directions for future research are given in Sect. 6.

2   Related Work

The most prominent related work relevant to learning disjointness axioms consists of the contributions by Johanna Völker and her collaborators [10,12,13]. In early work, Völker developed supervised classifiers from LOD incorporated in the *LeDA* tool [12]. However, the learning algorithms need a set of labeled data for training that may demand expensive work by domain experts. In contrast to *LeDA*, statistical schema induction via associative rule mining [10] was given in the tool *GoldMiner*, where association rules are representations of implicit patterns extracted from large amount of data and no training data is required. Association rules are compiled based on a transaction table, which is built from the results of SPARQL queries. That research only focused on generating axioms involving atomic classes, i.e., classes that do not consist of logical expressions, but only of a single class identifier.

Another relevant research is the one by Lorenz Bühmann and Jens Lehmann, whose proposed methodology is implemented in the DL-Learner system [11] for learning general class descriptions (including disjointness) from training data. Their work relies on the capabilities of a reasoning component, but suffers from scalability problems for the application to large datasets like LOD. In [9], they tried to overcome these obstacles by obtaining predefined data queries, i.e., SPARQL queries to detect specific axioms hidden within relevant data in datasets for the purpose of ontology enrichment. That approach is very similar to ours in that it uses an evolutionary algorithm for learning concepts. Bühmann and Lehmann also developed methods for generating more complex axiom types [14] by using frequent terminological axiom patterns from several data repositories. One important limitation of their method is the time-consuming and computationally expensive process of learning frequent axioms patterns and converting them into SPARQL queries before generating actual axioms from instance data. Also, the most frequent patterns refer to inclusion and equivalence axioms like $A \equiv B \sqcap \exists r.C$ or $A \sqsubseteq B \sqcap \exists r.C$.

Our solution is based on an evolutionary approach deriving from previous work, but concentrating on a specific algorithm, namely Grammatical Evolution (GE) [15] to generate class disjointness axioms from an existing RDF repository which is different from the use of Genertic Algorithm as in the approach of Bühmann and Lehmann [14]. GE aims at overcoming one shortcoming of GP, which is the growth of redundant code, also known as *bloat*. Furthermore, instead of using probability theory, we applied a possibilistic approach to assess the fitness of axioms.

3   Background

This section provides a few background notions required to understand the application domain of our contribution.

3.1   RDF Datasets

The Semantic Web[2] (SW) is an extension of the World Wide Web and it can be considered as the Web of data, which aims to make Web contents machine-readable. The Linked Open Data[3] (LOD) is a collection of linked RDF data. The LOD covers the data layer of the SW, where RDF plays the roles of its standard data model.

RDF uses as statements triples of the form (Subject, Predicate, Object). According to the World Wide Web Consortium (W3C), RDF[4] has features that facilitate data merging even if the underlying schemas differ, and it specifically supports the evolution of schemas over time without requiring all the data consumers to be changed. RDF data may be viewed as an oriented, labeled multi-graph. The query language for RDF is SPARQL,[5] which can be used to express queries across diverse data sources, whether the data is stored natively as RDF or viewed as RDF via some middleware.

One of the prominent examples of LOD is DBpedia,[6] which comprises a rather rich collection of facts extracted from Wikipedia. DBpedia covers a broad variety of topics, which makes it a fascinating object of study for a knowledge extraction method. DBpedia owes to the collaborative nature of Wikipedia the characteristic of being incomplete and ridden with inconsistencies and errors. Also, the facts in DBpedia are dynamic, because they can change in time. DBpedia has become a giant repository of RDF triples and, therefore, it looks like a perfect testing ground for the automatic extraction of new knowledge.

---

[2] https://www.w3.org/standards/semanticweb/.
[3] https://www.w3.org/egov/wiki/Linked_Open_Data.
[4] https://www.w3.org/RDF/.
[5] https://www.w3.org/TR/rdf-sparql-query/.
[6] https://wiki.dbpedia.org/.

3.2   OWL 2 Axioms

We are interested not only in extracting new knowledge from an existing knowledge base expressed in RDF, but also in being able to inject such extracted knowledge into an ontology in order to be able to exploit it to infer new logical consequences.

While the former objective calls for a target language, used to express the extracted knowledge, which is as expressive as possible, lest we throttle our method, the latter objective requires using at most a decidable fragment of first-order logic and, possibly, a language which makes inference problems tractable.

OWL 2[7] is an ontology language for the Semantic Web with formally defined meaning which strikes a good compromise between these two objectives. In addition, OWL 2 is standardized and promotes interoperability with different applications. Furthermore, depending on the applications, it will be possible to select an appropriate *profile* (corresponding to a different language fragment) exhibiting the desired trade-off between expressiveness and computational complexity.

4   A Grammatical Evolution Approach to Discovering OWL 2 Axioms

This section introduces a method based on Grammatical Evolution (GE) to mine an RDF repository for class disjointness axioms. GE is similar to GP in automatically generating variable-length programs or expressions in any language. In the context of OWL 2 axiom discovery, the "programs" are axioms. A population of individual axioms is maintained by the algorithm and iteratively refined to find the axioms with the highest level of credibility (one key measure of quality for discovered knowledge). In each iteration, known as a *generation*, the fitness of each individual in the population is evaluated using a possibilistic approach and is the base for the parent selection mechanism. The offspring of each generation is bred by applying genetic operators on the selected parents. The overall flow of such GE algorithm is shown in Algorithm 1.

4.1   Representation

As in O'Neill et al. [15] and unlike GP, GE applies the evolutionary process on variable length binary strings instead of on the actual programs. GE has a clear distinction in representation between genotype and phenotype. The genotype to phenotype mapping is employed to generate axioms considered as phenotypic programs by using the Backus-Naur form (BNF) grammar [15–17].

---

[7] https://www.w3.org/TR/owl2-overview/.

Algorithm 1 - GE for discovering axioms from a set of RDF datasets

Input: T: RDF Triples data; Gr: BNF grammar; $popSize$: the size of the population;
 $\quad\quad initlenChrom$: the initialized length of chromosome;
 $\quad\quad maxWrap$: the maximum number of wrapping; $pElite$: elitism propotion
 $\quad\quad pselectSize$: parent selection propotion; $pCross$: the probability of crossover;
 $\quad\quad pMut$: the probability of mutation;
Output: Pop: a set of axioms discovered based on Gr

1: Initialize a list of chromosomes L of length $initlenChrom$.
 $\quad$ Each codon value in chromosome are integer.
2: Create a population P of size $popSize$ mapped from list of chromosomes L
 $\quad$ on grammar Gr by performing $popSize$ times **CreateNewAxiom()**
3: Compute the fitness values for all axioms in Pop.
4: Initialize current generation number ($currentGeneration = 0$)
5: **while**($currentGeneration < maxGenerations$) **do**
6: $\quad$ Sort Pop by descending fitness values
7: $\quad$ Create a list of elite axioms $listElites$ with the propotion $pElite$ of the number
 $\quad$ of the fittest axioms in Pop
8: $\quad$ Add all axioms of $listElites$ to a new population $newPop$
9: $\quad$ Select the remaining part of population after elitism selection
 $\quad\quad Lr \leftarrow Pop \backslash listElites$
10: $\quad$ Eliminate the duplicates in Lr
 $\quad\quad Lr \leftarrow$ Distinct (Lr)
11: $\quad$ Create a a list of axioms $listCrossover$ used for crossover operation
 $\quad$ with the propotion $pselectSize$ of the number of
 $\quad$ the fittest individuals in Lr
11: $\quad$ Shuffle($listCrossover$)
12: $\quad$ **for** (i=0,1....$listCrossover.length$-2) **do**
10: $\quad\quad parent1 \leftarrow listCrossover[i]$
13: $\quad\quad parent2 \leftarrow listCrossover[i+1]$
14: $\quad\quad child1, child2 \leftarrow CROSSOVER(parent1,parent2)$ with the probability $pCross$
15: $\quad\quad$ **for each** $offspring$ $\{child1,child2\}$ **do** $MUTATION(offspring)$
16: $\quad\quad$ Compute fitness values for $child1,\ child2$
17: $\quad\quad$ Select $w1,\ w2$ - winners of competition between parents and offsprings
 $\quad\quad w1,w2 \leftarrow CROWDING((parent1,\ parent2,\ child1,\ child2)$
18: $\quad\quad$ Add $w1$, $w2$ to new population $newPop$
19: $\quad Pop= newPop$
20: $\quad$ Increase the number of generation $curGeneration$ by 1
21: return Pop

Structure of BNF Grammar. We applied the extended BNF grammar consisting of the production rules extracted from the normative grammar[8] of OWL 2 in the format used in W3C documentation for constructing different types of OWL 2 axioms. The noteworthy thing is that the use of a BNF grammar here does not focus on defining what a well-formed axiom may be, but on generating well-formed axioms which may express the facts contained in a given RDF triple store. Hence, resources, literals, properties, and other elements of the language that actually occur in the RDF repository could be generated. We organized our BNF grammar in two main parts (namely static and dynamic) as follows:

- the static part contains production rules defining the structure of the axioms loaded from the text file. Different grammars will generate different kinds of axioms.

---

[8] https://www.w3.org/TR/owl2-syntax/.

– the dynamic part contains production rules for the low-level non-terminals, which we will call *primitives*. These production rules are automatically built at runtime by querying the SPARQL endpoint of the RDF repository at hand.

This approach to organizing the structure of BNF grammar ensures the changes in the contents of RDF repositories will not require to rewrite the grammar.

In the following, we will refer to generating class disjointness axioms containing atomic expression such as DisjointClasses(Film, WrittenWork) or complex expression in the cases of relational operators, i.e., intersection and union, such as DisjointClasses(Film, ObjectIntersectionOf(Book, ObjectUnionOf(Comics, MusicalWork))). We built the following pattern of the grammar structured for generating class disjointness axioms:

% Static Part

```
(r1) Axiom := ClassAxiom
(r2) ClassAxiom := DisjointClasses
(r3) DisjointClasses := 'DisjointClasses' '(' ClassExpression ' 'ClassExpression ')'
(r4) ClassExpression :=   Class                (0)
                        | ObjectUnionOf        (1)
                        | ObjectIntersectionOf (2)
(r5) ObjectUnionOf := 'ObjectUnionOf' '(' ClassExpression ' ' ClassExpression ')'
(r6) ObjectIntersectionOf := 'ObjectIntersectionOf' '(' ClassExpression' 'ClassExpression ')'
```

% Dynamic Part - Primitives

```
(r7) Class := % production rules are constructed by using SPARQL queries
```

This produces rules of the primitive Class, which will be filled by using SPARQL queries to extract the IRI of a class mentioned in the RDF store. An example representing a small excerpt of an RDF triple repository is the following:

```
PREFIX dbr: http://DBpedia.org/resource/
PREFIX dbo: http://DBpedia.org/ontology/
PREFIX rdf: http://www.w3.org/1999/02/22\-rdf-syntax-ns\#

dbr:Quiet_City_(film)     rdf:type    dbo:Film.
dbr:Cantata               rdf:type    dbo:MusicalWork.
dbr:The_Times             rdf:type    dbo:WrittenWork.
dbr:The_Hobbit            rdf:type    dbo:Book.
dbr:Fright_Night_(comics) rdf:type    dbo:Comic
```

and options for the nonterminal Class are represented as follows:

```
(r7) Class := dbo:Film            (0)
            | dbo:MusicalWork     (1)
            | dbo:WrittenWork     (2)
            | dbo:Book            (3)
            | dbo:Comic           (4)
```

Encoding and Decoding. Individual axioms are encoded as variable-length binary strings with numerical chromosomes. The binary string consists of a sequence of 8-bit words referred to as *codons*.

According to the structure of the above BNF grammar, chromosomes are then decoded into OWL 2 axioms in different OWL syntaxes through the mapping process according to the function:

$$\text{Rule} = \text{Codon value } \textbf{modulo} \text{ Number of Rules for the current terminal} \qquad (1)$$

In the advantageous cases, axioms are generated before the end of the genome is reached; otherwise, a *wrapping* operator [15,16] is applied and the reading of codons will continue from the beginning of the chromosome, until the maximum allowed number of wrapping events is reached. An unsuccessful mapping will happen if the threshold on the number of wrapping events is reached but the individual is still not completely mapped; in this case, the individual is assigned the lowest possible fitness. The production rule for `ClassExpression` is recursive and may lead to a large fan-out; to alleviate this problem and promote "reasonable" axioms, we increase the probability of obtaining a successful mapping to complex axiom expressions, we double the appearance probability of non-terminal `ClassExpression`. Rule (r4) in the grammar is modified to

```
(r4) ClassExpression :=   Class                     (0)
                        | Class                     (1)
                        | ObjectUnionOf             (2)
                        | ObjectIntersectionOf      (3)
```

4.2   Initialization

In the beginning of the evolutionary process, a set of chromosomes, i.e., genotypic individuals, are randomly initialized once and for all. Each chromosome is a set of integers with the initialized length *initlenChrom*. Its length can be extended to *maxlenChrom* in the scope of the threshold of *maxWrap* in the wrapping process. The next step is the transformation of genotypes into phenotypic individuals, i.e., axioms according to grammar *Gr*, by means of the mapping process based on the input grammar called *CreateNewAxiom()* operator. The population of *popSize* class disjointness axioms is created by iterating *popSize* times *CreateNewAxiom()* operator described in Algorithm 2.

4.3   Parent Selection

Before executing the selection operator, the axioms in the populations are evaluated and ranked in descending order of their fitness. To combat the loss of fittest axioms as a result of the application of the variation operators, elitism selection is applied to copy the small proportion *pElite* of the best axioms into the next generation (line 7–8 of Algorithm 1). In the remaining part of the population, the elimination of duplicates is carried out to ensure only distinct individuals will be included in the *candidate list* for parent selection. The parent selection

---

Algorithm 2 - CreateNewAxiom()

Input: *Chr*: Chromosome - a set of integers; *Gr*: BNF grammar
Output: *A*: a new axiom individual

1: *maxlenChrom ← initlenChrom * maxWrap*
2: *ValCodon ← random(maxValCodon)*.
3: Set up *Chr* as input genotype *gp* used in mapping proccess to axiom individual *A*
4: *while* (*Chr.length <maxlenChrom*) && (incomplete mapping) *do*
5:     mapping from input genotype *gp* to output phenotype of axiom individual *A*
       according to grammar *Gr*
6: return *A*

---

mechanism amounts to choosing the fittest individuals from this list for reproduction. Figure 1 illustrates the process of selecting potential candidate solutions for recombination, i.e., a list of parents. The top proportion *pselectSize* of distinct individuals in the *candidate list* is selected and it is replicated to maintain the size *popSize* of population. The list of parents is shuffled and the individuals are paired in order from the beginning to the end of the list.

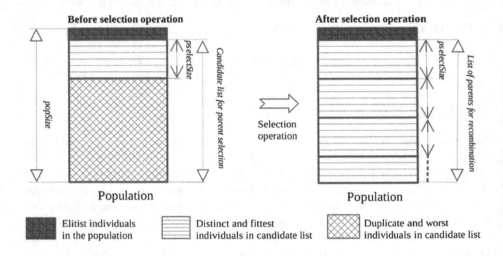

Fig. 1. An illustration of the parent selection mechanism.

4.4 Variation Operators

The purpose of these operators is to create new axioms from old ones. The standard genetic operators of crossover and mutation in the Evolutionary Algorithms (EA) are applied in the search space of genotypes. Well-formed individuals will then be generated syntactically from the new genotypes in the genotype-to-phenotype mapping process.

Algorithm 3 - Crowding(parent1, parent2, offspring1, offspring2)
Input: *parent1, parent2, child 1, child 2*: a crowd of individual axioms; **Output:** *A*: *ListWinners*- a list containing two winners of individual axioms

1: *d1* ← *DISTANCE(parent1,child1)* +*DISTANCE (parent2,child2)*
    d2 ← *DISTANCE(parent1, child2)* + *DISTANCE(parent2, child1)*
    in which *DISTANCE(parent, child)* - the number of distinct codons between
    parent and child.
2: **if**(*d1* >*d2*)
    ListWinners[0]← *COMPARE(parent1,child1)*
    ListWinners[1]← *COMPARE(parent2,child2)*
    else
    ListWinners[0]← *COMPARE(parent1,child2)*
    ListWinners[1]← *COMPARE(parent2,child1)*
    in which *COMPARE(parent, child)* - defines which individual in *(parent,child)*
    having higher fitness value.
3: **return** *ListWinners*

Crossover. A standard one-point crossover is employed whereby a single crossover point on the chromosomes of both parents is chosen randomly. The sets of codons beyond those points are exchanged between the two parents with probability *pCross*. The result of this exchange is two offspring genotypes. The mapping of these genotype into phenotypic axioms is performed by executing the *CreateNewAxiom()* operator (Algorithm 2) again with the offspring chromosomes as input.

Mutation. The mutation is applied to the offspring genotypes of crossover with probability *pMut*. A selected individual undergoes single-point mutation, i.e. a codon is selected at random, and this codon is replaced with a new randomly generated codon.

4.5   Survival Selection

In order to preserve population diversity, we used the *Deterministic Crowding* approach developed by Mahfoud [18]. Each offspring competes with its most similar peers, based on a genotypic comparison, to be selected for inclusion in the population of the next generation. Algorithm 3 describes this approach in detail. Even though we are aware that computing distance at the phenotypic level would yield more accurate results, we chose to use genotypic distance because it is much faster and easier to compute.

4.6   Fitness Evaluation

As a consequence of the heterogeneous and collaborative character of the linked open data, some facts (instances) in the RDF repository may be missing or erroneous. This incompleteness and noise determines a sort of *epistemic* uncertainty in the evaluation of the quality of a candidate axiom. In order to properly capture this type of uncertainty, typical of an open world, which contrasts with the *ontic* uncertainty typical of random processes, we adopt an axiom scoring heuristics based on possibility theory, proposed in [19], which is suitable for dealing with incomplete knowledge. This is a justified choice for assessing knowledge extracted from an RDF repository. We now provide a summary of this scoring heuristics.

Possibility theory [20] is a mathematical theory of epistemic uncertainty. Given a finite universe of discourse Ω, whose elements $\omega \in \Omega$ may be regarded as events, values of a variable, possible worlds, or states of affairs, a possibility distribution is a mapping $\pi : \Omega \to [0,1]$, which assigns to each ω a degree of possibility ranging from 0 (impossible, excluded) to 1 (completely possible, normal). A possibility distribution π for which there exists a completely possible state of affairs ($\exists \omega \in \Omega : \pi(\omega) = 1$) is said to be *normalized*.

A possibility distribution π induces a *possibility measure* and its dual *necessity measure*, denoted by Π and N respectively. Both measures apply to a set $A \subseteq \Omega$ (or to a formula ϕ, by way of the set of its models, $A = \{\omega : \omega \models \phi\}$), and are usually defined as follows:

$$\Pi(A) = \max_{\omega \in A} \pi(\omega); \tag{2}$$

$$N(A) = 1 - \Pi(\bar{A}) = \min_{\omega \in \bar{A}}\{1 - \pi(\omega)\}. \tag{3}$$

In other words, the possibility measure of A corresponds to the greatest of the possibilities associated to its elements; conversely, the necessity measure of A is equivalent to the impossibility of its complement $\bar{A}$.

A generalization of the above definition can be obtained by replacing the min and the max operators with any dual pair of triangular norm and co-norm.

In the case of possibilistic axiom scoring, the basic principle for establishing the possibility of a formula ϕ should be that the absence of counterexamples to ϕ in the RDF repository means $\Pi([\phi]) = 1$, i.e., that ϕ is completely possible. Let ϕ be an axiom that we wish to evaluate (i.e., a theory). The *content* of an axiom ϕ that we wish to evaluate is defined as a set of logical consequences

$$\text{content}(\phi) = \{\psi : \phi \models \psi\}, \tag{4}$$

obtained through the instatiation of ϕ to the vocabulary of the RDF repository; the cardinality of content(ϕ) is finite and every formula $\psi \in$ content(ϕ) may be readily tested by means of a SPARQL ASK query. Let us define $u_\phi = \|\text{content}(\phi)\|$ as the *support* of ϕ. Let then u_ϕ^+ be the number of confirmations (basic statements ψ that are satisfied by the RDF repository) and u_ϕ^- the number of counterexamples (basic statements ψ that are falsified by the RDF repository).

The possibility measure $\Pi(\phi)$ and the necessity measure $N(\phi)$ of an axiom have been defined as follows in [19]: for $u_\phi > 0$,

$$\Pi(\phi) = 1 - \sqrt{1 - \left(\frac{u_\phi - u_\phi^-}{u_\phi}\right)^2} \; ; \tag{5}$$

$$N(\phi) = \sqrt{1 - \left(\frac{u_\phi - u_\phi^+}{u_\phi}\right)^2} \; , \quad \text{if } \Pi(\phi) = 1, \text{ 0 otherwise.} \tag{6}$$

The cardinality of the sets of the facts in the RDF repository reflects the generality of each axiom. An axiom is all the more necessary as it is explicitly supported by facts, i.e., confirmations, and not contradicted by any fact, i.e., counterexamples, while it is the more possible as it is not contradicted by any fact. These numbers of confirmations and counterexamples are counted by executing corresponding SPARQL queries via an accessible SPARQL endpoint.

In principle, the fitness of axiom ϕ should be directly proportional to its necessity $N(\phi)$, its possibility $\Pi(\phi)$, and its support u_ϕ, which is an indicator of its generality. In other words, what we are looking for is not only credible axioms, but also general ones. A definition of the fitness function that satisfies such requirement is

$$f(\phi) = u_\phi \cdot \frac{\Pi(\phi) + N(\phi)}{2}, \tag{7}$$

which we adopted for our method.

5 Experiment and Result

We applied our approach to mine the classes disjointness axioms relevant to the topic *Work* in DBpedia. The statistical data of classes and instances about this topic in DBpedia 2015-04 is given in Table 1.

Table 1. Statistical data in the topic Work in DBpedia

Total number of classes	62
Total number of classes having instances	53
Total number of instances	519,5019

All data used in this experiment is represented in the form of RDF triples, as explained in Sect. 3. In order to assess its ability to discover axioms, we ran the GE indicated in Sect. 4 by repeating the sample procedure of Algorithm 1 for each run with the same parameters indicated in Table 2. The chart in Fig. 2 illustrates the average diversity of the population of axioms over the generations

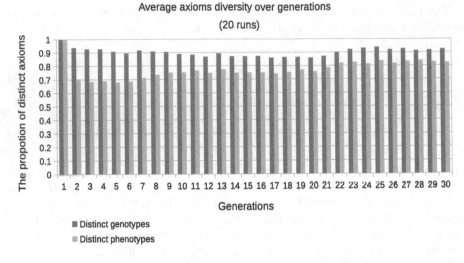

Fig. 2. The diversity of axioms over generations

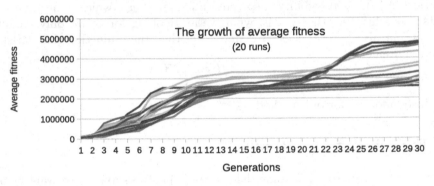

Fig. 3. The growth of average fitness over generations

of the evolutionary process. It shows how many different "species" of axioms are contained in the population, i.e., axioms that cover different aspects of the known facts. One of the remarkable points here is that there is a more rapid loss of diversity in the phenotype axioms compared with this decrease in the genotype ones. The use of *Crowding method* on genotypes instead of phenotypes can be the reason of this difference. Likewise, a set of codons of two parent chromosomes which are used for the mapping to phenotypes can fail to be swapped in the single-point crossover operator.

From the chart in Fig. 3, we can observe a gradual increase in the quality of discovered axioms over generations.

In order to evaluate the effectiveness of our method in discovering disjointness class axioms of the *Work* on RDF datasets of DBpedia, a benchmark of class disjointness axioms about this topic was manually created, which we called *Gold Standard*. The process of creating the *Gold Standard* was carried out by

Table 2. Input parameter values for GE

Parameter	Value
popSize	500
numGenerations	30
initlenChrom	20
maxWrap	2
pCross	80%
pMut	1%
pselectSize	70%
pElite	2%

Table 3. Experimental results

	Our approach		GoldMiner
	Complex axioms	Atomic axioms	Atomic axioms
Precision (per run)	0.867 ± 0.03	0.95 ± 0.02	0.95
Recall (per run)	N/A	0.15 ± 0.017	0.38
Recall (over 20 runs)	N/A	0.54	0.38

knowledge engineers and consisted of two phases. In the first phase, the disjointness of the top-most classes to their siblings was assessed manually. Therefrom, two sibling classes being disjoint will infer automatically the disjointness of their corresponding pair of subclasses. This process is repeated in the same way on the next level of concepts. The second phase of *Gold Standard* creation consisted in manually annotating the disjointness for the not yet noted pairs of classes which did not belong to the cases given in the previous phase. The result of the completion of the *Gold Standard* is the disjointness evaluation between 1,891 pairs of distinct classes relevant to the chosen topic. Table 3 summarizes the performance of our approach in discovering axioms with the parameters setting in Table 2 over 20 runs. The precision and recall are computed by comparison to the *Gold Standard*. Although the main purpose in our research is to focus on exploring the more complex disjointness axioms which contain logical relationship—intersection and union expression, we also performed experiments to generate axioms involving atomic classes only, for comparison purpose. We carry out the comparison with the results of *GoldMiner* [10] in generating class disjointness axioms about the topic *Work*. The precision and recall are computed by comparison to the *Gold Standard*. The results in Table 3 confirm the high accuracy of our approach in discovering class disjointness axioms in the case of atomic expressions (Precision = 0.95 ± 0.02). Also, the recall value is higher than the value in *GoldMiner*. There are a number of class disjointness axioms generated in our experiments which are absent in the result of *Gold-Miner*. For example, there are no any axioms relevant to class Archive in the axioms generated by *GoldMiner*. In the case of more complex axioms, there is a smaller degree of precision (Precision $= 0.867 \pm 0.03$). The reason may stem from the complexity in the content of generated axioms which is relevant to more different classes. We do not present the recall for the case of complex axioms, since the discovery process of this type of axioms cannot define how many of the complex axioms should have been generated. After 20 runs, from 10,000 candidate individual axioms, we got 5,728 qualified distinct complex axioms. We performed an analysis of the discovered axioms and found some noticeable points. Almost all generated axioms have high fitness values with millions of support instances from the DBpedia dataset, which witness the generality of the discovered axioms. However, we found some deficiencies in determining the disjointness of classes. As in the case of axiom DisjointClasses(MovingImage, ObjectUnionOf(Article,ObjectUnionOf(Image, MusicalWork))), 4,839,992 triples in DBpedia confirm that this class disjointness axiom is valid. However, according to the *Gold Standard*, these two classes should not be disjoint *a priori*. Indeed, the class MovingImage can be assessed as a subclass of Image, which makes the disjointness between class MovingImage and any complex class expression involving relational operator union of class Image altogether impossible. Another similar case is the axiom DisjointClasses(ObjectUnionOf(Article, ObjectUnionOf(ObjectUnionOf(ObjectUnionOf (ObjectUnionOf(TelevisionShow, WrittenWork), MusicalWork), Image), Film)), UndergroundJournal), having 5,037,468 triples in its support. However, according to the *Gold Standard*, these two classes should not be disjoint.

From the above examples we can infer that the main reason for such erroneous axioms may lie in the inconsistencies and errors in the DBpedia dataset. Therefore, a necessary direction to improve the quality of the knowledge base is to use the results of our mining algorithms to pinpoint errors and inconsistencies and thus aim at tighter constraints by excluding problematic triples from the RDF repository.

6   Conclusion and Future Work

We proposed an algorithm based on GE to discover class disjointness axioms from the DBpedia dataset relevant to the topic "Work". The experiment results have allowed us to evaluate the effectiveness of the model and analyze some of its shortcomings.

In future work, we will focus on three main directions:

1. improving the diversity of generated axioms by applying the crowding method at the level of phenotypes;
2. mining different types of axioms like identity axioms, equivalence axioms and relevant to broader topics;
3. enhancing the performance of the algorithm on parallel hardware in order to be able to carry out bigger data analytics.

References

1. Guarino, N., Oberle, D., Staab, S.: What is an ontology? In: Staab, S., Studer, R. (eds.) Handbook on Ontologies. IHIS, pp. 1–17. Springer, Heidelberg (2009). https://doi.org/10.1007/978-3-540-92673-3_0

2. Maedche, A., Staab, S.: Ontology learning for the semantic web. IEEE Intell. Syst. **16**(2), 72–79 (2001)

3. Lehmann, J., Völker, J.: Perspectives on Ontology Learning, Studies on the Semantic Web, vol. 18. IOS Press, Amsterdam (2014)

4. Drumond, L., Girardi, R.: A survey of ontology learning procedures. In: WONTO. CEUR Workshop Proceedings, vol. 427. CEUR-WS.org (2008)

5. Hazman, M., El-Beltagy, S.R., Rafea, A.: Article: a survey of ontology learning approaches. Int. J. Comput. Appl. **22**(8), 36–43 (2011)

6. Zhao, L., Ichise, R.: Mid-ontology learning from linked data. In: Pan, J.Z., et al. (eds.) JIST 2011. LNCS, vol. 7185, pp. 112–127. Springer, Heidelberg (2012). https://doi.org/10.1007/978-3-642-29923-0_8

7. Tiddi, I., Mustapha, N.B., Vanrompay, Y., Aufaure, M.-A.: Ontology learning from open linked data and web snippets. In: Herrero, P., Panetto, H., Meersman, R., Dillon, T. (eds.) OTM 2012. LNCS, vol. 7567, pp. 434–443. Springer, Heidelberg (2012). https://doi.org/10.1007/978-3-642-33618-8_59

8. Zhu, M.: DC proposal: ontology learning from noisy linked data. In: Aroyo, L., et al. (eds.) ISWC 2011. LNCS, vol. 7032, pp. 373–380. Springer, Heidelberg (2011). https://doi.org/10.1007/978-3-642-25093-4_31

9. Bühmann, L., Lehmann, J.: Universal OWL axiom enrichment for large knowledge bases. In: ten Teije, A., et al. (eds.) EKAW 2012. LNCS (LNAI), vol. 7603, pp. 57–71. Springer, Heidelberg (2012). https://doi.org/10.1007/978-3-642-33876-2_8

10. Völker, J., Fleischhacker, D., Stuckenschmidt, H.: Automatic acquisition of class disjointness. J. Web Sem. **35**, 124–139 (2015)

11. Lehmann, J.: Dl-learner: learning concepts in description logics. J. Mach. Learn. Res. **10**, 2639–2642 (2009)

12. Völker, J., Vrandečić, D., Sure, Y., Hotho, A.: Learning disjointness. In: Franconi, E., Kifer, M., May, W. (eds.) ESWC 2007. LNCS, vol. 4519, pp. 175–189. Springer, Heidelberg (2007). https://doi.org/10.1007/978-3-540-72667-8_14

13. Fleischhacker, D., Völker, J.: Inductive learning of disjointness axioms. In: Meersman, R., et al. (eds.) OTM 2011. LNCS, vol. 7045, pp. 680–697. Springer, Heidelberg (2011). https://doi.org/10.1007/978-3-642-25106-1_20

14. Bühmann, L., Lehmann, J.: Pattern based knowledge base enrichment. In: Alani, H., et al. (eds.) ISWC 2013. LNCS, vol. 8218, pp. 33–48. Springer, Heidelberg (2013). https://doi.org/10.1007/978-3-642-41335-3_3

15. O'Neill, M., Ryan, C.: Grammatical evolution. Trans. Evol. Comput. **5**(4), 349–358 (2001). https://doi.org/10.1109/4235.942529

16. Dempsey, I., O'Neill, M., Brabazon, A.: Foundations in Grammatical Evolution for Dynamic Environments - Chapter 2 Grammatical Evolution. Studies in Computational Intelligence, vol. 194. Springer, Heidelberg (2009). https://doi.org/10.1007/978-3-642-00314-1

17. Ryan, C., Collins, J.J., Neill, M.O.: Grammatical evolution: evolving programs for an arbitrary language. In: Banzhaf, W., Poli, R., Schoenauer, M., Fogarty, T.C. (eds.) EuroGP 1998. LNCS, vol. 1391, pp. 83–96. Springer, Heidelberg (1998). https://doi.org/10.1007/BFb0055930

18. Mahfoud, S.W.: Crowding and preselection revisited. In: PPSN, pp. 27–36. Elsevier (1992)
19. Tettamanzi, A.G.B., Faron-Zucker, C., Gandon, F.: Testing OWL axioms against RDF facts: a possibilistic approach. In: Janowicz, K., Schlobach, S., Lambrix, P., Hyvönen, E. (eds.) EKAW 2014. LNCS (LNAI), vol. 8876, pp. 519–530. Springer, Cham (2014). https://doi.org/10.1007/978-3-319-13704-9_39
20. Zadeh, L.A.: Fuzzy sets as a basis for a theory of possibility. Fuzzy Sets Syst. 1, 3–28 (1978)

Author Index

Ali, Mohammad Zawad 178
Anjum, Muhammad Sheraz 3
Assunção, Filipe 197
Atkinson, Timothy 19
Azzali, Irene 35, 213

Bakurov, Illya 213
Banzhaf, Wolfgang 49
Bertolotti, Luigi 35
Bisanzio, Donal 35

Drake, John 19

Gervasi, Riccardo 35
Giacobini, Mario 35, 213

Hemberg, Erik 64
Heywood, Malcolm I. 162
Hu, Ting 49, 178
Husa, Jakub 228

Jakobovic, Domagoj 262

Karsa, Athena 19
Kelly, Jonathan 64
Kocnova, Jitka 81
Koncal, Ondrej 98

Langdon, William B. 245
Lensen, Andrew 114
Leporati, Alberto 262
Liang, Xiaodong 178
Lorenz, Ronny 245
Lourenço, Nuno 197

Machado, Penousal 197
Mariot, Luca 262
McDermott, James 131
Moore, Jason H. 146
Mosca, Andrea 35

Nguyen, Thu Huong 278

O'Reilly, Una-May 64

Picek, Stjepan 262

Ribeiro, Bernardete 197
Ryan, Conor 3

Sekanina, Lukas 98
Shabbir, Md. Nasmus Sakib Khan 178
Silva, Sara 213
Sipper, Moshe 146
Smith, Robert J. 162
Swan, Jerry 19

Tettamanzi, Andrea G. B. 278
Tomassini, Marco 49

Urbanowicz, Ryan J. 146

Vanneschi, Leonardo 213
Vasicek, Zdenek 81

Xue, Bing 114

Zhang, Mengjie 114
Zhang, Yu 178

Printed in the United States
By Bookmasters